PTEROSAURS
FROM DEEP TIME

DAVID M. UNWIN
CURATOR, FOSSIL REPTILES AND BIRDS
MUSEUM OF NATURAL HISTORY
HUMBOLDT UNIVERSITY, BERLIN

A PETER N. NÉVRAUMONT BOOK

PI PRESS ♦ NEW YORK

PI PRESS

An imprint of Pearson Education, Inc.
1185 Avenue of the Americas,
New York, New York 10036

Pi Press offers discounts for bulk purchases. For more information,
please contact U.S. Corporate and Government Sales, 1-800-382-3419,
corpsales@pearsontechgroup.com. For sales outside the U.S., please
contact International Sales at international@pearsoned.com.

Printed in the United States of America
First Printing

Library of Congress Catalog Number: 2005906801

Pi Press books are listed at www.pipress.net.
ISBN 0-13-146308-X

Pearson Education LTD.
Pearson Education Australia PTY, Limited.
Pearson Education Singapore, Pte. Ltd.
Pearson Education North Asia, Ltd.
Pearson Education Canada, Ltd.
Pearson Educatión de Mexico, S.A. de C.V.
Pearson Education—Japan
Pearson Education Malaysia, Pte. Ltd.

Produced by Névraumont Publishing Company,
New York, New York 10006

A Peter N. Névraumont Book

PTEROSAURS

CONTENTS

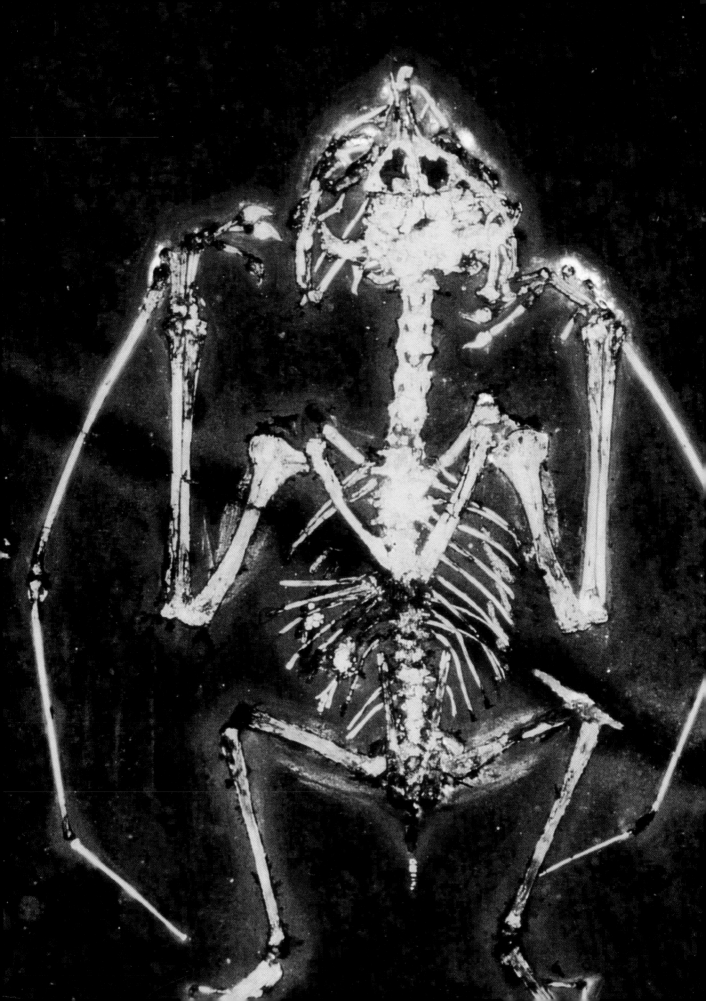

I
DRAGONS OF THE AIR

The dragon blinked in the fierce light of the sun as it emerged from the clouds and banked hard, its tremendous wings arching under the load. Reflected in a massive dark eye, the world below slowly tilted into view. Vast herds of dinosaurs were strung out across a dusty yellow-orange plain, occasionally gathered in knots where they had stopped to feed on patches of stunted vegetation. Then came marshes and—the dragon focused—a long, still, clear-blue lake. In its glassy depths, rainbow-hued fish hung almost motionless in the warm water, fins undulating, gills slowly pulsing. Suddenly, a huge dark shadow swept across the water's surface and, before it could dive into the safety of the weeds, a fish was gripped by long, sharp-pointed jaws and jerked violently into the air. As its consciousness faded, it gazed up into a limpid eye, set in a reptilian skull framed by fine, straggly hair. Slender, powerful wings beat smoothly up and down, membranes tensed and relaxed as the animal rose swiftly. At the top of its climb the Quetzalcoatlus paused, swallowed, then dived down again toward the shimmering lake far below.[1]

FIGURE 1.1 A beautiful new fossil of the Upper Jurassic (approximately 148 million years old) Solnhofen Limestone pterosaur *Anurognathus*, only the second example yet known for this species. This pterosaur had a wingspan of about 16 inches (40 centimeters). (Photograph courtesy of Helmut Tischlinger.)

Around 215 million years ago, at about the same time as dinosaurs first spread across the continents, an altogether different group of reptiles took to the air. Reptiles had tried this before, on several occasions, but this time was different. This time they got beyond gliding, where all previous groups had given up, and, as bats and birds were to do millions of years later, they evolved a rare and complex adaptation: true flapping flight. Going boldly where no reptile had gone before, these intrepid aeronauts entered a new realm—the open sky—and developed into a riotous multitude of species. Some, such as the frog-headed beetle-cruncher *Anurognathus*, shown in *Figure 1.1*, were as small as a starling, but others, like *Quetzalcoatlus*, became as large as an airplane. They thrived for 150 million years, only to disappear forever in the events that also killed off many of their contemporaries, including, most famously, the dinosaurs, and brought evolution's finest hour, the Mesozoic, to a close. This extraordinary group of animals—nature's real dragons—was the pterosaurs.[2]

The Trouble with Pterosaurs Pterosaurs, "winged reptiles," as their Greek name puts it, are familiar to most people as rather fierce-looking, leather-winged monsters featured in classic novels, such as Arthur Conan-Doyle's *Lost World*, or flit across the screen in TV documentaries[3] and in movies from *King Kong* to *Jurassic Park*.[4] No dinosaur scene is complete without them—usually a *Pteranodon*, the most famous pterosaur of them all, with large teeth to give it added fierceness, even though this species was actually completely toothless.[5] Trying to see beyond these superficial images and get a glimpse of the real thing triggers all sorts of questions: What were pterosaurs really like? How big did they get? Could they fly as well as birds? And one of the first questions that is always asked and hardest to answer: Why did they become extinct?

Driven more than most by curiosity, scientists find these strange creatures absolutely fascinating (pterosaur talks always fill the hall at scientific meetings) and some of the sharpest paleontological minds ever to ponder a fossil were so beguiled by these extraordinary animals that they made them the centerpiece of their researches. The first thing they discovered was that pterosaurs are really hard to understand. Even the term embodied in their name—winged reptile—seems contradictory. The word reptile, from the French "repere," meaning to creep, is not especially flattering, but describes living reptiles rather well. Wings, on the other hand, are about the last thing one might expect such a creature to have, and yet for pterosaurs, this defined them and their very existence.

As soon as pterosaurs were discovered, at the end of the 18th century, naturalists started quarrelling about them. Disagreements came thick and fast: Pterosaur origins, for instance, were hotly debated on several occasions, and, in one particular case detailed in Chapter 4, led to a protracted and acrimonious feud, while several other disagreements, for example, over wing shape and walking posture, continue even now. It was a long time before scientists could even agree as to what kind of animal pterosaurs might be—reptiles, as it turned out, but others argued that they were birds, and several eminent naturalists became convinced that they were bats. Debates also raged over the lifestyle of pterosaurs. Were they some kind of water creature? Or could they fly? And what did they do on the ground—strut around on their hind limbs like birds, or clump around on all fours? Even their likely metabolism became a scientific battleground: cool-blooded like crocodiles and lizards, insisted several authorities, warm-blooded like birds and mammals, countered others.

Pterosaurs became a *cause célèbre*, one of the great paleontological mysteries. Each generation of scientists "had a go" at them, and everyone had an opinion that, almost without exception, differed from that held by everyone else. The arguments, the confusion and the misunderstandings continued right up until a decade ago, and a few persist even today, but before we get into that, we should look a little more closely at why these ancient fliers caused so much controversy in the first place.

A Fossil Problem Pterosaurs have successfully defied more than two centuries' worth of scientific probing for several reasons. The most obvious is the problem of trying to understand animals that are known only from fossils. Just a tiny proportion of all the pterosaurs that ever existed, probably less than one individual in a million, has actually made it into the fossil record. The processes by which their cadavers became fossils, normally survived only by the hardest parts of the body—bones and teeth[6]—mean that most of the important information about anatomy, movement and behavior, how pterosaurs were colored, what noises they made, was lost forever.

Compounding the problem, even pterosaurs' hard parts were not well-suited for the rigors of fossilization. Pterosaurs were creatures of the air, with a relatively light and delicate skeleton constructed from slender, hollow bones whose walls were often little thicker than a credit card. This is not a good design if you want to become a fossil. To begin with, it meant that even pterosaurs' skeletons were relatively easily destroyed, so, compared with other backboned animals, their fossil remains are rare. Worse still, if you pull open a museum drawer, you find that most of their fossilized remains consist of isolated, often broken, bones—dumb witnesses that tell us little more than "here be pterosaurs." To cap it all, most of the decent pterosaur fossils that we do have, whole skeletons and, very occasionally, fragments of fossilized soft tissues, come from just a few locations scattered across the world and are separated by vast, barren, pterosaur-less gaps of thousands of miles and millions of years.

Thin-walled bones also mean that the complete skeleton of *Anurognathus*, shown in *Figure 1.1*, buried at the bottom of a Bavarian lagoon 148 million years ago, is rather less helpful than one might expect. Like many other beautiful-looking pterosaurs, several of which are featured in this book, it is absolutely flat—a "picture" fossil—its hollow-tube bones unable to resist the inexorable crushing weight applied over countless millennia by the overlying rock. Without the three-dimensional, sticking-out-here, dimpled-in-there form of the skull, the pelvis, or any of the 300 or so bones that made up a pterosaur skeleton, and unable to measure the exact shape, size and position of the joints, paleontology is robbed of critical data.

Confronted with a row of these "road-kills," it is often hard for an observer to establish even basic facts, such as: How many species are there? Two? Or more? Or just one, its representatives flattened in different ways? Trying to go further and find out, for example, how these pterosaurs might have stood, walked, or flown, is even harder. Strictly speaking, these are relatively simple questions (some of the really tough ones, for example, about physiology and breathing, will pop up later), but, enigmatic as the Mona Lisa, and sometimes just as smiley, the "picture" pterosaurs rarely give an answer.

More Problems: Analogy and Chauvinism   As if fossils themselves were not difficult enough to interpret, the ways in which scientists have gone about studying them also have their pitfalls. One trap that pterosaur researchers seem to have queued up to throw themselves into is a method much used, and no less often abused, by paleontologists: analogy. Confronted with the fossilized remains of an organism from deep time—be it a tiddly little ammonite, a 10-ton dinosaur or a toothless *Pteranodon*, all extinct for half an eternity and tragically bereft of any living descendants—it is terribly tempting to reach for analogy. Essentially, this means choosing a living organism that, although completely unrelated to ammonites or dinosaurs or even pterosaurs, seems to be sufficiently similar to them in external appearance or supposed lifestyle that we can use it as a mental vehicle for trying to better understand those long-expired denizens of the past.

In the case of pterosaurs, one does not have to look far (just upward) to spot their living analogies: birds and bats. Often bolstered by the mistaken belief that pterosaurs were somehow related to one or the other, scientists have repeatedly tried to use these modern fliers as models for understanding the flying reptiles. Yet, time and again, birds and bats proved to be treacherous allies. Similarities apparent in all three aeronauts—compact bodies, large wings, lightly built skeletons—were not inherited from a common ancestor, but result from convergent evolution. That is, the features that these fliers share evolved quite independently in each group as a response to the same difficult and highly demanding activity—flight.

Put them under the spotlight, though, and these three groups are found to be quite different. Pterosaurs, as the fossil in *Figure 1.2* shows, had a membrane wing, the outer part of which was supported by a single extraordinarily long and robustly built finger (hence, the vernacular name "pterodactyl," meaning wing-finger). Bats also have a membrane wing, but its outer part is supported by not one, but four fingers that, while long, are relatively thin and spindly. Birds, by contrast, have feathered wings, and their fingers (compared with pterosaurs and bats, at least) are strongly reduced.

The message is clear: pterosaurs were *not* birds or bats, nor were they related to them. Bitter experience has taught researchers to avoid the beguiling analogies offered by living fliers and to rely on the fossil remains of pterosaurs themselves as the best means for unraveling the mysteries of these animals.

A second pitfall, chauvinism, is rather more subtle, but no less dangerous. A widespread misperception of the living world is that organisms alive today are somehow "better" than those that existed in the past. Bombarded as we are by advertisements to buy the latest, fastest, shiniest whatever, it's easy to see why this notion is so pervasive. This "temporal chauvinism" is closely bound up with another misleading idea: that organisms can be ranked as if on a ladder, with the simplest forms of life on the bottom rung and the most complex and important—man, obviously (or even more meretriciously, certain races of man)—on the top rung. This is all complete hogwash, as that late great dispeller of such myths Stephen J. Gould has so effectively

FIGURE 1.2 "Dark-Wing" *Rhamphorhynchus*. This fantastically well-preserved Upper Jurassic fossil, seen here in ultraviolet light, has one of the best preserved pterosaur wing membranes ever found. With a wingspan of about 3 foot (1 meter) in life, this specimen was first described by Helmut Tischlinger and Dino Frey (2001). (Photograph courtesy of Helmut Tischlinger.)

shown in his writings, but nineteenth-century science was riddled with such ideas and, though now more subtle, they survive even today, lurking in the much debated notion that evolution is always progressive.[7]

Pterosaurs, like many other groups, such as dinosaurs and, closer to home, Neanderthals, have been innocent victims of this chauvinism. Obviously, such primitive-looking leather-winged lizards were an early but doomed attempt at flight and clearly inferior to birds and bats. Otherwise they would still be around, wouldn't they? A double whammy—temporal and biological chauvinism ganging up together. Generally speaking, of course, such a caricature, though it still exists and can even be found in the scientific literature, is a relatively trivial problem. The real difficulty is that such ideas can hinder scientists from grasping the true nature of the fossil organisms with which they are dealing, be it pterosaurs or any other extinct group, because, right from the start, they ensnare the victims (both object and observer) within a false and misleading perspective.

In fact, contrary to our often unthinkingly biased expectations, the latest scientific findings suggest that pterosaurs living 100 million years ago may have been more efficient fliers than the birds and bats that fill our skies today[8]—a poke in the eye for our self-centered human chauvinism.

Solving the Pterosaur Puzzle Fettered by a patchy fossil record, misleading analogies with birds and bats and a chauvinistic milieu, it's not surprising that the true nature of these Mesozoic dragons has proven so elusive—until now. A glut of new finds, including some extraordinarily well-preserved wing membranes (illustrated in *Figure 1.2*), thousands of fossilized tracks of pterosaurs (shown in *Figure 1.3*) and, for my money the most thrilling of all, an embryonic pterosaur in an egg, are finally laying bare some of the greatest mysteries of these enigmatic creatures. No little help has been provided by novel techniques and technologies such as CAT scanning,[9] which permits researchers to look inside pterosaur skulls (a view reproduced in *Figure 1.4*), allowing them to squeeze the last drops of information out of fossil finds, both old and new.

Rather like cresting the top of a hill and seeing a marvelous panorama for the first time, when all these new discoveries are put together, they reveal a startlingly new but also remarkably coherent and convincing picture of

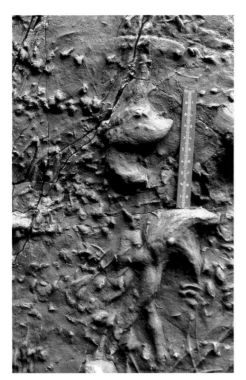

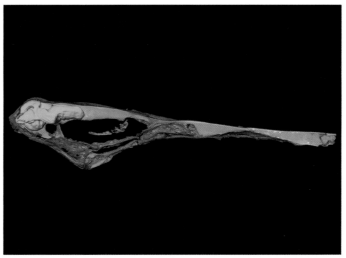

FIGURE 1.4 Inside a pterosaurs' head. This image, prepared from a Computer Automated Tomographic (CAT) scan of the skull of *Rhamphorhynchus*, shows a vertical slice about 4 inches (10 centimeters) long running through the middle of the skull, from the front (right) to the back (left), with bones colored purple and a reconstruction of the brain in green. The large opening in the rear half of the skull housed the eyeball. (Image courtesy of Larry Witmer.)

FIGURE 1.3 Scuttling around on a beach about 150 million years ago, pterosaurs left impressions of their hands, feet, and what seem to be beak marks in sands that subsequently became a layer of stone, now broken up into boulders to be found lying on the sea shore in the Asturias region of northern Spain. (Image courtesy of José Carlos Martínez García-Ramos and Laura Piñuela.)

pterosaurs. The best way to explain how this breakthrough was achieved is with a metaphor.

Imagine that you have been given a large box containing a jigsaw puzzle with many thousands of pieces. When completed, the picture reveals how pterosaurs were constructed, how they walked, flew, fed and grew, even how they evolved over their 150-million-year history. Now, without looking at them, take most of the pieces out of the box and throw them in the fire. Those were all the species, anatomy, behaviors and events that didn't make it into the fossil record.

The handful of pieces that remains represents pterosaur fossils buried in rocks scattered across the seven continents and, in some cases, under the sea. The first problem for paleontologists has been to get a few of those pieces out of the box, that is, collect some fossils—not always easy when they lie buried in the ground in remote regions of Earth, such as western Mongolia, for example. The next challenge has been to try to examine most of the available puzzle pieces, now to be found in fossil collections all over the world, decipher what it is that they seem to show, be it a feature of the brain or a trail of footprints, and then fit them together in a way that, we hope, matches up to some part of the original, true picture.

Using the puzzle metaphor, we can readily grasp how extraordinarily difficult it must have been 200 years ago for the first naturalists who tried to comprehend pterosaurs with the equivalent of just one piece of the puzzle in their hands. Even so, the legendary French anatomist Baron Georges Cuvier, the great-grandpére of all pterosaur researchers, was spot on when he proposed, in 1801, that pterosaurs were flying reptiles.[10] Slowly, over the decades, more and more pieces of the puzzle—fossil finds in Europe and then in the Americas, Africa and Middle Asia, and finally in China and even Antarctica—came to light.

Still, even by the late 20th century, several critical puzzle pieces—such as the design of pterosaurs' wings: broad or narrow? Their walking ability: on two legs or four? Their physiology: like reptiles or birds?—remained, well, puzzling. Just as perplexing, no matter how much one rearranged them, there were always points where the pieces that we had already collected just would not match up. So, for example, while one study concluded that pterosaurs could not have used their arms for walking, another described clear, well-preserved handprints in the tracks of these animals (this paradox is resolved in Chapter 9).

Then, in just a few years, beginning in the mid-1990s, the puzzle suddenly came together and revealed a consistent and convincing picture of pterosaurs, their lives and their fate. But why then and not before? If we step back a couple of decades to the time in the 1970s when modern vertebrate paleontology (the study of extinct backboned animals, including pterosaurs), was just getting under way, we find that among its many new interests was an age-old problem: the pterosaurs. Ground-breaking work in those early years by the Munich-based paleontologist Peter Wellnhofer, the world's leading authority on these animals, eventually attracted the attention of a whole new generation of researchers and triggered a tidal wave of research that "broke" in the 1990s but even now, in mid-2005, shows no sign of abating. In the past two decades, "pterosaurology," as we might now refer to the study of pterosaurs, has seen more fossils collected, more techniques brought to bear and a greater volume of scientific studies and publications than ever before.

The jigsaw puzzle is still far from complete, but we have enough pieces to make out the picture, and it seems at last to make some sense. This book puts that picture on display for the first time and, as with any good exhibition, there are lots of surprises.

Rulers of the Mesozoic Skies Pterosaurs, as the sketch in *Figure 1.5* shows and the following chapters will reveal in glorious detail, were completely different from any other animal, living or extinct. They were reptiles, but, unlike lizards or crocodiles, they were not scaly (except perhaps on the legs and soles of the feet), but furry,[11] and, unlike their "cold-blooded" reptilian relatives, they seem to have been capable of strenuous and protracted exercise, such as flapping their wings, for hours on end, something that in the modern world only warm-blooded animals can do. Equipped with relatively large bird-like brains, they also appear to have been much more intelligent than living reptiles and, if the spectacular and astonishingly diverse range of "look at me" head crests is anything to judge by, they also had complex social behaviors.

Above all else, pterosaurs were creatures of the air. Their entire bodies were highly modified for flight, powered by a warm-blooded physiological engine and lungs that may have been as efficient as those of birds. Pterosaurs also appear to have had highly sophisticated flight membranes that were directly connected to the brain—clever wings that may have been better than

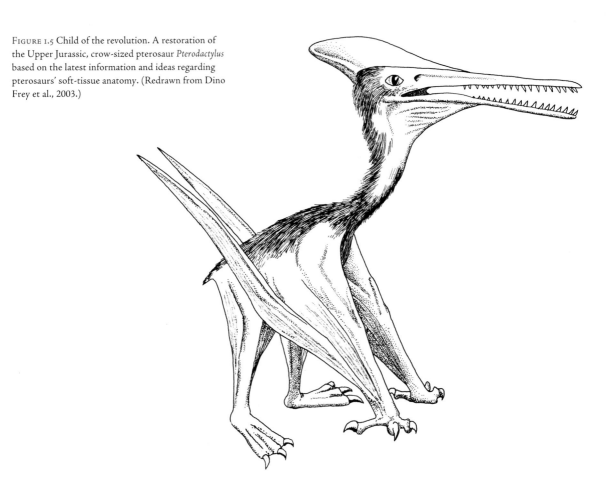

FIGURE 1.5 Child of the revolution. A restoration of the Upper Jurassic, crow-sized pterosaur *Pterodactylus* based on the latest information and ideas regarding pterosaurs' soft-tissue anatomy. (Redrawn from Dino Frey et al., 2003.)

anything to be found in modern aeronauts, natural or man-made. Consequently, these Mesozoic dragons are now envisioned as highly competent and vigorous fliers, capable of snatching up their prey while on the wing and of staying aloft for hours or even days at a time. Indeed, the latest studies (reported in Chapter 9) suggest that some aspects of flight performance, such as the soaring ability of *Pteranodon*, a highly specialized albatross-like pterosaur, may have outstripped that of any living bird.

The ground, however, was quite a different matter. With their legs fastened to each other and to their arms by the flight membranes, long-tailed rhamphorhynchoid pterosaurs were rather hampered in their movements and probably kept out of harm's way by clinging to trees or cliffs using long, strongly curved claws on their fingers and toes. By contrast, short-tailed pterodactyloid pterosaurs evolved somewhat narrower flight membranes that gave their legs more freedom and endowed them with the ability to scamper around rather more adroitly. Having "conquered" the ground, pterodactyloids were able to exploit all kinds of new life styles. Many of these involved

FIGURE 1.6 A tiny pterosaur tragedy. This individual of *Pterodactylus*, from the Upper Jurassic Solnhofen Limestone of Bavaria, at only 8 inches (20 centimeters) in wingspan, was probably only a few days or weeks old when it died, perhaps after an accident on one of its maiden flights. (Photograph courtesy of Helmut Tischlinger.)

wading in ponds and streams and, as they stumped around in a uniquely pterosaurian fashion, rather like a saddle-worn cowboy on crutches, they left behind their peculiar "feet before the hands" tracks imprinted on mud flats and seashores all over the world.

Not content with merely looking different, pterosaurs also appear to have been unique among flighted animals in that they adopted a "hands-off" approach to bringing up their young. Almost without exception, hatchling birds and baby bats are looked after by their parents and must be almost fully grown before they can take to the air. Astonishing as it may seem, baby pterosaurs like the one shown in *Figure 1.6* could fly soon after hatching and may not have needed or received any assistance from mom or dad. Interestingly, this remarkable ability might help explain another unique and spectacular feature of pterosaurs: gigantism. They were equipped with the unique property of being able to grow and fly, and it would seem that pterosaurs were not restricted to a particular size but, in some cases, just continued growing until they were as big as an airplane.

The unique construction, abilities and behaviors of pterosaurs paid off—big time—and the group became tremendously successful. This is seen, for example, in the remarkable diversity of these animals, with forms ranging from agile, aerial insect hunters, through flamingo-like filter-feeders with thousands of teeth, to highly evolved ocean-going soarers. They also achieved an incredible range of sizes: While the smallest was only about 20 centimeters (8 inches) in wingspan and probably weighed less than a starling, the largest had wings more than 10 meters (almost 40 feet) from tip to tip and probably tipped the scales at around 50 kilograms (110 pounds) or more. Pterosaurs were successful in other ways, too. After they first appeared about 215 million years ago, it did not take them long to spread around the world, after which they dominated the skies for the next 150 million years. That is more than twice the length of the known history of bats and rivals the evolutionary longevity of birds.

Why Pterosaurs? Prehistoric reptiles are big business. Every day thousands of books detailing the lives and deaths of *Tyrannosaurus*, *Diplodocus* and a host of their dinosaurian relatives are sold across the world, and literally millions of viewers tune in to the latest TV documentaries to catch the breaking news from the Mesozoic. Huge, weird, dangerous-looking monsters from the deep past excite, fascinate and entrance people of all ages everywhere and, thanks to their extinction long ago, these creatures are a "safe" thrill. But, just as in the past, dinosaurs rule. Other denizens of the dinosaurian world such as ichthyosaurs and plesiosaurs—toothy killers that swam the Mesozoic seas—are occasionally allowed to show their faces, but whole TV programs or entire books devoted to these animals are still rare.

It is the same with pterosaurs. There is no doubting the extraordinary appeal these incredible creatures exert on the imagination of anyone who glimpses them on a TV screen or billboard, but until now, if you wanted to read more about them, it wasn't easy. The first book on pterosaurs, *Dragons of the Air*, written by the English paleontologist Harry Seeley, appeared in 1901, more than 100 years after pterosaurs had first come to light. Rare, expensive and completely out of date, this marvelously idiosyncratic work is still a literary delight. Having read it, though, readers had to wait another 90 years for a second book on pterosaurs—Peter Wellnhofer's magnum opus *The Illustrated Encyclopedia of Pterosaurs*. Chock-full of information on every pterosaur species known to science in 1991, this was the principal reference volume for the group but, like *Dragons of the Air*, it, too, is now out of print, rare and expensive.

Much has changed since the *Encyclopedia* first appeared. The many critical ideas about pterosaur biology that were fought over in the 1990s—Were they two- or four-footed on the ground? Did the flight membranes attach to the legs? Did they leave tracks—have been resolved into a convincing and (among pterosaurologists) widely agreed-upon picture. At the same time, a stream of new fossil discoveries (more than 30 pterosaurs previously unknown to science have been described since the *Encyclopedia* was published) and the application of modern techniques, for example, with regard to discovering pterosaur genealogy, have dramatically improved our knowledge of the evolutionary history of these animals.

The pages that follow contain the first comprehensive account of our new understanding of how pterosaurs were constructed and how they lived their lives: how they flew, walked, breathed and grew. What this book also reveals, for the first time, is how the design and function of these animals launched their successful invasion of the skies and also shaped their final doom.

2 PTEROSAUR PLANET

Picture a world with a warm, humid climate extending as far north as present-day Alaska. In this Mesozoic Eden, luxuriant groves of cycads and tree ferns and seemingly endless forests of conifers rang to the crash of ever-browsing herds of truck-heavy, bus-sized sauropod dinosaurs. Wreaking swathes of destruction, these mountains of flesh were trailed by quick, bright-eyed, hook-clawed theropods, ceaselessly eyeing the sick, the young and the unwary. Offshore, schools of Leedsichthys, each fish the size of a whale, slowly sieved the waters of deep blue ammonite-filled seas, while refracted shadows of snake-necked plesiosaurs and spear-snouted ichthyosaurs played on the sides of their titanic bulk.[1]

FIGURE 2.1 Predator and prey. Collected from the Karatau region of Kazakhstan in the 1960s by a team of Russian paleontologists led by Alexander Sharov, most of this slab is occupied by a superbly preserved skeleton of the pigeon-sized Late Jurassic pterosaur *Sordes pilosus* (a restoration of the same pterosaur adorns the cover of this book), surrounded by patches of fossilized skin and hair. This pterosaur was buried in mud at the bottom of a freshwater lake rimmed by dense stands of conifer trees, fragments of which also ended up in the sediment and are now preserved as fine black flecks scattered across the slab. The lakes were also home to *Sordes'* prey—palaeoniscoid fish—one example of which was, by chance, preserved alongside its predator.

The world in which pterosaurs lived was very different from our own. It was not completely different, though; some things would have looked rather familiar: mosses, ferns, conifers, insects, even turtles and lizards haven't changed all that much, at least in external appearance, since they were perched on, eaten, or flown over by pterosaurs. But in other ways, things were very alien. Life on land was dominated not by mammals, as it is today, but by a remarkable panoply of dinosaurs. Many of these were herbivores that, in turn, were preyed upon by an astonishing array of theropods, not least, the king of them all, *Tyrannosaurus rex*. Mammals, including the line that eventually culminated in you, me and other members of *Homo sapiens*, were mostly mouse-sized, while familiar plants such as grasses lay unimaginably far in the future.

This chapter considers how scientists discovered, mapped and reconstructed the world in which pterosaurs lived, and how that world changed, sometimes dramatically, during their 140-million-year tenure. The journey takes in global geography and climate, and along the way, we catch some brief glimpses of the animals and plants that formed the backdrop to the everyday life of pterosaurs. First, though, we will attempt to comprehend the incomprehensible: deep time.

Long, Long Ago Deep time, like quantum physics and cricket,[2] is really hard to grasp, and generations of writers have expended considerable amounts of this commodity trying to devise metaphors to help you, the reader, come to terms with the concept. One of the most popular metaphors invites us to consider Earth's entire history, all 4 billion years of it, as a 24-hour clock, a scale within which our own species appears at about 4 seconds before midnight. Impressive, but it doesn't quite convey the sheer ineffableness of deep time, so let's try something else.

If we assume that, on average, humans produce a new generation approximately every 20 years,[3] then five human generations equate to a century, 50 generations to a millennium and 500 generations, or 10 millennia, takes us back to the end of the last Ice Age and the dawn of civilization. If we step up a scale and use the entire length of human civilization as our basic unit, then we need at least 100 of these (50,000 human generations) just to reach back a million years. This means that to return to the last moment in time when pterosaurs existed—the final days of the Mesozoic, 65 millions years ago—one would have to experience a time span equivalent to the whole of human civilization repeated 6,500 times, which is equivalent to about 3 million human generations.

For pterosaurs, however, this was the end of what had already been a stupendously long history. At this, their final moment on Earth, they had been around (or rather above) for more than 150 million years, having first appeared at least 215 million years ago. Or, to try to frame this in a human context, pterosaurs lived, died and evolved for a period equivalent to more than 15,000 human civilizations. This also means, rather surprisingly, that the last pterosaurs existed much closer to us in time than to their earliest ancestors.

The time interval in which pterosaurs lived is referred to as the Mesozoic (*Figure 2.2*) and divided by geologists into three periods: the Triassic, toward the end of which pterosaurs first appeared; the Jurassic, during which, apart from insects and the original early bird, *Archaeopteryx*, pterosaurs were practically the only creatures to be seen in the skies; and the Cretaceous, when the heavens must have thrummed with multitudes of pterosaurs and birds. Each period is subdivided into Early, Middle and Late (apart from the Cretaceous, which has no Middle), and each of these is more finely divided into units of time called stages, usually named after a region or location where the stage was first defined.[4]

Stages, and the longer intervals to which they belong, such as periods, are firmly embedded in a relative time scale tied to the fossils of what were abundant, rapidly evolving creatures such as ammonites (which floated by the millions in Mesozoic seas) and an absolute time scale based primarily on the slow decay of certain radioactive isotopes.[5] The significance of the stage here is that the most accurate geological date we can obtain for most pterosaur fossils is only to the stage level.[6] Since, typically, stages are about 6 million years in length (equivalent to about 600 human civilizations), this means that, as a rule, it is rather difficult to resolve any "short-term" events in pterosaur history, unless they lasted at least a few million years.

An Ever-Changing World By pooling information from a wide range of disciplines, scientists have been able to develop a surprisingly detailed picture of the geography, climate, vegetation and faunas of the Mesozoic worlds in which pterosaurs lived.[6] Matching the fit of coastlines and using other types of information such as paleomagnetic data,[7] which give away the positions of continents in the past, scientists are able to track the drift of land masses and reconstruct the geographical history of our world. Features of sediments that make up the rock record, such as the size, shape and composition of the individual grains from which they are formed and the fossils they contain, give geologists clues about the local geography and climate.

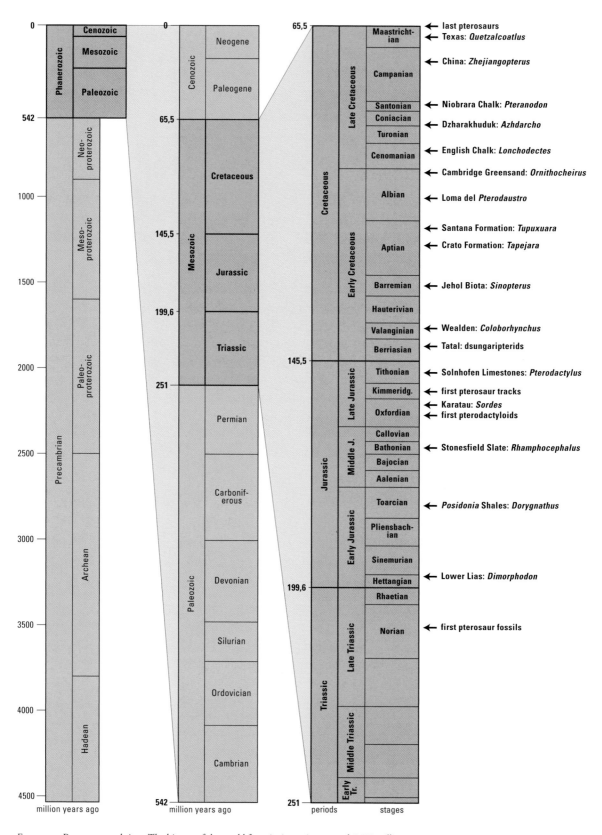

FIGURE 2.2 Pterosaurs and time. The history of the world from its inception around 4,500 million years ago to the present is depicted in the left column. Life is thought to have first appeared some 700 million to 800 million years later, but only really got going at the beginning of the Phanerozoic, shown in the middle column, about 542 million years ago. The 150 million year long history of pterosaurs, together with some of the more important pterosaur fossil localities, is illustrated in the right column.

Slice up samples of fossilized wood into sections so thin that you can see right through them and, under the microscope, variations in cell size can be used to detect wet and dry seasons and their relative intensity. Keep collecting more plant fossils, and eventually paleobotanists (paleontologists who specialize in the study of fossil plants) will be able to reconstruct much of the ancient flora of the region.

If we could go back 215 million years to the Late Triassic, a time when the first rather primitive-looking rhamphorhynchoid pterosaurs were beginning to flap through the skies, we would encounter a world very different from that of today. Rather than being spread all over the globe, as they are now, most of the continents were fused together in a single huge land mass called Pangaea (*Figure 2.3, top*). This had a profound effect on the climate: There were no ice caps, average global temperatures were much higher than today, and in general, the differences between seasons were far less marked. As a consequence, much of the interior of Pangaea was dry and arid and seems to have been dominated by extensive deserts.

Out on the coastal plains, conditions were more humid, especially in the vicinity of the main waterways. Here, plants could grow in abundance, but the Late Triassic flora over which early pterosaurs flitted would have looked quite strange to our eyes. Tree ferns, ginkgoes, cycads and conifers dominated the vegetation, forming extensive forests whose shady understory was carpeted by myriad ferns and mosses. Elsewhere, immense drifts of reed ferns and horsetails fringed swampy regions and colonized the edges of lakes and rivers.

The start of the Jurassic, approximately 199 million years ago, saw several well-established groups of long-tailed pterosaurs living around the margins of a supercontinent that was beginning to break up and whose coastal regions had been flooded by rising sea levels. By the end of this period (*Figure 2.3, middle*), Pangaea had split right through the middle, forming two land masses: Laurasia in the north, composed of modern-day North America, Europe and Asia, and Gondwanaland, formed from Antarctica, South America, Africa, Australasia and India, in the south. This new configuration also brought about important climatic changes. Tropical or subtropical conditions extended across both land masses but, crucially, it was more humid than before the split and the differences between the seasons were more pronounced, with hot, dry summers and cooler, wetter winters.

The vegetation was rather like that in the Late Triassic and, as the humidity increased, ferns, in particular, prospered in the lowland areas, forming vast, dense tracts. Drier upland environments were dominated by

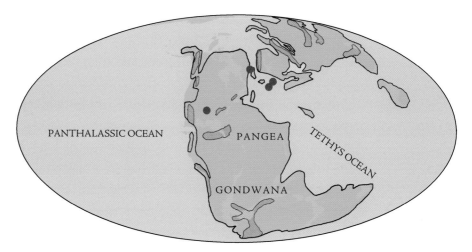

TRIASSIC

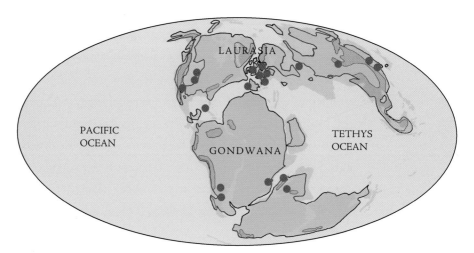

JURASSIC

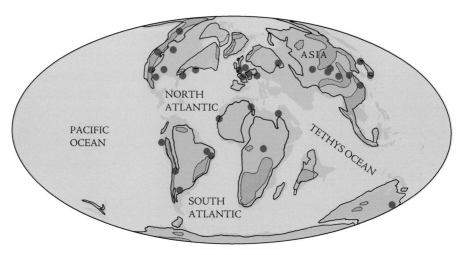

CRETACEOUS

FIGURE 2.3 Pterosaur world. The disposition of the
main continental land masses and oceans are shown for
the Upper Triassic (top), Upper Jurassic (middle) and
Upper Cretaceous (bottom). The red dots pinpoint
locations where pterosaurs were fossilized.

cycads and conifers, some reaching 30 or 40 meters (120 feet) high. Among them was a new type of cycad, the bennettitalean, with a short trunk and a well-developed bush-like crown. This particular cycad was so extraordinarily successful that it ended up completely dominating the flora and waxed so luxuriant that vast accumulations of its rotting remains eventually led to the formation of Jurassic coals.

As new kinds of short-tailed pterosaurs spread across the world at the start of the Cretaceous, the breakup of Pangaea progressed even further. Australasia and Antarctica broke away from the rest of Gondwanaland, and the remaining land mass, South America and Africa, had also completely separated by the Late Cretaceous (*Figure 2.3, bottom*). The rift that split them apart continued northward and had sundered North America and Greenland from Europe by the end of the Mesozoic. In addition to all this continental drift, sea levels rose to some of their highest levels ever in the early Late Cretaceous, flooding over the continents and creating a *Pteranodon*-haunted midcontinental seaway down through North America and an archipelago of islands where Europe stands today.

Initially as warm as in the Jurassic, temperatures cooled toward the middle of this period and seem to have become more variable toward the end of the Cretaceous. The vegetation also underwent some major changes. Early Cretaceous floras were similar to those of the Jurassic, except for the appearance of a new group: angiosperms, or flowering plants. At first, these newcomers remained small and shrubby, but in the Late Cretaceous they grew much bigger and evolved into many new types of plants, prominent among them magnolias, sycamores and oaks, which formed forests that largely replaced those of cycads and ginkgoes.

Meet the Neighbors Like birds today, pterosaurs were widespread and lived as members of many different communities, not only in mountains, forests or on the plains, but also in coastal regions and even far out to sea. Many of the simpler types of animals that formed the bulk of these communities—worms, spiders, crabs, shellfish, corals and sponges—are little different from their living descendants that surround us now. Indeed, the single most important group by almost any measure, insects, was already well-established,[8] and the pterosaur world hummed, buzzed and whirred with dragonflies, beetles, bugs and cockroaches, just as our world does today. Not everything was the same, though. One type of shellfish common in the Mesozoic, the brachiopod, though still with us, has dwindled to but a shadow of its former glory, and the molluscan epitome of the Mesozoic, the ammonites, probably one of the commonest animals in the seas over which pterosaurs soared, died out, together with the dinosaurs.

Some of the backboned animals that roamed the Mesozoic lands or swam in the seas would also have seemed rather familiar. Modern sharks and rays and the so called ray-finned fish (actinopterygians), which include early forms such as sturgeon and paddle fish, were common in the later Mesozoic. Actinopterygians also include the greatest of all the finny races, teleosts—bony fish. The bewildering variety of modern teleosts—ranging from seahorses to sail fish—is only the latest development of a group that first rose to dominance during the Cretaceous (*Figure 2.4*). In the Jurassic and earlier, the seas, lakes and rivers that teleosts later filled were occupied by older, more ancient groups of fish. Prominent among them were hybodontids, a group of early, spiny-finned sharks, and a large heterogeneous association of fish called the paleonisciforms, often of rather small size, that were common in the Triassic and Jurassic and probably figured largely in the diet of many pterosaurs.

The top predators in the seas were several groups of reptiles: ichthyosaurs, plesiosaurs, and arriving somewhat later, mosasaurs.[9] Originally descended from small lizard-like animals that lived on land, these creatures took to the water very seriously. Ichthyosaurs became completely adapted for a life under the ocean waves. Like modern whales and dolphins, even their young were born in the water, as revealed by several fossilized remains of pregnant mothers that died while giving birth. Generally rather dolphin-shaped, but with vertical fish-like tails, ichthyosaurs were typically about 2 to 3 meters (6 to 9 feet) long, though some whale-sized forms reached lengths of 15 meters (about 50 feet) or more. Most species had long jaws that were crammed with simple, sharp-pointed teeth that enabled ichthyosaurs to get a good grip on their prey, which, if their fossilized stomach contents are anything to go by, was primarily squid.

Plesiosaurs first appeared in the Late Triassic, some time after the ichthyosaurs, but whereas the latter seem to have died out at the start of the Late Cretaceous, plesiosaurs, like ammonites, dinosaurs and pterosaurs, survived to the very end of the Mesozoic (*Figure 2.4*). Famous for their peculiar appearance "like a snake threaded through a barrel," these animals had a long neck, a short tail and two pairs of flippers, with which they propelled

FIGURE 2.4 Meet the neighbors. A selection of the most important groups of backboned animals that shared the Mesozoic world with pterosaurs. The dagger symbol indicates the point in time when a particular group is thought to have become extinct.

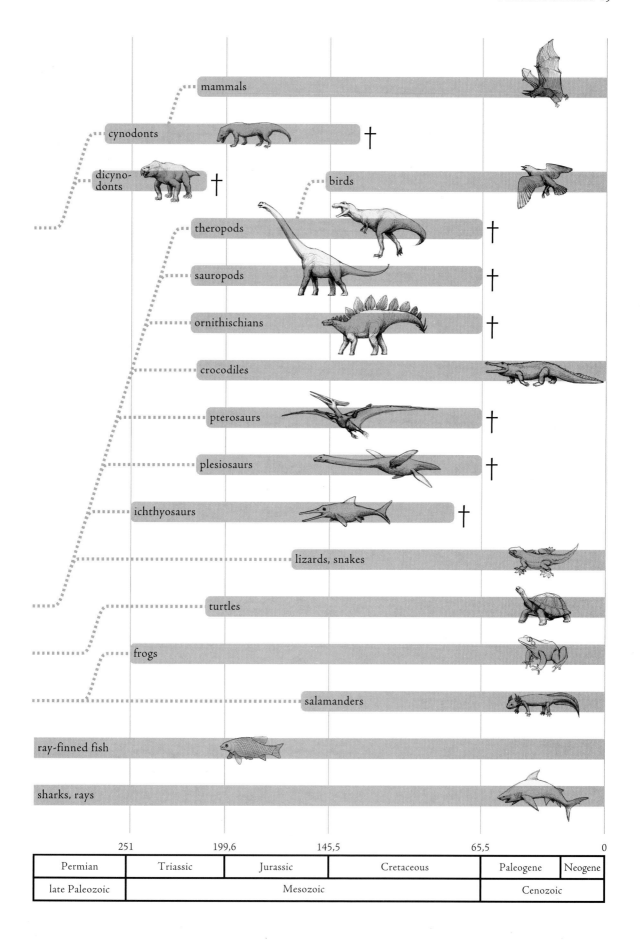

Permian	Triassic	Jurassic	Cretaceous	Paleogene	Neogene
late Paleozoic	Mesozoic			Cenozoic	

themselves through the water. Such an arrangement does not seem well-designed for high-speed swimming, especially because it is not at all clear how the flippers were deployed so that they did not interfere with one another. Consequently, it is thought that plesiosaurs are more likely to have been ambush predators, lurking in the murky depths waiting for their prey to swim by, rather than pursuit predators like the ichthyosaurs, which chased down their dinners. Smaller forms probably lived on fish, but the big plesiosaurs, such as *Liopleurodon*, with a skull 2 meters (6 feet) long, a total body length of 15 meters (45 feet), and some of the biggest teeth in the animal kingdom, were probably capable of killing and eating anything else in the sea, including sharks, ichthyosaurs and other plesiosaurs.

Propelled by a powerful tail and stabilized by paddle-shaped limbs, mosasaurs, like ichthyosaurs and plesiosaurs, were superbly adapted to a life in the water, even giving birth there. Descended from true lizards that took to the seas in the early Late Cretaceous, mosasaurs evolved several different lifestyles: Some developed large, globular teeth and went in for feeding on shellfish while others, among them huge 15 meter-long (50 foot-long) leviathans such as *Mosasaurus*, preyed on fish or perhaps even other mosasaurs.

Several important groups of land-living backboned animals, still alive today, also first appeared at about the same time as pterosaurs. Mesozoic lakes and ponds were home to the earliest frogs and salamanders and many, but not all, crocodiles. Some small, agile crocodiles roamed the land, while several Cretaceous species seem to have become vegetarian and taken to chomping the undergrowth.[10] Yet another group of crocodiles, the thalattosuchians, evolved paddle-like limbs and became completely adapted to a life in the seas, presumably only returning to land to breed. Plodding along patiently, protected by their shells, turtles were there, too. Among the most common of vertebrate fossils, they were to be found worldwide in the Jurassic and Cretaceous. They also took to the seas, where some species became giants, reaching the size of a family car—4 meters (13 feet) in length and up to a ton in weight.

Not all groups opted for gigantism, though. Mammals were highly successful (literally so in some cases, where the design of the limbs shows that they must have been completely at home up in the cycads and tree ferns), spreading across the entire Mesozoic world and evolving into many different types and kinds, but with one proviso—they always stayed small.[11] Rarely larger than a rabbit, and usually smaller than a rat, Mesozoic mammals lived in the shadows, feasting on insects, shoots or seeds, and we, their descendants, might still be there today if all but one lineage of dinosaurs (the birds), had not eventually become extinct.[12]

Dinosaurs, of course, are what made the Mesozoic really different from our modern mammal-dominated world. They seem to have risen to prominence at about the same time as pterosaurs; they rapidly became the dominant land animal and stayed on top for the next 150 million years.[13] Often large, or even gigantic when compared with today's mammals, their complex communities developed on every land mass and must have had a profound impact on both the flora and fauna of the Mesozoic world.

The majority of dinosaurs were plant-eaters. The most successful, and from the flora's point of view the most dangerous, were the sauropods. These huge, long-necked, four-legged behemoths, typified by *Diplodocus*, a dinosaur familiar to all, must have wreaked havoc on the vegetation, either by consuming it or by trampling it underfoot. Worse still for the landscape, fossilized tracks suggest that sauropods congregated in herds and migrated long distances to find fresh fodder.

Other vegetarians included the armored dinosaurs, such as stegosaurs, whose spikes and plates attest to the need for a really effective defense. In the Cretaceous, other groups of herbivorous dinosaurs came to the fore: hadrosaurs and their relatives the iguanodontians, the ceratopsians (led by the king of the spiky heads, *Triceratops*) and the pachycephalosaurs—the thick heads. Apart from some of the ceratopsians, who returned to a life on all fours, most of these dinosaurs ran around on long powerful hind limbs, only supporting themselves with their arms when they needed to feed, rest or move slowly.

Inevitably, where there are herbivores there are carnivores—in this case, theropods—some of the biggest and most dangerous predators that ever lived. Theropods were killers. Their jaws were filled, as a rule, with dagger-shaped teeth equipped with sharp cutting edges; many had wickedly hooked claws on their fingers; and some even bore large killing claws on their toes. Their long, powerful hind limbs tucked under the body enabled them to move quickly and efficiently. To cap it off, they were at least as intelligent as any of their potential victims, big or small, and might even have ganged up on their prey by hunting in groups.

Pterosaurs were never free of the menace posed by these predators, as some dramatic fossils emphasize—pterosaur bones with theropod teeth still embedded in them.[14] Theropods of all kinds and sizes from tiny kill-ers no bigger than a chicken up to 4- or 5-ton, 12-meter-long (39 feet) "you name it and we've eaten it" monsters such as *Tyrannosaurus rex* were at the top of the food chain throughout the Jurassic and Cretaceous. They even did something that only pterosaurs had ever done before: took to the air, disguised as birds.

Feathered Friends, or Foes? The earliest birds, represented by the most famous fossil in the world, *Archaeopteryx*,[15] first appeared on the scene in the Late Jurassic, approximately halfway through the reign of the ptero-saurs. The question of the origin of birds, one of the most important and fiercely debated scientific issues of modern times, was recently answered in a most dramatic fashion. Complete, undisturbed skeletons of small theropods from Lower Cretaceous lake sediments of northeast China, surrounded by halos of beautifully preserved feathers, indistinguishable from the feathers of crows, ducks or pigeons living today, show beyond any possible doubt that birds are the direct descendants of meat-eating dinosaurs.[16] In fact, so many fossils have been found that the pathway from small, ground-living, fast-running *Velociraptor*-like theropods to the early birds that swooped and screeched through the Cretaceous skies can now be traced in astonishing detail, even down to the origin and evolution of one of nature's greatest in-ventions: the feather.[17]

Although sharing some basic similarities in body design, birds differ in many important ways from pterosaurs. Most obviously, the wings were com-posed of feathers, not membranes, and, perhaps even more significantly, the wings of birds were only supported by the arms and had no connection to the legs at all. Freed from any major role in flight, the legs and feet were able to evolve and adapt to doing many other things, among them perching, run-ning and grabbing prey.

These and many other features of birds, such as their warm-blooded phys-iology and care of the young, seem to have ensured the extraordinary suc-cess of this group. After their origin in the Jurassic, many different lineages of birds seem to have appeared quite rapidly, and by the Early Cretaceous, birds had become firmly established worldwide.[18] Specialized, flightless div-ing birds, completely unrelated but remarkably similar to today's loons, fol-lowed,[19] and by the Late Cretaceous, the ancestors of at least some modern groups of birds, such as gulls and ducks, had also appeared.

So, it seems that the Cretaceous world was full of birds, which raises an interesting question: How did they achieve this, if pterosaurs were already there? After all, the two living groups of fliers, birds and bats, are pretty well segregated between day and night, with just a few specialists such as owls or fruit bats that trespass in each other's realm. I believe that the answers to this question and to its logical follow-up (Were birds responsible for pterosaurs' extinction?) lie in the different design, construction and function of these two animal aeronauts. This issue is explored in later chapters, the last of which (Chapter 11) returns to the main question: Was the Cretaceous a long, slow, showdown between birds and pterosaurs?

3 CONSIDERING MEDUSA

The storm had been coming for days, and when it hit, its ferocity was overwhelming. The normally placid waters of the lagoon were whipped into a maelstrom, while the trees and shrubs that clothed the archipelago of low islands were blown this way and that. Some, torn from their roots, whirled off toward the horizon. Pterosaurs, who normally rode out storms on the wing, fought vainly for control and found themselves being rudely tossed around the sky. A big, old male seeking escape by climbing to a higher altitude was flipped on his back by a sudden eddy and then blasted by a tremendous gust. With a dull crack, his wing-finger broke and, simultaneously, a flight membrane tore away from his leg. Crippled beyond hope, torn wings fluttering like forlorn pennants, he spiraled down toward the hundreds of other pterosaurs whose bodies littered the surface of the lagoon far below. Dropping into the sea with a crumpling splash, the pterosaur gradually lost consciousness as water flooded his lungs, and within a few minutes, he was dead. Soon, waterlogged by the surge of the waves, the carcass slipped beneath the surface and slowly began to sink down through the blue waters into the dark, lifeless depths and the pristine mud far below.[1]

FIGURE 3.1 Fragments of the Upper Cretaceous pterosaur *Azhdarcho* rest on the hand of their discoverer, the late Lev Alexandrovich Nesov. This photograph was taken at the fossil locality of Dzharakhuduk in Uzbekistan in the early 1980s. (Photograph courtesy of Lev Nesov.)

The Fossil Factory The fate of those who looked upon Medusa was to be turned to stone—forever—which is generally considered to be a bad thing, especially if one had other plans for the evening. Pterosaurs, along with millions of other animals and plants, suffered a similar destiny at the hands of geological processes, but in this case, it was a good thing. This goes both for the paleontologist, who now has something to work with, and the original victims, which have achieved an enviable degree of immortality, and, if they get very lucky, a meeting with someone who would, in a sense, like to bring them back to life.

Fossilization, which is primarily about replacing or replicating biological tissues with relatively inert minerals (the basic stuff of stone) that can last practically forever, is the process by which pterosaurs traveled across millions of years from their dinosaur-filled Mesozoic world to our modern mammal-run planet.[2] Most of what we know about pterosaurs is founded on fossils that survived this journey. Inevitably, not everything was fossilized. Usually only the hardest, toughest materials, such as teeth and bone, were capable of hanging around long enough in just the right sorts of places, such as the bottom of tranquil lagoons, to stand any real chance of becoming fossilized. This means that in order to be able to extract the most from what remains of pterosaurs, which, because of the relatively delicate nature of their construction, are rare as fossils in any case, it is vitally important to try to understand how they were fossilized in the first place: what survived, what did not, and how the conversion to stone modified what was originally there to what we see now.

In this chapter, then, we explore the transformation of a pterosaur from a living animal to a fossil and its subsequent journey into the hands of a paleontologist, such as Lev Nesov, shown in *Figure 3.1*, intent on understanding what this animal originally looked like, how it functioned during life and what its role was in the long-extinct communities of the Mesozoic. The road from a living pterosaur to a reconstruction on a computer screen is long (*Figure 3.2*). It begins with a corpse, a burial, fossilization and a few tens of millions of years underground. It continues with discovery, preparation (the freeing of fossils from their rocky tomb) and identification, and ends beneath the microscope of a paleontologist intent on wringing as much information as possible from his petrified subject. In order to start this process, however, another process must end: life.

When the Grim Reaper Calls As is the case for wild animals today, it was probably rare for a pterosaur to die of old age, and certainly all, or nearly all, of the fossils collected so far probably represent individuals that succumbed to disease or accidents or were killed in some local, or perhaps more widespread, catastrophe. We can be absolutely sure, for example, that several hatchling-size individuals of the Upper Jurassic pterosaur *Pterodactylus* (one of which is shown in *Figure 1.6*) were probably only a few days or weeks old when they died and had their brief lives dramatically cut short by some kind of accident—probably related to their inexperience of flight.

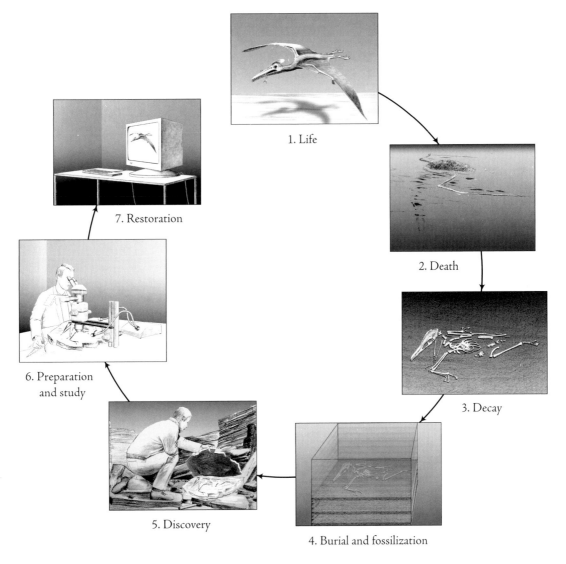

1. Life

2. Death

3. Decay

4. Burial and fossilization

5. Discovery

6. Preparation and study

7. Restoration

FIGURE 3.2 Pterosaurology begins with a living pterosaur and proceeds through death, decay, burial and fossilization, discovery, preparation and study to arrive back at a restoration of a living pterosaur.

Generally speaking it is not clear exactly how most pterosaurs met their doom, but there are one or two cases where the cause of death can be pinpointed. Returning to *Pterodactylus*, among the more than 100 specimens already found, there is one pigeon-sized adult individual with a wing-finger bone that is clearly snapped in two: a major accident that, irrespective of its true cause, perhaps a storm, was undoubtedly fatal.[3] The discovery of the victim, preserved in the Upper Jurassic Solnhofen Limestones, a thick sequence of limey muds that accumulated at the bottom of lagoons about 148 million years ago in the region of what is now Bavaria[4] and is still the world's single most important source of pterosaur fossils, certainly supports this idea, because these sediments are considered, by some, to have been generated as a result of powerful storms. If this is true, it might be that many of the more than 1,000 individual pterosaurs thought to have been recovered from this deposit,[5] including the hatchling *Pterodactylus* just mentioned, perished in similar deadly events.

Perhaps the most spectacular example of a pterosaur for which cause of death seems clear is the so-called "tree-biter," a large pterodactyloid from Lower Cretaceous rocks of the Araripe Plateau in Brazil. Described by Dino Frey and colleagues from the Staatliches Museum für Naturkunde in Karlsruhe, Germany, and named *Ludodactylus*,[6] one of the several surprising features of this pterosaur is a large spike-like leaf, similar to those of the yucca plant, wedged between its mandibles (see *Figure 3.3*). Attractive as the idea is, this accident probably did not happen as a result of *Ludodactylus* flying into, or even attacking, a tree, but most probably occurred while this pterosaur was fishing. *Ludodactylus* had large teeth at the front of the jaws that formed a grab-like structure that it used to snatch up its prey as it flew low over the water surface. It seems that on this occasion this particular pterosaur may have mistaken the leaf for a fish and, after it snapped it up, the point speared through its throat sac[7] and became stuck between the mandible and the tongue. Hindered by this encumbrance and unable to close its jaws or feed properly, the pterosaur must have slowly starved to death. As Dino Frey and colleagues point out, the same accidents occasionally befall pelicans today: Victims may change, but death, even in its strangest of forms, is always waiting.

Not all pterosaurs seem to have died alone, though. There are several fossil localities in Argentina, Mongolia and China, for example, where large numbers of pterosaur bones, and sometimes whole skeletons, have been found preserved together in just a single or several closely spaced rock layers. Although this cannot yet be demonstrated with any certainty, it is possible that these accumulations reflect the results of natural disasters such as droughts, volcanic eruptions, hurricanes or perhaps even long-term events such as major changes in weather patterns. Such catastrophes may have been the direct cause of mass mortalities among pterosaurs. Or, more subtly, they may have led to temporary breakdowns in the food web, wreaking havoc on animals such as pterosaurs that were near the top of the pile, in much the same fashion that aberrant weather conditions such as El Niño devastate bird populations today.[8]

FIGURE 3.3 "The tree-biter" *Ludodactylus*, from the Lower Cretaceous Crato Limestone Formation of Brazil, and its deadly cargo, a yucca leaf lodged between its mandibles. Unable to close its beak fully, or dislodge the leaf by rubbing it against the ground (resulting in its frayed end), the pterosaur either died of starvation, or an illness or accident, brought on by its half starved state. The main part of the skull of this pterosaur was almost 19 inches (a half meter) long. (Photograph courtesy of Dino Frey.)

Burial Irrespective of how pterosaurs died, the chances of them becoming fossilized were vanishingly small—on the order of winning the main lottery prize twice in the same month. The reason is that in the Mesozoic, as they do today, after death, almost all organisms, including pterosaurs, immediately began to decay and were broken down or devoured by bacteria, scavengers or even predators long before they had any chance of becoming entombed in sediment. Then, as now, this was generally a good thing, because it ensured that many elements vital for life, such as carbon, nitrogen, potassium and phosphorus, were recycled. It also saved the world from being buried beneath an ever-deepening layer of insect corpses interspersed with the odd dead dinosaur.

The secret to immortality through fossilization is to make sure that after death, one's carcass is buried as rapidly as possible in a place where conditions are so extreme that they prevent living organisms from reaching the carcass and breaking it down, or even accidentally dismembering it, merely by ploughing through the sediment on which it sits or in which it was buried. For the lucky few who get that far, the next step is to ensure that the key process of fossilization, the replacement or replication of organic tissues by minerals, actually takes place. Then, if not already buried, the body must be interred in sediment that, over the millennia, slowly becomes rock. Barring the odd geological accident, such as disturbances by movements of the land masses or volcanism, or the exposure and destruction of the rock at Earth's surface, the enclosed and protected fossil should last almost indefinitely.

This then, very briefly, is the typical path of fossilization along which pterosaurs and all other fossils traveled. Now, we need to take a closer look at this sequence of events to see exactly how they led to the different kinds of pterosaur fossils that we have today: some flattened, some not, a few with soft parts, the vast majority with only their bones and teeth.

As we have seen, the best spots for getting fossilized should, if possible, have a complete absence of living creatures of any kind and, preferably, a plentiful supply of sediment, the finer-grained, the better. It also helps if the water is still, or nearly so, because strong currents can damage the cadaver or wash it away altogether. Such fossil "traps" are not that common, but they do exist. The bottom of stagnant lakes, or very salty lagoons and even shallow land-locked seas are perhaps obvious examples, but "events" such as underwater mud flows, volcanic eruptions, or even a sediment-laden river in flood could also do the job, although there is a much greater risk that the carcass will be damaged or destroyed. All these and many other kinds of fossil "traps" also existed in the Mesozoic. The problem was getting dead pterosaurs into them. Fortunately (at least from the pterosaurologist's point of view), pterosaurs' main means of locomotion—flight—meant that occasionally they found themselves over such "traps" into which they fell, or were blown, from the air. Indeed, at one locality in Zhejiang, China,[9] they may even have been "downed" by volcanic eruptions. Aside from an aerial delivery, most pterosaurs probably reached fossilization traps by floating in, carried by currents.

The Solnhofen Limestones "trap" that we first met earlier in this chapter provides a good example. Recall that these rocks formed as a result of very fine-grained limey muds settling out at the bottom of lagoons. Conditions on the lagoon floor seem to have been extremely unpleasant, possibly because the stagnant water contained little or no oxygen and had become very salty, and no organism larger than bacteria could live there. Consequently, nothing disturbed the sediments, which thus retained their fine lamination, and any animals that did accidentally wander in did not last long, as the bodies of horseshoe crabs preserved at the end of their tracks (so-called death marches) eloquently show.

So many pterosaur remains have been recovered from the Solnhofen Limestone that it seems reasonable to conclude they must have lived in the vicinity of these lagoons of death, but the speed at which their bodies arrived in these cemeteries seems to have varied. Many, including the broken-winged pterosaur mentioned earlier, may have been killed in storms and sent to the bottom almost immediately—the rapidity of their arrival and entombment reflected in the condition of their skeletons: complete and often undisturbed, as can be seen from the examples in *Figures 1.1* and *1.2*. Other individuals may have floated for days and weeks, buoyed up by their light, air-filled skeletons and shedding odd pieces such as head, wings, legs or even feet, until finally the water-logged carcass, now lacking most of its soft parts, sank to the bottom. Once on the lagoon floor, most, but not all, pterosaurs were quickly buried in limy mud. Some carcasses, such as that shown in *Figure 3.4*, seem to have lain uncovered for months or perhaps even years, slowly decaying,[10] the skeleton becoming increasingly jumbled up by water currents, until everything was buried by the next storm-generated influx of mud.

FIGURE 3.4 Death and decay in the Solnhofen lagoons. Above left: the snapped wing-finger bone, seen in the lower left region of this photograph, and in greater detail in the photograph to the right, must have been almost instantly fatal for this individual of *Pterodactylus* from the Solnhofen Limestone. In another skeleton of a similar sized 20 inch (50 centimeter) wingspan *Pterodactylus* (below) many of the original bones have been dissolved away over the millennia, leaving empty cavities. The jumbled-up nature of the skeleton suggests that the carcass of this individual had been decaying for months or years on the lagoon floor before it was finally buried. (Photographs courtesy of Peter Wellnhofer, above, and Carola Radke, below.)

From Bone to Stone The fine details of the actual process of fossilization, converting organic material to stone, are still not fully understood, but the preservation of bones and teeth—which form 99 percent of the pterosaur fossil record—is fairly straightforward. Pterosaur hard parts (like our own skeletons) were largely composed of the relatively inert mineral apatite (calcium phosphate), and thus were already well on the way to being fossils even before their owner was dead. The main event during fossilization, as these hard parts lay shrouded in sediment, appears to have been an enrichment of their mineral component by the addition of further calcium phosphate or a similar substance, such as calcium carbonate. Both these and other minerals could have crystallized out from the water that percolated through the sediment.

The preservation of soft parts is more complicated and can occur in different ways. Internal soft parts, such as major organs (heart, liver, lungs), the blood system or nerves, were literally soft and decayed and degraded extremely rapidly. Not surprisingly, they are almost unknown in pterosaurs. External soft parts, that is, the skin and its various derivatives, such as "hair," wing membranes and foot webs, all discussed in more detail in Chapter 6, are a lot tougher and, on one or two rare occasions, were preserved, although usually only in small patches.

The most common type of pterosaur soft-part preservation, illustrated for two different species in *Figure 3.5*, consists of impressions. If they survived long enough, patches of skin or wing membranes, for example, could leave indentations (forming a negative image of the imprinting surface) on the sediment that over- or underlay them. Ideally, the sediment should have been extremely fine-grained (i.e., mud) and of the right consistency. The Solnhofen Limestone corresponded exactly to these requirements (for which generations of paleontologists have been eternally thankful) and has yielded numerous examples of impressions with superbly preserved copies of pterosaur wing membranes and other structures, some showing incredibly fine detail, such as thread-like lineations of individual "hairs."

Sometimes, rather than an impression, the actual soft parts themselves are preserved. Usually, in this case, the result is a fine black film that consists of partially decayed organic remains that have reacted with minerals in the surrounding sediments or groundwater to form a complex but relatively inert substance. The effect is rather like a photograph or a painting, essentially two-dimensional but, as in the Karatau pterosaur depicted in *Figure 3.6*, subtle variations in the color and texture of the fossilized soft parts can pick out fine details only fractions of a millimeter in width.

FIGURE 3.5 Above: superb wing impression of the
so-called Zittel wing specimen of *Rhamphorhynchus*
(about 40 inches [1 meter] in wingspan), preserved
in the Upper Jurassic Solnhofen Limestone of
southern Bavaria. Below: *Pterodactylus* (about 16 inches
[40 centimeter] in wingspan) from the same rock
sequence, also with well-preserved wing impressions
partially picked out by the orange-red mineral goethite.
(Photographs courtesy of Peter Wellnhofer.)

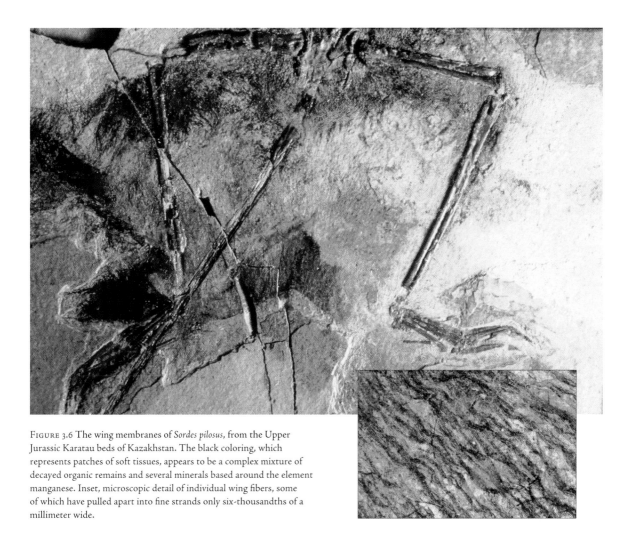

FIGURE 3.6 The wing membranes of *Sordes pilosus*, from the Upper Jurassic Karatau beds of Kazakhstan. The black coloring, which represents patches of soft tissues, appears to be a complex mixture of decayed organic remains and several minerals based around the element manganese. Inset, microscopic detail of individual wing fibers, some of which have pulled apart into fine strands only six-thousandths of a millimeter wide.

Going a step further, there are one or two pterosaurs in which small patches of soft parts were mineralized in their original condition, before any significant decay could take place, where even the three-dimensional details are fossilized. So far, this so-called exceptional preservation has been reported in only a single specimen,[11] part of which is illustrated in *Figure 3.7*, showing one of many pterosaur fossils from the Lower Cretaceous Santana Formation, which crops out around the edges of the Araripe Plateau in Brazil.

Two critical steps fostered the extraordinary preservation seen in this Santana fossil. First, within hours or even minutes of death, the pterosaur cadaver, now sinking toward the bottom of a largely land-locked Early Cretaceous lagoon, encountered a region that was saturated in phosphate.[12] This mineral precipitated out on the bacteria that were by now furiously breaking down the pterosaur soft parts, resulting in a film of mineralized bacteria that replicated even the finest details, such as individual muscle fibers. This explanation was proposed by David Martill, a paleontologist at Portsmouth University in England, and he coined for it the delightfully appropriate term "The Medusa Effect."[13] The second step, ensuring the long-term survival of this apparently instantaneously petrified pterosaur, was the development of a hard stony casing, termed a "concretion," around the pterosaur cadaver as it lay in the sediment on the floor of the lagoon. Critically, this protected the three-dimensionally preserved soft parts from getting crushed or from other damage due to geological perturbations, such as earthquakes.

The processes we have dealt with so far concern only the original tissues, hard and soft, from which pterosaurs were constructed and result in what

FIGURE 3.7 The extraordinarily well-preserved wing membrane of an Early Cretaceous Brazilian pterosaur from the Santana Formation of the Araripe Plateau. Several different layers, including a sheet of muscle fibers (lowermost), are seen in this cross-sectional view of the membrane which, as preserved, is about 1 millimeter in thickness. (Photograph courtesy of David Martill.)

are generally referred to as "body fossils." This is not, however, the only fossil evidence we have of these creatures. Handprints and footprints left by pterosaurs when they walked or ran over a soft surface, such as the mud on a seashore, could ultimately become trace fossils. Normally, of course, such traces were erased by the next tide or the destruction of the track surface by erosion, but occasionally, events conspired to bring about their preservation. Prolonged drying out, followed by sudden burial under another layer of sediment and then further layers, as a sea gradually flooded over a coastal plain, is just one plausible scenario among many, all of which removed this fleeting moment to the depths.

In Pluto's Realm This brings us neatly to the next stage in our fossil journey, a sojourn lasting many millions of years in the geological underworld. Once petrified and buried, one might expect that little else could happen to a pterosaur fossil, snug in its rocky tomb. But even here, it wasn't safe.

Over the millennia, as the weight of overlying sediment built up, the underlying rock layers, in one of which lay our pterosaur fossil, were slowly squashed down and down until, in many cases, they were only one-tenth or less of their original thickness. Three-dimensional parts of the pterosaur skeleton, such as the skull, shoulder girdle and pelvis, and even the individual, hollow, thin-walled bones, were usually quite incapable of resisting such compression and, as a consequence, the vast majority of pterosaurs ended up as picture-fossils, crushed completely flat. The most memorable example that I have encountered consisted of an incomplete skeleton of a small Upper Triassic pterosaur from Austria that had been reduced to a vanishingly thin film of bones probably less than a tenth of a millimeter thick.[14]

Not all pterosaurs suffered this fate, however. Sometimes, the sediments seem to have been compressed at an early stage when they were still very soft, so that the relatively hard skeletons "floated" within them and were not crushed. This appears to have happened in several Solnhofen Limestone pterosaurs. A similar process also seems to have occurred in the Santana Formation, although in this case, the fossils were encased in concretions, which then "floated" in the surrounding sediment.

Another major geological danger to fossils was chemical in origin. In particular, variations in the acidity or alkalinity (pH), or other chemical properties of the groundwater percolating through the sediments could lead to the fossil being dissolved away, leaving mere holes in the rock where the bones formerly lay, as the example in *Figure 3.4* demonstrates. Alternatively, the precipitation of minerals around the fossil skeleton can lead to its encrustation and, in some extreme cases, even its destruction. A common feature of Solnhofen Limestone pterosaurs, for example, is the presence of calcite crystals, which look a bit like granulated sugar, around the ends of limb bones. Sometimes they are so profuse that they completely obscure and even obliterate parts of the skeleton and, because they often merge into the bone itself, they are damned difficult to remove without damaging the fossil.

Escape From the Underworld Having survived the rigors of fossilization and several eons of entombment in rock, our pterosaur fossil now approached one of the most dangerous moments in its journey. In order to be found and collected, a fossil must be on or very near Earth's surface, but as soon as it is exposed, for example, by natural erosion or by quarrying, it and the rock in which it is embedded immediately begin to weather away. If not rescued quickly, the fossil can be lost forever. Hard as it is for a pterosaurologist to cope with, this is what happens to the vast majority of fossils. Let us pause here for a minute's silence, dedicated to the remembrance of all those pterosaurs that, having survived the almost impossible journey to our modern world, were reduced to dust in some remote, ever-windy, Mongolian landscape or, bitterer still, were fed into the maw of a colossal earth-moving machine as it clawed its way along the bottom of an Oxfordshire clay pit.

Surprising as it may seem, most of the pterosaur fossils now housed in museum collections, from Brighton, England, to Beijing, China, were not found as a result of paleontological expeditions or searches specially sent out to look for them, but were accidental discoveries made during other activities, usually quarrying for stone or minerals. The fine, platy Solnhofen Limestone, which splits in a most satisfying way into sheet-like slabs, was, and still is, used both for printing and for building and decorating. Ubiquitous fossils such as *Saccocoma*, a floating crinoid,[15] and much rarer items—crabs, fish, pterosaurs, even *Archaeopteryx*—found while working the stone, were traditionally sold by the quarrymen for "beer money," although the high prices commanded by rarities such as pterosaurs mean that most are now traded for large sums by the quarry owners and fossil dealers. Many other "classic" locations that have yielded pterosaurs, for example, quarries in the *Posidonia* shales at Holzmaden, southeast of Stuttgart, Germany,[16] and strip mines in the Cambridge Greensand around Cambridge, England,[17] also developed for purely commercial reasons.

Some locations that began as stone quarries produced fossils in such quantity and of such value that work eventually switched largely to fossil collection as the main source of income. Two of the most important sources for Lower Cretaceous pterosaurs developed in this way. The Santana Formation of Brazil,[18] whose nodules were originally burnt to produce lime, now generates its wealth and fame by producing thousands of superb fossils. Among these are some of the best pterosaur skeletons ever found and crateloads of fossil fish, occasionally with astonishingly well-preserved soft parts, such as eyes, muscles and guts.

Similarly, rocks in northeast China that belong to the Jehol Group, long exploited by local farmers for stone and originally formed from sediments that accumulated in large freshwater lakes, have recently achieved worldwide fame by yielding huge numbers of exquisitely preserved remains of the animals and plants—the so-called Jehol Biota—that lived in, over and around the lakes in the Early Cretaceous.[19] The fossils range from early flowering plants to complete skeletons of early mammals and, most sensational of all, feathered dinosaurs. These ancient lake beds have also yielded a jaw-dropping array of new pterosaurs, several with fossilized soft parts, including one of the most astounding pterosaurian discoveries of all time—eggs containing embryos. More on these can be found in Chapter 7.

The general rarity of pterosaur fossils, even in rocks of the kind suitable for preserving pterosaurs, means that, usually, they are remarkably difficult to find out in the field, and even highly experienced collectors may make only one or two discoveries in a whole lifetime of work.[20] For this reason, very few scientific expeditions have set out with the specific intention of finding pterosaurs, and those that have went to locations that had already produced at least one or two remains and were thought to have at least some chance of finding more.

One of the most arduous but ultimately successful expeditions ever made to collect pterosaurs set out from Ulanbataar, Mongolia, in the summer of 1981. Spurred by just a handful of peculiar-looking wrist bones from the collections of the Paleontological Institute in Moscow that Natasha Bakhurina, the leader of the expedition, felt sure were pterosaurian despite her colleagues' doubts, a small team of Russian and Mongolian paleontologists made for the remote region of Tatal, five days west of Ulanbataar. Enduring stormy weather, a daily drive of 30 kilometers (nearly 19 miles) to obtain fresh water, a brush with the Black Death, and no means of contacting the outside world,

Bakhurina and her crew (pictured in *Figure 3.8*) eventually found the locality. At first, their searches were fruitless, but then, as they were about to give up in despair and move on to another region, they found a tiny fragment of pterosaur bone, then another and another—and then they hit the jackpot: a rock layer full of bones. Weeks of painstaking and backbreaking work collecting the fragile fossils and transporting them back to civilization were eventually rewarded with a fabulous prize: an almost complete set of skulls, vertebrae and limb bones of a brand new heron-sized, Lower Cretaceous, lake-dwelling, clam-eating dsungaripterid pterosaur.[21]

Released From the Rock Getting the fossil back to the museum or research laboratory is still far from the end of the story. Except for those rare occasions when the sediment is very soft and can be blown or brushed away when first found, the fossil remains of most pterosaurs are deeply embedded in the millennium-hardened sediment within which they were originally interred. Sometimes, a lucky split might have exposed much of the fossil, but usually, it has to be carefully freed (the technical term is "prepared") from the rock that still surrounds it in order to reveal anatomical details and render the fossil suitable for study or display. More often than not, this is a

FIGURE 3.8 Russian paleontologists in the summer of 1981 searching Lower Cretaceous beds in the Tatal region of Western Mongolia for signs of pterosaurs. (Photograph courtesy of Natasha Bakhurina.)

difficult, time-consuming and very labor-intensive process that has to be carried out in specially equipped laboratories.

Most pterosaurs are prepared using extremely sharp needles, usually mounted in a chuck. Dental picks are also very effective, and various power tools that vibrate needles or chisels at high speed or blast away rock using abrasive powders carried in an airstream can help to winkle out details or clear away large areas of stone rapidly. Acid preparation, which involves immersing the whole specimen in a bath of very weak acid for hours or days, followed by prolonged washing in pure water, then painting over exposed bone to protect it from the next acid bath, is slow, but can be very effective. Indeed, it can be too effective. A beautiful pterosaur skull from the Santana Formation prepared in this way was absolutely fabulous to behold, but so fragile that it was almost impossible to handle.

The main aim of preparation, of course, is to expose the fossil as far as possible. In the case of pterosaurs, this generally means revealing as much of the skeleton as one can. Problems arise, however, when soft parts are preserved alongside the bones. Recent investigations of older specimens, mostly from the Solnhofen Limestones, collected and prepared mainly in the 19th century, reveal that fossilized soft parts were much more common than previously realized. Unfortunately, the existence of impressions of wing membranes, skin or claw sheaths often seems to have been overlooked and, as a specimen of *Pterodactylus* in the collections of the American Museum of Natural History in New York dramatically shows, in several cases they were partially or completely destroyed during work to expose the skeleton.[22]

Even when fossilized soft tissues were recognized, they occasionally had to suffer the indignity of being "cleaned up," probably to improve the fossil's appearance so that it could be sold for a higher price. A classic example of this practice can be seen in the so-called "Zittel wing," pictured in *Figure 3.5*, which originally belonged to an individual of *Rhamphorhynchus*, one of the long-tailed pterosaurs from the Solnhofen Limestone.[23] This fossil, collected in the mid-1800s, is famous for having some of the best-preserved impressions of the flight membranes of any pterosaur. In particular, the rear edge of the main flight membrane, the so-called cheiropatagium (see Chapter 8) is remarkably (one might say suspiciously) straight and even—almost certainly not because this is how it was in a living *Rhamphorhynchus*, but because at some point a scalpel or knife blade was run along this edge to tidy it up. Now, sadly, we will never know the exact shape of the main wing membrane in this particular fossil.

Pencils, Paper, and Pixels Much of the basic information that we have for pterosaurs—anatomical details, measurements, drawings—was collected by previous generations of pterosaurologists using just pencils, paper, sharp eyes and, if they could gain access to one, a good binocular microscope. This was more than adequate when the first pterosaur was found 200 years ago and pretty much remained so until quite recently. Nowadays, laptop computers and digital cameras have largely replaced the paper and pencil, but sharp eyes and a microscope are still obligatory.

Even in the brightest daylight, however, you cannot always see everything. Fortunately for pterosaurologists there are other kinds of light; the most useful of these is ultraviolet. Bathe bones or patches of fossilized soft tissues in its skin-burning rays and, if certain minerals are there, the fossil will fluoresce. With the judicious use of color filters and plenty of patience, one can produce high-contrast photographs that show a fluorescing fossil in brilliant detail. Helmut Tischlinger, a pterosaurologist from Stammham in Bavaria, who has spent a lifetime studying, preparing and photographing Solnhofen Limestone fossils, is the master of this technique.[24] Several of his superb photographs adorn this book (for example, the first and last illustrations in Chapter 1) and reveal the incredible sharpness of detail visible in ultraviolet light. Features that normally are barely discernible suddenly leap out at the viewer. In another of Tischlinger's photos (*Figure 8.1*), the magic of ultraviolet light brings forth the delicate tracery of blood vessels in a pterosaur wing membrane.

Those who would like to see even further can turn to several different pieces of modern equipment. Ultrafine anatomical details, such as the minutiae of bone cells and the microstructure of the protective layer of enamel that coated pterosaur teeth, can be viewed with a scanning electron microscope. Another heavy-duty piece of equipment, the CAT scanner, has also been of much aid to pterosaurologists. Larry Witmer and his group from Ohio State University linked up with the CAT scanning team at Austin, Texas, to peer inside the skulls of two different pterosaurs.[25] Using computers to render the internal volume of the braincase as a "solid" three-dimensional image revealed some previously unsuspected details of the brain, not least the remarkably large size of the organs for balance—which as Chapter 8 explains, led to some surprisingly new ideas about how a pterosaur's wings worked.

While scrutinizing fossil remains in any and every possible way is the starting point for investigating pterosaurs, many other techniques and approaches have enabled pterosaurologists to take this understanding further.

There is a long tradition of model making, much of which has been aimed at trying to determine how pterosaurs flew. One of the earliest attempts in this direction was made by Erich von Holst, a renowned German scientist who was passionately interested in animal flight. At the German Paleontological Society meeting in Wilhelmshaven in 1956, he demonstrated a rubberband-powered model of *Rhamphorhynchus* that, according to eyewitness accounts,[26] flew most elegantly. More recently, pterosaurologists have turned, with some success, to testing models in wind tunnels, investigating the behavior of the extraordinary head crest of *Pteranodon* and experimenting with different configurations of the wings.[27]

All the tools we have seen so far, however, are overshadowed by a technological development that has dramatically stepped up the pace of research on pterosaurs (and other fossils), and has reshaped the way we do that work and communicate it to one another: It is also the same tool on which I am doggedly tapping out these words—the desktop computer.[28] Computers provide an extremely effective way of organizing information about pterosaurs, whether it concerns their anatomy, the dimensions of their bones, where they were found, or the age of the rocks from which they were collected. These machines also save huge amounts of time when it comes to analyzing information; in a few microseconds, they can identify and illustrate growth patterns that in the past took weeks to calculate and graph with a pencil and paper. Perhaps even more importantly, computers now permit us to communicate our findings far and wide (and very fast), and to participate in virtual research teams whose members live on different continents and might actually meet each other only once or twice in a lifetime.

Computers have already had a huge impact on pterosaur research, but this is only the start. Try to trace in your mind's eye the exact three-dimensional trajectory of each and every bone in a pterosaur leg as it extends and flexes through a single step. Now try to do this for all four limbs at once—a walking pterosaur. Not easy, is it? But it can be with a computer. Using measurements from a superbly preserved, uncrushed skeleton of a Santana pterosaur and a piece of software downloaded for free from the World Wide Web, Don Henderson, a computer-literate physicist turned paleobiologist now working at the University of Calgary, Canada, developed a computer model that could be used to test pterosaurs' walking ability. How the first virtual pterosaur, which Henderson nicknamed "Robodactylus," fell over, flew apart and eventually performed is revealed in Chapter 9.

Computers have also been busy elsewhere—right at the very heart of 21st century pterosaurology. Modern methods of discovering how species are related to one another—and this goes for all organisms, not just pterosaurs—use a basketful of techniques that are collectively known as phylogenetic systematics.[29] This work is founded on tables of data that consist of tens or even hundreds of species that form the tables' rows, and the hundreds or even thousands of characters—shape of the teeth, number of toes—that vary among these species and that make up the tables' columns. Where each row and column intersect lies a number, a single data point, that tells you exactly what kind of character was present in a particular species— the large crest on the skull of *Pteranodon*, or the spiked prow on the lower jaw of *Rhamphorhynchus*.[30] Each data point provides a tiny clue as to how species were originally related to one another, so phylogenetic systematics takes advantage of this and tries to fit all the thousands of bits of data together in as harmonious a fashion as possible, thereby revealing the pattern of relationships among species.[31] The problem is that even with small tables that have relatively few rows and columns there can be millions or even billions of ways of fitting together the data points, many of them only very slightly less harmonious than the most harmonious solution. It would take humans several lifetimes to do this work by hand, which is where computers come in: the bigger and faster, the better.

Tables of phylogenetic data now exist for pterosaurs, too.[32] They are still modest, with rows of species numbering in the tens and columns of characters at only 100 or so, but they still need a computer to search for the most harmonious combination and reveal how pterosaurs are related to one another. This type of research only really got under way in the 1990s, but most of the main branches of the pterosaur evolutionary tree (we will meet them in the next chapter) have already been sketched out, although, just as for many other groups of organisms, the exact arrangement of many of the twigs continues to be hotly disputed. Still, the importance of this tree cannot be overemphasized: It forms the fundamental framework upon which our understanding of pterosaur evolutionary history is being built, and it also lies behind much of what is written in this book—yet without computers, most of this tree would still be invisible.

On the Record Our journey from living pterosaur, via death, fossilization, burial, discovery, preparation and study, back to a pterosaur that lives, at least in the human intellect, is done. But that still leaves one important question: What, in fact, do we have, in terms of pterosaur fossils? The answer would fill an entire book. That book was written by Peter Wellnhofer and is called *The Illustrated Encyclopedia of Pterosaurs*. All we require here is an up-to-date summary of the pterosaur fossil record, while those who desire a fuller account should turn to the appendix at the back of this book or get a copy of Wellnhofer's *Encyclopedia*.

As you will have already deduced from this chapter, pterosaurs have what can only be described as a modest fossil record, at best.[33] In terms of sheer numbers of individuals in museum collections, common fossils such as ammonites can be counted in the millions, and even groups such as fish number in the tens of thousands. By contrast, only about 5,000 to 6,000 pterosaur individuals have been collected so far (*Figure 3.9*). These fossils range from a few hundred complete skeletons, through all possible combinations of incompleteness, to several thousand single, isolated bones, or even just bone fragments. Unsurprisingly, fossilized soft parts of pterosaurs are rare. To date, they have been reported in just over 100 individuals, are often rather patchy, and, as a rule, are usually associated with relatively complete skeletons. In addition to these body fossils (where original remains are preserved), thousands of footprints and trackways made by pterosaurs wandering along the edges of rivers, lakes and seashores have recently come to light at several sites around the world.[34]

As *Figure 3.9* illustrates, most of the "good" fossils, by which I mean those that are sufficiently complete and well enough preserved to tell us something about the biology of pterosaurs, have been recovered from just a small number of fossil localities. These sources, such as the Solnhofen Limestones of Bavaria, the Jehol Group of northeast China and the Santana Formation of Brazil, are separated by long periods of time and large geographical distances. Thus, while some evidence of pterosaurs has been reported from almost all the 24 stages (those 6-million year or so blocks of time that geologists use to slice up the past) that lie between the oldest records in the Upper Triassic and the youngest at the end of the Cretaceous, the quality of the evidence is very uneven.

	Location	Original environment	Fossil type	Nb.	Principal pterosaurs
Upper Jurassic (−145 to −161)	Solnhofen Limestones, Germany	lagoons		1000+ (F,J,A)	*Scaphognathus, Rhamphorhynchus Pterodactylus, Germanodactylus*
	Villaviciosa, Spain	shoreline		70+	*Pteraichnus*
	Crayssac, France	shoreline		1000's	*Pteraichnus*
	Morrison Formation, USA	floodplain		10+ (A)	*Harpactognathus, Kepodactylus, Mesadactylus*
	Alcova Lake, Wyoming	shoreline		many	*Pteraichnus*
	Karatau Ridge, Kazakhstan	lake		9 (J,A)	*Sordes, Batrachognathus*
	Vinales, Cuba	shallow sea		3 (A)	*Nesodactylus, Cacibupteryx*
Middle Jurassic (−161 to −175)	Canadon Asfalto, Argentina	river/lakes		3+ (A)	Unnamed scaphognathine
	Stonesfield Slate, England	coast line		100+ (J,A)	*Rhamphocephalus*
	Dashanpu, Zigong, China	rivers/lakes		1 (A)	*Angustinaripterus*
	Bakhar, central Mongolia	lake		1 (A)	anurognathid
Lower Jurassic (−175 to −199)	Posidonia Shales, Germany	shallow sea		30+ (J,A)	*Dorygnathus, Campylognathoides*
	Lower Lias, Dorset, England	sea		1 (A)	Unnamed dimorphodontid
	Lower Lias, Dorset, England	sea		30+ (J,A)	*Dimorphodon*
Up. Trias. (−199)	Fleming Ford Fm., Greenland	mudlfats, lakes		1 (J)	*Eudimorphodon*
	Dolomites, N. Italy, Austria	coastal lagoons		8+ (J,A)	*Preondactylus, Peteinosaurus, Eudimorphodon, Austriadactylus*

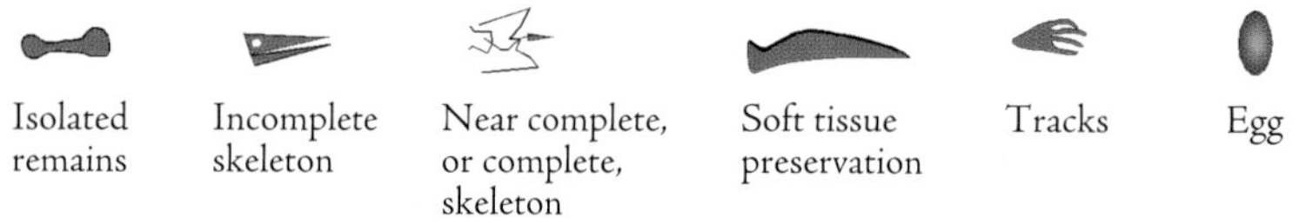

Isolated remains | Incomplete skeleton | Near complete, or complete, skeleton | Soft tissue preservation | Tracks | Egg

FIGURE 3.9 An overview of the fossil record of pterosaurs. (F, flapling; J, juvenile; A, adult)

Location	Original environment	Fossil type	Nb.	Principal pterosaurs
Javelina Fm., Texas, USA	inland plain		10+ (A)	Quetzalcoatlus
Roseifa, Jordan	sea		5+ (A)	Arambourgiania
Dinosaur Park Fm., Canada	plain		10+ (A)	Quetzalcoatlus
Two Medicine Fm., Montana, USA	coastal plain		5+ (A)	Montanazhdarcho, Quetzalcoatlus
Tangshan Formation, Linhai, China	lake adjacent to volcanoes		5 (J,A)	Zhejiangopterus
Haenam, S. Korea	shore line		many	Haenamichnus
Niobrara Chalk, Kansas, USA	continental sea way		1200+ (J,A)	Pteranodon, Nyctosaurus
Muzquiz, Coahuila, Mexico	open sea		1 (A)	Unnamed nyctosaurid
Dzharakhuduk, Uzbekistan	river delta		50+ (F,J,A)	Azhdarcho
Lower Chalk, Southern England	sea		40+ (A)	Coloborhynchus, Lonchodectes
Kem Kem, Morocco	shallow sea		5+ (A)	Coloborhynchus, Tapejaridae
Cambridge Greensand, England	shallow sea		100's (J,A)	Ornithocheirus, Coloborhynchus, Lonchodectes
Toolebuc Formation, Australia	open sea		4+ (A)	Anhanguera
Loma del Pterodaustro, Argentina	lake		100's (F, J, A)	Pterodaustro
Santana Formation, Brazil	inland sea		30+ (J,A)	Ornithocheirus, Cearadactylus Tapejara, Tupuxuara
Crato Limestones, Brazil	lagoon		20+ (J,A)	Tapejara, Ludodactylus, Arthurdactylus
Wealden, Isle of Wight, England	costal lagoons		5+ (A)	Istiodactylus
Jehol Biota, China	lakes		100+ (J,A)	Jeholopterus, Pterorhynchus, Eosipterus, Haopterus, Boreopterus, Sinopterus
Tugulu Group, Xinjiang, China	lakes		25+ (J,A)	Dsungaripterus, Noripterus
Hastings Sand, England	coastal plain		5+ (A)	Coloborhynchus
Tsagaantsav Svita, Tatal, Monglia	shallow lake		45+ (J,A)	Tatal dsungaripterid
Purbeck Limestone England	coast		10+ (A)	Gnathosaurus, Purbeckopus

Left margin (geological time scale):
65 — Upper Cretaceous — 99 — Lower Cretaceous — 145

Pterosaurs are practically unknown from some stages, for example, the Bajocian in the Middle Jurassic, but abundant (relatively speaking) in others, such as the Barremian in the Lower Cretaceous. This unevenness is also encountered in their geographic distribution. As the maps in *Figure 2.2* show, pterosaur body fossils have now been found on every continent, even Antarctica, but the vast majority have emerged from sites in Europe and North America, mainly because that is where most of the effort to collect them has been made so far.

As we saw earlier, most of the pterosaur fossil record consists of petrified bones and teeth. Virtually all of the components of the skeleton are fully known, many of them in fine detail, thanks to several beautifully preserved complete, uncrushed fossil pterosaurs from the Santana Formation, in which even the subtlest features, such as muscle scars and openings for the passage of nerves and blood vessels, can be identified. How skeletal anatomy varies among pterosaurs is also quite well-understood, because skulls and a substantial part of the rest of the skeleton are known for approximately half of the more than 100 species found so far.

The fossil record of pterosaur soft parts is largely confined to relatively tough materials, such as the skin and its various derivatives. Thus, apart from the skin itself, we also have evidence of several structures that were constructed from modified skin—the horny covering of the beak, throat sacs, skull crests, claw sheaths, tail flaps, webs of skin between the toes and wing membranes. Other soft parts, such as stiffening rings in the wind pipe and what might be sections of the gut and patches of muscle, are exceptionally rare, and their exact identity is often disputed. An easily overlooked but important aspect of fossilized soft parts is that they have been found in several different pterosaurs that, collectively, represent most of the main evolutionary lines. This means that not only have we gained some information about pterosaurs' soft parts but, occasionally, it is even possible to see how these structures, such as the wing membranes, varied.

The pterosaur fossil record is not extensive, especially compared with that of most other groups of backboned animals, but with the steady improvement in our knowledge of where to find the fossils and how to prepare them more accurately, it is slowly but surely getting bigger and better. Although scattered in collections all over the world, the fossils that we already have are sufficient for us to be able to begin to establish some fundamental aspects of pterosaur biology. The trick to this has been not to rely on single "Rosetta Stone" specimens, but to combine knowledge from as many specimens and as many different types of fossils—skeletons, soft tissues and tracks—as possible. Thanks to a few pterosaurs who gazed upon Medusa, we now have fossils that reveal how these animals were constructed, how they flew, how they moved on the ground, even how they managed to evolve into flying giants. But, before we begin that part of the story, we first need to meet some of the principal actors.

4
A Tree for Pterosaurs

Shaded from the glaring sun beneath the canopy of several tree ferns, insects whined and hummed in the stifling heat of yet another cloudless Triassic day. A large, metallic green dragonfly, wings whirring, detached itself from the tip of a dead twig and began its patrol back and forth, occasionally darting from its path to grab some slow-flying victim. Returning to its perch, it dismembered its prey, then swooped back into the air—lord of the skies, the biggest thing on the wing. At least, until now. Higher up, above the dragonfly, deep in the shadows of the massive fronds, hung something quite a lot bigger. It had watched the green hunter on its sentinel beat, and now it was unfolding short, broad wings, its legs were tensing, and down it came, in a rush of membranes, a swish of a tail and a lot, such a lot, of needle-sharp teeth. The dragonfly flickered its wings and sped away from the path of this tumbling threat toward the edge of the canopy shade and the safety of the open sky. Except that safety was now full of more membrane-winged fliers, one of which bit the dragonfly in half as the others fanned out in search of more victims. Powered by a rich diet of insects, protopterosaurs had really taken to the air.[1]

FIGURE 4.1 Could this early, sparrow-sized gliding reptile, *Sharovipteryx* from Upper Triassic rocks of the Fergana Valley in Kirgizia, be ancestral to pterosaurs? Probably not, because it is almost the same geological age as early pterosaurs and, with its remarkably long neck, already highly specialized. Still, it might have something important to tell us about how pterosaurs first took to the air.

Naming Names Before plunging into the heart of pterosaur biology, a little familiarity with the family tree is required. Pterosaurologists have toiled long and hard to discover the genealogy of these animals, identifying and naming species, linking them together into their natural clans and tracing out the relationships between one clan and the next, until all are linked together in a complete family tree.

Before such a grand classification scheme could be drawn up, however, came the task of identifying and naming the basic components from which this tree was made—species. According to the generally accepted definition, a species, including the one to which we belong, *Homo sapiens*,[2] consists of a group of individuals (often totaling several million or more) that can successfully interbreed with one another, but not with members of other species, to produce offspring that are themselves capable of reproducing.[3] While, in theory, one could test the members of living species to see if they fit this definition, it is impossible to observe the breeding behavior of fossils, so paleontologists rely on another aspect of extinct species in order to identify them: Their members look more like each other than members of other species.[4] And that, basically, is how pterosaurologists have recognized, defined and named the approximately 100 species of pterosaur discovered so far.[5]

Among pterosaurs (and many other backboned animals), features that best distinguish species—their hallmarks—are generally to be found in the anatomically most complicated part of the body: the skull. The basic design of this structure and, in toothed pterosaurs, details of the teeth, such as their shape, relative size, spacing and arrangement, are usually sufficient to distinguish species at a glance. Features of the limb bones, especially their relative lengths, can also be helpful, although they tend to be typical of larger clans (genera and families, for example) rather than particular species.

As each new fossil comes to light, it is compared with those found previously to see whether it belongs to an existing species of pterosaur. Usually it does, but when it doesn't, a new species must be defined, named and described—a process that lies at the core of the science known as taxonomy.[6] Often, the formal scientific names given to pterosaurs refer to distinctive parts of their owners' anatomy. In the case of *Pteranodon longiceps*, first proposed in 1876 by Professor Othniel C. Marsh of Yale University for a pterosaur that had been found in the Upper Cretaceous chalk bluffs of Kansas, the genus name *Pteranodon*, meaning winged and toothless, is remarkably apt, while the specific epithet *longiceps* perfectly describes the extraordinary development of the jaws.[7]

Many pterosaur names contain a reference to the location where the fossil on which they are founded was first discovered. Dsungaria and Zhejiang, both in China, gave us *Dsungaripterus* and *Zhejiangopterus*. These names also illustrate another tradition of pterosaur taxonomy, the inclusion, in modified form, of the Greek term "pteron," which means wing, alluding to the aerial mode of life of these animals. My favorite tradition with regard to the concoction of pterosaur names is the references made to dragons, spirits and gods. Take, for instance, *Azhdarcho*, from the Upper Cretaceous of middle Asia, a memorable moniker that stems from the Uzbek word for dragon, or *Tapejara*, an ancient Brazilian spirit, and perhaps the most evocative pterosaur name of all, *Quetzalcoatlus*, derived from the Mexican deity Quetzalcoatl, the plumed serpent. (A complete list of all valid pterosaur names can be found at the back of this book.)

Pterosaur taxonomy never sleeps. As new, more complete, or better-preserved fossils are found, they are compared with previously named species, while, in turn, these "older" species are reassessed in light of the new finds. Sometimes, one or more fossils may be split away from their original species to form a new species. Much more commonly, taxonomists will take several supposedly different species and lump them together under a single name.[8] The difficulty with pterosaurs, as indeed with most animals and plants, is that while members of a species are supposed to look like one another, they may, in fact, appear quite different. This can happen for several reasons. Natural variation, especially in size, is common in adult reptiles, and pterosaurs are no exception. Differences between the sexes can also be striking, especially when one or the other is ornately decorated, as in many birds, for example, most spectacularly, the peacock. Age can also have a profound impact on appearance. With their relatively large eyes and short limbs, youngsters may look quite different from their parents and, as has happened on several occasions for pterosaurs, if their immaturity is not recognized, they could be misidentified as members of a "small" species.

This problem of variability has long plagued pterosaur taxonomy and, as recent studies have shown, youngsters and adults, males and females have often found themselves in completely different species. Painstaking taxonomic work has helped to reunite many of these strays with other members of their own species. As *Figure 4.2* illustrates, what for many years were thought to be five distinct species of *Rhamphorhynchus* have now been recognized by the American pterosaurologist Chris Bennett,[9] based at Fort Hays University in Kansas, as five stages in the growth of just a single species: *Rhamphorhynchus*

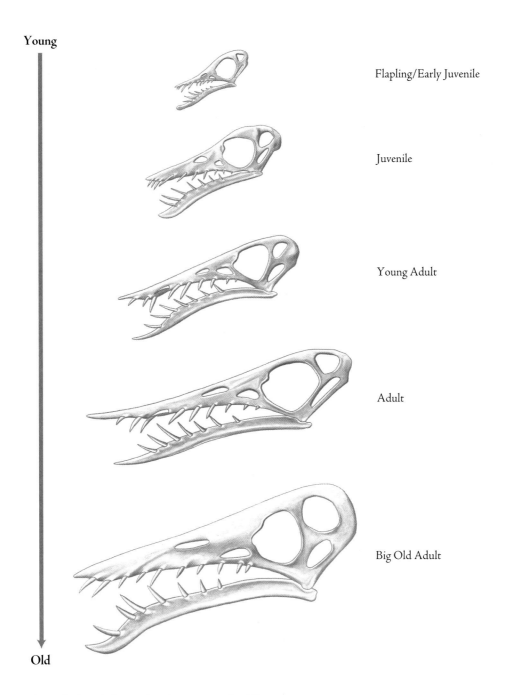

Young

Flapling/Early Juvenile

Juvenile

Young Adult

Adult

Big Old Adult

Old

FIGURE 4.2 Until recently, these skulls were thought to represent five different species of the long-tailed Late Jurassic Solnhofen Limestone pterosaur *Rhamphorhynchus*. Work by Chris Bennett (1995) has demonstrated that they are in fact just different growth stages of a single species: *Rhamphorhynchus muensteri*. (Redrawn from Bennett, 1995.)

muensteri. Features such as the degree of unification of the several bones contributing to the shoulder or to the pelvis (ranging from completely separate through partially united to fully fused, without even a trace of the suture) that were originally believed to distinguish the various species, are now recognized as natural changes that occurred as individuals grew into adults. Bennett's taxonomic "welfare" work has also been felt elsewhere.[10] Under his aegis, different forms of *Pteranodon*, cruelly divorced from one another for more than a century, are now happily reunited—tall-crested males with short-crested females, in the single species *Pteranodon longiceps*.

Establishing how different species of pterosaur were related to one another was for many years primarily based on their overall degree of similarity and their geological age. Because the Early Jurassic prow-jawed pterosaur *Dorygnathus* appeared to be quite similar to the Late Jurassic prow-jawed *Rhamphorhynchus*, it was not only assumed that they were more closely related to each other than to any other pterosaur, but that *Rhamphorhynchus* was directly descended from *Dorygnathus*.[11]

Modern phylogenetic systematics (which we first met at the end of the previous chapter) largely ignores geological age and employs a more refined technique, whereby rather than utilizing any and every characteristic, be it skull shape, the length of the neck or a detail of the foot, only characteristics unique to particular groups are used to establish genealogy. Moreover, in what was for some researchers a painful break from tradition, species, or the larger groups into which they clustered, were not generally considered to be descended from one another, as you can see in the family tree that appears later in this chapter.

Pterosaurs have recently been subjected to several "phylogenetic analyses"[12] among the results of which is the discovery that while *Dorygnathus* and *Rhamphorhynchus* still belong in the same prow-jawed clan—the rhamphorhynchines—they have been joined by two or three relatives and are no longer thought to be directly related to one another. More importantly for pterosaurologists, the main conclusions of these new phylogenetic studies match well with one another, although discrepancies here and there inevitably continue to fuel debates and squabbles.

That, then, is how pterosaurologists built their pterosaur family tree, but before we set off to explore its various branches and meet some of the inhabitants, an even more general question must be addressed. How are pterosaurs related to other backboned animals? The pterosaur tree is, itself, just a single branch among the many on the great tree of life—but where is that branch upon which pterosaurs sit?

Bird or Bat? Pterosaurs belonged to a large group of vertebrates called the tetrapods—four-footed beasts (*Figure 4.3*). That they were amniotes,[13] a particular group of tetrapods whose members laid eggs with a waterproof membrane, is shown by several tell-tale features of the skeleton, such as a single rounded knob of bone on the back of the skull (the occipital condyle), which connects it with the spinal column. Although their amniote credentials have never been doubted, the recent discovery of several fossilized pterosaur embryos *"in ovum"* in Lower Cretaceous rocks of China and Argentina demonstrates beyond any question that they could produce the ultimate proof of membership of the amniote club: a desiccation-proof egg.[14]

So far, so good, but now it gets more complicated. In the past, pterosaurs were allied, on different occasions, with each of the three main amniote groups: reptiles, birds and mammals. Pterosaurs certainly belonged within one of these, but which?

One of the biggest blunders to be made by a pterosaurologist was perpetrated by Samuel Thomas von Soemmering, professor of anatomy and surgery at the University of Munich in the early 1800s. Soemmering mistakenly decided that one of the first pterosaurs to be found, a young *Pterodactylus* from the Solnhofen Limestone, was some kind of bat[15] and thereby landed pterosaurs among the mammals. Compounding the problem, this error was enthusiastically embraced in some quarters, the English zoologist Edward Newman even going so far as to depict pterosaurs as marsupial bats,[16] adorned with fur and sporting a pair of large and rather cute-looking ears. Soemmering was quite wrong, however, as his French contemporary, Baron Georges Cuvier, the father of comparative anatomy, showed in the pages of his 10-volume magnum opus *Recherches sur les Ossemens Fossiles*.[17] Cuvier pointed out numerous features that disqualified pterosaurs from any possible kinship with mammals, including their simple, single-crowned teeth, quite unlike our multi-cusped molars, and the presence of a distinctive "quadrate" bone in the skull, upon which the lower jaw hinged, which is typical of reptiles, but reduced in modern mammals to a tiny element in the ear.[18]

Another serious misunderstanding of pterosaurs, with ramifications that are still being felt today, was made by Harry Govier Seeley,[19] the author of *Dragons of the Air*. At an early stage in his career, Seeley became quite convinced that pterosaurs were the ancestors of birds, an idea that, according to the English naturalist Richard Lydekker in his review of *Dragons*, had impressed itself with "peculiar force" upon Seeley's mind.[20] Later, as this notion drew increasingly sharp criticism, he shifted, albeit reluctantly, to a slightly

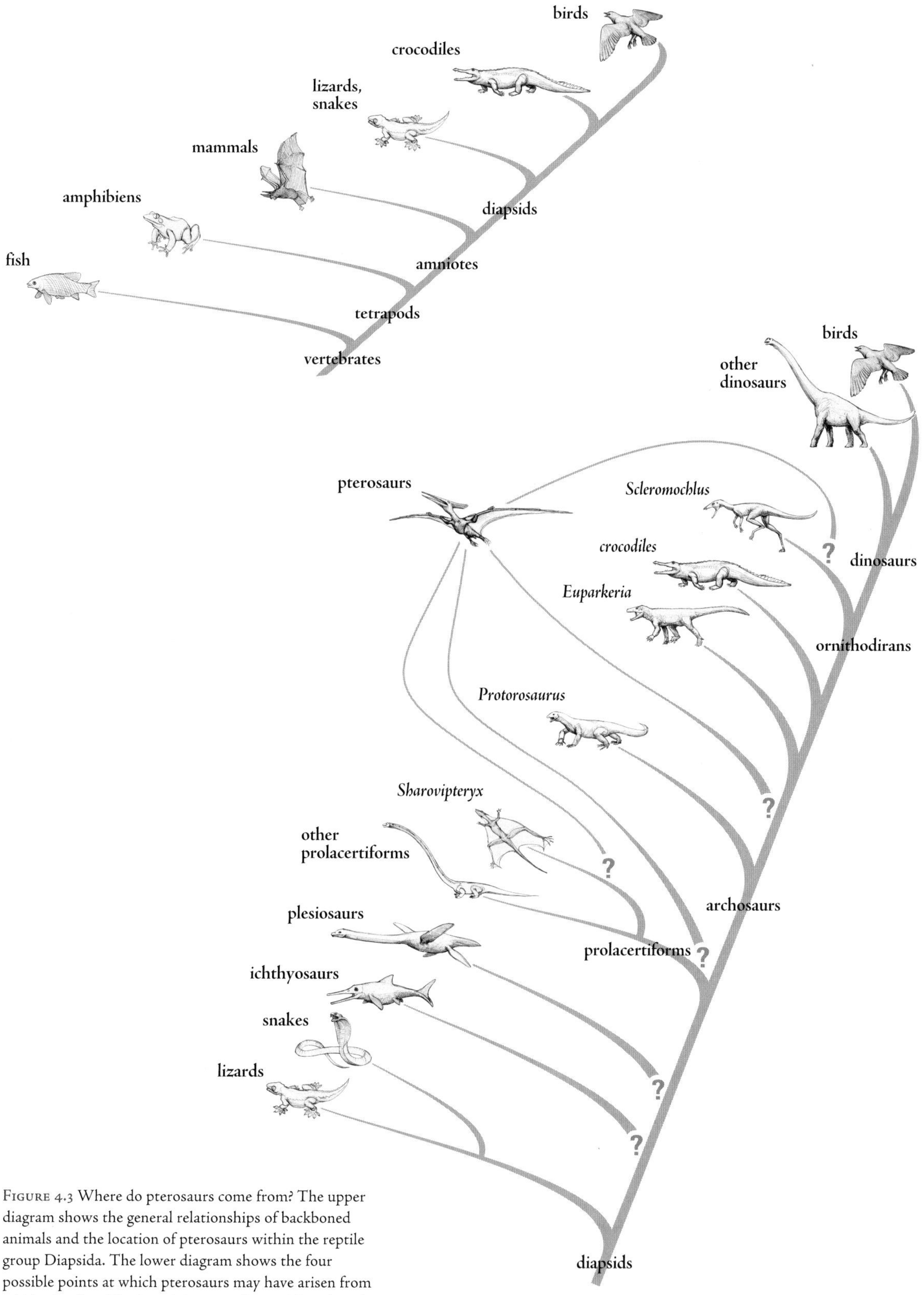

FIGURE 4.3 Where do pterosaurs come from? The upper diagram shows the general relationships of backboned animals and the location of pterosaurs within the reptile group Diapsida. The lower diagram shows the four possible points at which pterosaurs may have arisen from within the diapsids. An origin somewhere within the split between archosauriforms and prolacertiforms seems most likely at present.

different position, in which birds shared a common origin with pterosaurs, rather than being directly descended from them. One of the stratagems that Seeley used to bolster his arguments was to make reference to birds in the many new names that he coined for pterosaurs. Thus, we have *Ornithocheirus* ("bird hand"), *Ornithostoma* ("bird mouth"), and *Ornithodesmus* ("bird link")[21]—and the crowning glory, "Ornithosauria," which he proposed as a replacement for Pterosauria. Despite his inventiveness, Seeley's big idea ultimately failed. His contemporaries, who included such illustrious scientists as Richard Owen, the British version of Cuvier, highlighted the many problems associated with this hypothesis. Pterosaurs, they said, have numerous features in the construction of the skull and design of the hands and feet that resemble the condition in reptiles, but are completely different from those of birds, ruling them out from any close relationship to these feathered fliers.

Reptiles Then, But Which? Cuvier was the first scientist to recognize pterosaurs for what they were—flying reptiles—but it wasn't until the beginning of the 20th century that their reptilian affinities were universally agreed upon. The general relationships of pterosaurs to other reptiles were also established at about this time. The presence just behind the eye of two openings, one above the other, shows beyond any doubt that pterosaurs were diapsids,[22] one of the most diverse and prominent of all the amniote groups.

The diapsid line first appeared more than 300 million years ago and subsequently sprouted many important branches, which, apart from pterosaurs, also culminated in lizards, snakes, ichthyosaurs, plesiosaurs, crocodiles, dinosaurs and birds.[23] The relationships of these different kinds of diapsids to one another are now fairly well understood, with one glaring exception: pterosaurs. As *Figure 4.3* illustrates, paleontologists don't really know where this group should sit within the diapsid family tree.

The reason for this confusion is simple—a complete lack of protopterosaurs that might link this group to other diapsids. Birds, by contrast, have an almost perfect intermediate—*Archaeopteryx*—that, with its mosaic of avian and reptilian characteristics, unites them in a most convincing fashion with their reptilian relatives, theropod dinosaurs such as *Velociraptor*. Pterosaurs have no equivalent of *Archaeopteryx* and so, at present, sit in splendid isolation, definitely related to, but somehow remote from, other diapsids. Even the earliest, most basic pterosaurs, such as *Dimorphodon*, which we will meet again later, are pterosaurian through and through, and little of what's known of their anatomy seems to have been left untouched by adaptations for flight, which seem to have had a profound and almost universal impact on their anatomy. This means that many of the features used by paleontologists to reconstruct the genealogy of diapsids, among them skull shape, ankle construction and numbers of fingers and toes, are so altered in pterosaurs that any messages they may contain regarding the origins of these animals are difficult to decipher.

Undaunted by this obstacle, specialists have come up with several different suggestions as to where pterosaurs might be lodged in the diapsid tree. The three main proposals, shown in *Figure 4.3*, are: perched next to the dinosaurs; squatting among dinosaur's relatives, the archosauriforms; or hanging out with another group altogether, the prolacertiforms.

Dinosaurian Bedfellows? The most popular current notion regarding pterosaur origins is that they were close relatives of dinosaurs. Several features, mostly to be found in the legs of pterosaurs, dinosaurs, birds and a few other reptiles, such as *Scleromochlus* (*Figure 4.4*), seem to support this idea and supposedly define a clan that has been named Ornithodira, which means literally "bird-like ankles." With the exception of sauropods and some other dinosaur groups that returned to all fours, perhaps because of their extremely large size, the hallmark of this clan was the ability of its members to move around on their back legs alone, skipping along on the tips of their toes as living ornithodirans—which we call birds—do today. This arrangement—which is quite different from that found in reptiles, where the legs sprawl out sideways as, for example, in lizards and crocodiles—required numerous modifications, such as an in-turning of the head of the thigh bone that allows the legs to be tucked in beneath the body, a relatively long shin and a simplification of the ankle joint—features that are, to some extent, found in pterosaurs.

Scleromochlus, a small reptile from Upper Triassic rocks of Scotland,[24] is especially important, because some scientists have suggested that it is the closest known relative of pterosaurs and could even be considered a pterosaurian *Archaeopteryx*.[25] But there are difficulties with this suggestion and with the more general idea that pterosaurs are ornithodirans. Significantly, the broad, shield-like pelvis of pterosaurs is quite different from that of dinosaurs or *Scleromochlus*, where it is constructed from bony spars that radiate forward and backward. Moreover, the design of the pterosaur foot, not

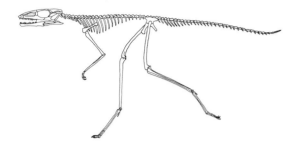

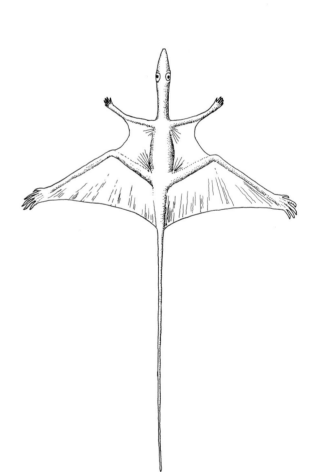

FIGURE 4.4 Meet the relatives? All these
diapsids—*Scleromochlus*, an ornithodiran about 8
inches (20 centimeters) long (above); *Euparkeria*,
an archosauriform (middle) about 20 inches
(50 centimeters) long; and the 10 inch (25 centimeter)
long *Sharovipteryx*, a prolacertiform (below)—have
been proposed as close relatives of pterosaurs.

to mention thousands of tracks (see Chapter 9), show that pterosaurs did not walk or run on their toes, but stamped along in a decidedly flat-footed fashion. So, what about the ornithodiran aspects of pterosaur legs? Nothing to do with ornithodirans at all, some have argued, just features that look superficially similar to those of *Scleromochlus* and its relatives, but that evolved for a quite different reason—for use in the flight apparatus.[26]

Further Down the Diapsid Tree? A second possibility is that pterosaurs branched off somewhat lower in the diapsid tree, from among several early lineages of an important diapsid group—the archosauriforms. As *Figure 4.3* shows, after lizards and snakes and several other groups, such as the sea-living ichthyosaurs and plesiosaurs, had separated off to follow their own evolutionary destiny, the main diapsid stem split into two major lines. They evolved in one case into prolacertiforms, which we shall return to shortly, and in the other into archosauriforms, a lineage of land-living carnivores that eventually gave rise to crocodiles, dinosaurs and birds.

Traditionally, pterosaurs were thought to have originated from somewhere near the base of the latter group, perhaps not far from *Euparkeria*, a medium-size, rather generalized archosauriform.[27] An in-depth study by Chris Bennett in the mid-1990s arrived at this conclusion and is supported by some archosauriform features found in pterosaurs, such as the presence of an antorbital fenestra—a bone-bounded window that pierced the skull between the openings for the nostril and the eye.

Again, however, there are difficulties with this idea, not least because if it is true, then, as *Figure 4.4* shows, pterosaurs' nearest known relatives were rather large, heavily built, superficially crocodile-like animals with short arms. Other features also speak against this relationship. Another opening, this time in the lower jaw and called the mandibular fenestra, is a particularly telling example. Most archosauriforms have this opening, but pterosaurs do not. This is surprising, because if pterosaurs are archosauriforms, their ancestors must have possessed this opening only for it to have been refilled with bone just as they were evolving into pterosaurs, even though at this point in their evolutionary history, they were also developing a flight ability and, judging by what's known of their skeletal design, losing weight wherever possible.

On Sharov's Wing? A third option is that pterosaurs originated from within another quite different diapsid clan: the prolacertiforms. These generally small, rather lizard-like reptiles[28] evolved along several different lines, one of which culminated in *Tanystropheus*, a bizarre-looking animal that had an incredibly long neck that looks well-suited for cleaning out drains, but is more likely to have been used as a means for reaching its fishy prey.[29] One of *Tanystropheus'* relatives, *Sharovipteryx*, a small, very lightly built, long-legged reptile from the Upper Triassic of Kirghizia,[30] illustrated in *Figures 4.1* and *4.4*, is of special interest because in some respects it really does fit the bill as a pterosaurian ancestor—not least because it seems that it could fly.

Pterosaurs and *Sharovipteryx* share several unusual features, such as hollow bones and a shin that was longer than the thigh. More importantly, however, *Sharovipteryx* has superbly preserved impressions of flight membranes that evidently attached to the back of each leg and ran out along the fifth toe—exactly as in pterosaurs. As if that were not enough, details of the wings of *Sharovipteryx*, shown in *Figure 4.4*, reveal a very distinctive pattern—numerous, closely packed fine lines running out toward the edges of the membranes. Among all backboned animals, this kind of lineation has only been found in pterosaurs, where, as detailed in Chapter 8, it appears as the external manifestation of an internal structure—long, thin fibers that helped stiffen the flight membranes.

Is this it, then? Is *Sharovipteryx* the pterosaurian *Archaeopteryx*? Well, probably not. The difficulty with *Sharovipteryx* is that while it does share some features in common with pterosaurs, it also has many characteristics that are quite unlike anything found in these animals. There is no antorbital opening between the nostril and the eye, the neck is extraordinarily long, unlike the short neck of early pterosaurs and, according to the latest studies of *Sharovipteryx*, the arms are very short and small,[31] exactly the opposite of the condition in pterosaurs. This Middle Asian Triassic aeronaut appears to be more closely related to other prolacertiforms than to pterosaurs, leading to the conclusion that the features of the skeleton and wings that they share in common must have evolved independently in response to the stringent demands of flight. Indeed, pterosaurs lack almost all the anatomical hallmarks of prolacertiforms and probably do not belong within this group at all.[32]

Outcasts? Pterosaurs certainly belong somewhere in the diapsid family tree, yet, at present, they do not sit comfortably in any of the positions on offer. "Does it really matter?" one might ask. The answer is, "yes, it does." If we had some idea of how pterosaurs were related to other diapsids, we could at least begin to understand what their ancestors looked like, how they evolved their most characteristic feature—wings—and under what circumstances this might have taken place.

We will return to this issue at the start of the "Pterosaur Story" (see Chapter 10), but for now, the best accommodation for these outcasts is a temporary dwelling in a cleft in the tree between the archosauriforms and the prolacertiforms (see *Figure 4.3*). It is not ideal, but it is the least uncomfortable fit for all the anatomical features that we have met so far. At this location, one would expect to find diapsids with an antorbital opening but no mandibular fenestra, along with hollow limb bones, a long shin, and a simplified ankle construction. All these particulars are, at least, found in pterosaurs, but I should point out that they form just a tiny part of the large and still expanding mass of data being used by paleontologists to map out the relationships of diapsids. Where pterosaurs will finally come to roost in this family tree is still quite unclear.

Pterosaur, or Not Pterosaur? Before we get to grips with the pterosaur family tree, we must briefly consider another important issue: How do we decide who actually belongs in the clan Pterosauria, to give it its formal name, and who does not? Happily, on this occasion, pterosaurs' highly distinctive anatomy and the large gap between them and other diapsids is very helpful. It means that usually it is quite easy to determine whether a fossil is pterosaurian.

Among the characteristics that proclaim a fossil to be pterosaurian, one of the most useful is the remarkable thinness of their bone walls. Typically, they are only 1 or 2 millimeters thick, even in giant species. This immediately distinguishes pterosaurs from other animals, even those with hollow bones, such as birds and some dinosaurs, where the walls are almost invariably thicker. Remarkable as it may seem, using this feature, even badly preserved, isolated bones that have been through so much that all their articular ends have broken off and only a piece of shaft remains can still be confidently identified as pterosaurian.

Other unique features, often sufficient to tell at a glance that one is dealing with a pterosaur, include the enormously enlarged fourth finger of the
hand—the wing-finger—composed of long, spar-like bones; the highly distinctive design of the wrist, which contains the pteroid, a rod-like element
only found in pterosaurs; and the peculiar design of the foot, which, in many
species, sports a long, clawless, fifth toe quite unlike anything found in other
reptiles, living or extinct.

Despite these and numerous other distinctive skeletal details, discussed
more fully in the next two chapters, pterosaurs have sometimes been mistaken for other animals, usually birds[33]—or other animals, usually birds,
have been mistaken for pterosaurs. The most notorious example concerns
a character we have met before, Richard Owen, one of his bitterest rivals,
Gideon Mantell,[34] and several small, delicate bones purchased by Mantell
in the early 1800s from quarrymen working the Lower Cretaceous rocks
of Tilgate Forest in Sussex, England. Mantell, a doctor from Lewes who
became famous for making some of the earliest discoveries of dinosaurs,
was sure that these bones belonged to birds, an idea that was supported by
Georges Cuvier and, initially, by Owen. Later, however, Owen changed his
mind and, without telling anyone, prepared a scientific paper that he sprung
on his contemporaries, including an outraged Mantell, in which he opined
that all the "Wealden 'birds' are pterodactylian."[35] Relations between these
two scientists reached a new low, with Mantell confiding to his diary, "It is
deeply to be deplored that this eminent and highly gifted man, can never act
with candor or liberality."[36] But, as is often the case, Owen was right, and
Mantell's birds proved to be pterosaurs after all.[37]

A reverse example, wherein a supposed pterosaur proved to be something
else, fell to my own experience in the early 1990s. During a visit to Beijing, I
was asked to inspect a small, headless, rather jumbled-up skeleton about the
size of that of a thrush, which had been collected from Lower Cretaceous
rocks in Inner Mongolia, China. Although identified as a pterosaur, it just
did not seem to have enough wing-finger bones, a problem that disappeared
when I came to the realization that it was a bird and not a pterosaur at all.[38]
To my chagrin, this new identity seemed to spark much more interest in the
fossil than had been the case when it was still a mere pterosaur.

Up the Tree The general layout of the pterosaur family tree, illustrated in *Figure 4.5*, has eight main branches. The arrangement of these branches and the species that belong to them, listed at the end of this book, are generally agreed upon by pterosaurologists, although, inevitably, disputes about some of the fine details remain. The lowermost four branches, consisting almost entirely of long-tailed pterosaurs, are traditionally grouped together as the rhamphorhynchoids, a name derived from one of the best-known members of this group—*Rhamphorhynchus*. The uppermost four main branches, forming the crown of the tree, are referred to collectively as the pterodactyloids, short-tailed pterosaurs whose common ancestry is signposted by a row of unique features, not least, their relatively short tails.

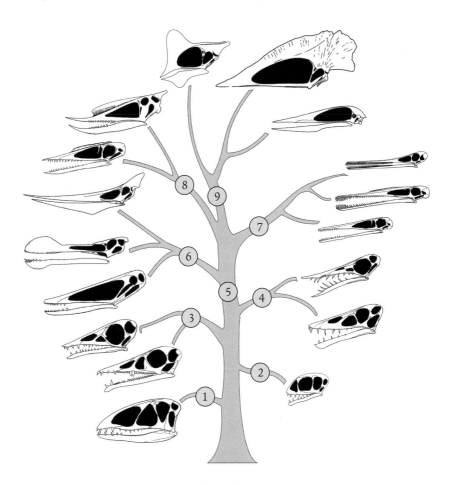

FIGURE 4.5 The pterosaur family tree, showing the main branches and some of the principal players. Clockwise, from left, with size given in centimeters (cm): *Dimorphodon* (22cm) *Eudimorphodon* (9cm), *Campyognathoides* (13cm), *Istiodactylus* (about 56cm), *Ornithocheirus* (67cm), *Pteranodon* (100cm), *Germanodactylus* (13cm), *Dsungaripterus* (41cm), *Tapejara* (20cm), *Tupuxuara* (85cm), *Zhejiangopterus* (29cm), *Ctenochasma* (10cm), *Gnathosaurus* (28cm), *Pterodactylus* (9cm), *Rhamphorhynchus* (10cm), *Scaphognathus* (12cm), *Anurognathus* (3.2cm). Branch names are as follows: 1, Dimorphodontidae; 2, Anurognathidae; 3, Campylognathoididae; 4, Rhamphorhynchidae; 5, Pterodactyloidea; 6, Ornithocheiroidea; 7, Ctenochasmatoidea; 8, Dsungaripteroidea; 9, Azhdarchoidea.

Dimorphodon and the Short-Winged Big-Heads Starting at the base of the tree, the first branch encountered belongs to *Dimorphodon* and its relatives (*Figure 4.6*). As might be expected from their basement location, dimorphodontids,[39] with their large, reptilian-looking skulls and relatively broad wings, are the least evolved of all pterosaurs known at present. Often reaching well over a meter in wingspan, the most distinctive feature of members of this clan is summarized in their name—dimorphodontids. This refers to the two strikingly different types of teeth found in their jaws: a few large fangs at the front (for grabbing the prey); and a row of tiny spikes at the back (for holding onto it afterward). Another instantly recognizable feature that distinguishes dimorphodontids from all other pterosaurs is the big, deep skull with its large openings framed by long, slender bars of bone. This, as one pterosaurologist has pointed out,[40] follows the design principles of a bicycle, combining the minimum amount of material with the maximum amount of effect.

Dimorphodon, from Lower Jurassic rocks of southern England, is easily the best-known dimorphodontid and is represented by several skeletons, the first of which was collected in the 1820s by one of paleontology's greatest legends, Mary Anning. Based in Lyme Regis, Anning, the inspiration for the nursery

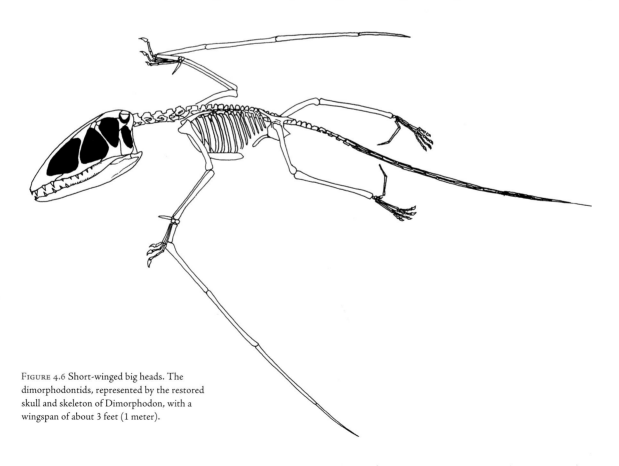

FIGURE 4.6 Short-winged big heads. The dimorphodontids, represented by the restored skull and skeleton of Dimorphodon, with a wingspan of about 3 feet (1 meter).

rhyme "she sells sea shells on the seashore," was perhaps the most famous and important British fossil collector of the early 19th century. Apart from *Dimorphodon*, she made many other astounding finds, including some of the first fossil remains of the so-called sea dragons, ichthyosaurs and plesiosaurs.

The only other dimorphodontid known at present, *Peteinosaurus*, is represented by just two incomplete skeletons that number among some of the oldest pterosaur fossils in the world, having been found in 220-million-year-old Triassic limestones that now form part of the beautiful and dramatic landscape of the Dolomites in northern Italy.[41] These same rocks have produced other fossil reptiles, among which are poorly preserved remains of another Triassic pterosaur called *Preondactylus*.[42] This small, rather enigmatic form might belong on a twig that branched off even earlier than the dimorphodontids, because in some respects, such as the construction of the lower jaw and the tail, *Preondactylus* is more similar to other diapsids than any other pterosaur found so far.

Anurognathus *and the Fabulous Flying Frog-Heads* Returning to the main trunk of pterosaur evolution, after dimorphodontids had branched off along their own path the principal development seems to have been linked to flight ability. The forelimbs became much longer, mainly through the lengthening of bones in the forearm and wing-finger, and the hind limbs were lightened and streamlined, mainly by the reduction of the outer bone of the lower leg (fibula) to a thin, slender, splint-like element. The first clan in which these features are to be seen also happens to be one of the most extraordinary and enigmatic of all pterosaur groups—the anurognathids, named for their remarkably bulbous and frog-like heads.

The first evidence of anurognathids, a "road kill with exploded head" was recovered from the Solnhofen Limestones in the 1920s and for many years was practically the only evidence of this particular clan. Another superb example of this pterosaur, *Anurognathus*, which came to light quite recently and is illustrated in *Figure 1.1*, consists of an entire skeleton laid out in a remarkably lifelike pose, with the added bonus of fossilized wing membranes. Apart from *Anurognathus*, several incomplete, rather crushed remains of anurognathids have been found in ancient lake sediments deposited in Kazakhstan in the Upper Jurassic and northeast China in the Lower Cretaceous.[43] One of these, a Chinese pterosaur called *Jeholopterus*, shown in *Figure 8.4*, has astonishingly complete and well-preserved wing membranes and even some patches of skin, replete with a covering of "hair."[44]

Anurognathids were small pterosaurs, usually only half a meter (20 inches) or less in wingspan, with a short, compact body and relatively broad wings. The most distinctive feature was the skull: Short, broad and deep, and perforated by large openings (*Figure 4.7*), it looks so frog-like that one anurognathid, *Batrachognathus*, literally "amphibian jaw," was named for this similarity. The teeth are also quite peculiar: rare, widely spaced, and resembling sharp little spikes.

Another feature that sets anurognathids off from all other rhamphorhynchoids is the reduction of the tail to a short little stub just like that found in pterodactyloids. Is this a hint that the two groups were related? Earlier it was thought that anurognathids, with their swollen heads, might have descended directly from dimorphodontids, which have similar, if rather more elongated, swollen heads, but this is now considered unlikely.[45] At the same time, anurognathids do not, at least superficially, look very much like pterodactyloids. Yet, in addition to the short tail, they share several other special features in common with this group, such as the loss of the neck ribs and a reduction in the number of vertebrae in the main part of the spinal column. It is doubtful that pterodactyloids descended directly from anurognathids, but the two clans might have shared a common ancestor.

Eudimorphodon—*First of the Long Snouts*

At some point soon after anurognathids had branched out on their own, pterosaurs experienced another major evolutionary event. It happened to the skull. Unlike dimorphodontids and anurognathids, whose skull shape harks back to pterosaurs' reptilian ancestry, all later clans have a longer, lower skull. This evolved mainly by lengthening of the snout region—almost as if someone had pinched a pterosaur by the nose and tugged very hard. Why the snout evolved in this way is not clear, but most likely it was driven by adaptations for feeding that endowed pterosaurs with a longer "prey-grabbing" tool. As the snout extended forward, several things happened to the skull: The bones upon which the lower jaw articulated—long, strong, rod-like elements called the quadrates—began to slope forward, rather than standing vertically, as they do in early pterosaurs. This opened up two possibilities: more room both for muscles that operated the jaw and for the bony capsule that enclosed the brain.

Campylognathoidids,[46] illustrated in *Figure 4.8*, were the first pterosaurs to show these important changes. The flag-bearer for this group, *Campylognathoides*, is best known from the Lower Jurassic *Posidonia* Shales of Germany,

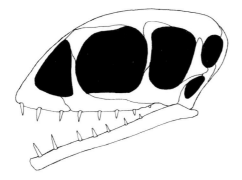

FIGURE 4.7 Anurognathids, the fabulous flying frog heads, represented by the restored skull of *Anurognathus*, with a wingspan of about 16 inches (40 centimeters).

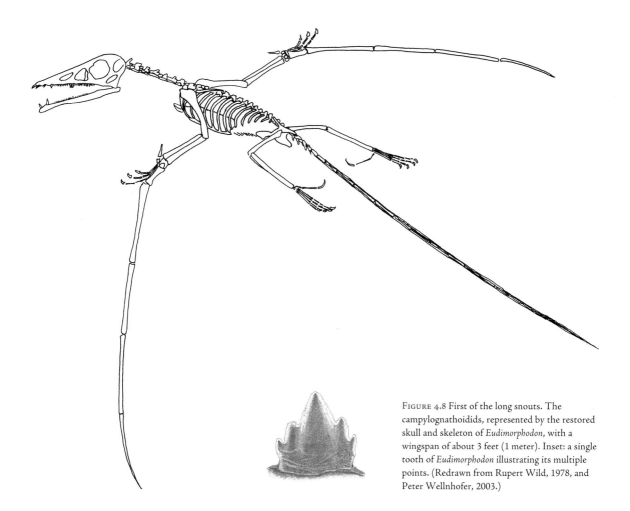

FIGURE 4.8 First of the long snouts. The campylognathoidids, represented by the restored skull and skeleton of *Eudimorphodon*, with a wingspan of about 3 feet (1 meter). Inset: a single tooth of *Eudimorphodon* illustrating its multiple points. (Redrawn from Rupert Wild, 1978, and Peter Wellnhofer, 2003.)

in which the remains of more than 20 individuals, generally a bit less than a meter (3 feet) in wingspan, have come to light over the last 100 years.[47] *Campylognathoides* demonstrates several typical features of the lineage, most noticeably the "drooped" tip of the lower jaw, from which erupt two large "prey-grabbing" fangs.

The other important member of this clan, *Eudimorphodon*, is mostly known from small individuals, though one or two also reached over a meter (3 feet) in wingspan. The first example of *Eudimorphodon*, an old individual with a most impressive-looking skull, was found quite by chance in 1973 by an Italian paleontologist, Rocco Zambelli, in debris from a rockfall at Cene near Bergamo in northern Italy. The first pterosaur to be described from Triassic rocks, one of the most surprising features of *Eudimorphodon* is its teeth: Unlike those of other pterosaurs, which have a single point, some of them have three or even five points, a shape that presumably helped them to grip their prey (slippery, writhing fish) more effectively. Several species of *Eudimorphodon* are now known, mainly from northern Italy, but also from elsewhere in Europe and even, in one case, from Greenland.[48] *Austriadactylus*, represented by a single, paper-thin, picture fossil from the Upper Triassic of Austria,[49] is probably a close relative of *Eudimorphodon*, because it too has multi-pointed teeth. But its crowning glory lies above: a magnificent crest that adorned the skull from the tip of the snout to the top of the head, present in a pterosaur right at the beginning of its 150-million-year history.

Rhamphorhynchus *and the True Prow Beaks*

The last and most evolved group of long-tailed pterosaurs, the rhamphorhynchids, was not only one of the longest-lived and most diverse clans, but also gave rise to several large species, individuals of which grew to almost 3 meters (about 10 feet) in wingspan. Early in its history, this clan split into two quite different lineages: On the one branch were rhamphorhynchines, distinguished by a formidable-looking array of fang-like teeth that splayed forward and outward from the jaws to form a fish-snagging tooth grab. On the other branch were the scaphognathines. They had fewer, but stronger, well-spaced, upright teeth mounted in a broader, more heavily built jaw that presumably allowed them to go after bigger, heavier prey than their rhamphorhynchine cousins.

Dorygnathus, a medium-size pterosaur from the same Lower Jurassic *Posidonia* shales of Germany as *Campylognathoides*, is the earliest known rhamphorhynchine and reached about 1 meter (3 feet) in wingspan. More than 20 skeletons of this pterosaur, some with skulls that positively bristle with fish-grabbing teeth, as *Figure 4.9* illustrates, have been recovered, one or two of them with patches of fossilized soft tissues.[50] Rhamphorhynchines have also been found in rocks from as far afield as England, Madagascar and China, often as fragmentary isolated remains, although the occasional gem, such as the beautifully preserved uncrushed skull of *Cacibupteryx* from Cuba, sometimes come to light.

The king of the prow beaks, and beakiest of them all,[51] was *Rhamphorhynchus*. At least 200 or 300 examples of this pterosaur (far more than for any other rhamphorhynchoid) have already been recovered from the Solnhofen Limestones, and more are found almost every year.[52] One happy consequence of this abundance is that the entire skeletal anatomy, from the point of the beak via every nook and cranny of the braincase to the tip of the toes, has been thoroughly pored over, drawn and described.[53] The same goes for the numerous examples of fossilized soft tissues: wing membranes, skin, "hair," claw sheaths, tail flaps and foot webs, which are better known in *Rhamphorhynchus* than for almost any other pterosaur.[54] *Rhamphorhynchus* is also one of those rare cases of a pterosaur for which much of the growth series is known, ranging from youngsters less than 30 centimeters (12 inches) from wing tip to wing tip, to big old adults well over a meter (3 feet) in wingspan.

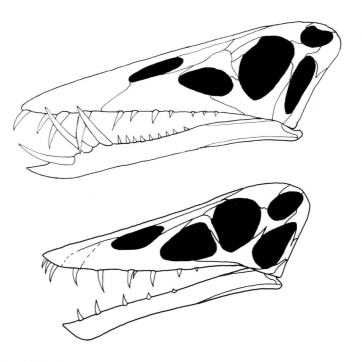

FIGURE 4.9 Rhamphorhynchids, the true prow beaks, represented by the skulls of the rhamphorhynchine *Dorygnathus*, 5 inches (13 centimeters) long, above; and the scaphognathine *Scaphognathus*, 4.5 inches (12 centimeters) long, below. (Redrawn from Wellnhofer, 1975.)

Until recently scaphognathines were far less well-known than rhamphorhynchines, but a row of new finds in China and the Americas is changing all that. A rich new Middle Jurassic pterosaur site, Cerro Cóndor in Patagonia, discovered by Ollie Rauhut, a paleontologist now based in Munich, Germany, has just yielded the earliest known scaphognathine,[55] while the well-preserved snout of *Harpactognathus*, found in 1996 in the Upper Jurassic of Wyoming, not only shows that scaphognathines lived in North America, too, but reveals that, like many other pterosaurs, some members of this group also sported large, well-developed crests on the skull.[56] This discovery has been dramatically confirmed by *Pterorhynchus*, a scaphognathine from the Lower Cretaceous of China, wherein the bony part of the cranial crest is continued upward by a large, sail-shaped skin-covered cartilaginous flap.[57]

Fossil remains of *Sordes* from Upper Jurassic lake sediments in the Karatau mountains of Kazakhstan, are the best-preserved of any scaphognathine. Nine individuals were collected in the early 1960s and are justly famous for the incredible fidelity and completeness of their fossilized soft tissues, which include rare examples of entire wing membranes and even "hair."[58] Last, but not quite least, the clan name bearer, *Scaphognathus*, illustrated in *Figure 4.9*, is known from just three skeletons, all from the Solnhofen Limestone.[59] The presence of a long tail in the second and third specimens to be found confirmed that *Scaphognathus* is a rhamphorhynchoid, although because the first example, which was found in the early 1800s, lacked the lower half of the body, it was thought, initially, that this pterosaur was a short-tailed form—a pterodactyloid, the group that we shall consider next.

Pterodactyloids: *A Long-Armed, Short-Tailed Revolution*

There is still some debate as to the exact relationship of pterodactyloids to rhamphorhynchoids. *Figure 4.5* shows the current consensus but, as mentioned earlier, it is also possible that they branched off, together with the anurognathids. Everyone agrees, however, that pterodactyloids themselves form a single, unique, clearly defined group. This is because on the line leading to pterodactyloids, pterosaurs underwent a profound reshaping and rebuilding of their anatomy that affected almost every part of their bodies.

In the skull, the nostril and the antorbital fenestra merged to form a single large opening and the braincase was considerably expanded, developments that hint at some significant, but as yet only poorly understood, physiological and neurological changes. Elsewhere, there was a sharp reduction in the length of the tail and various modifications to the limbs, most notably the lengthening of the metacarpals, the bones between the wrist and the wing-finger, and the extreme reduction and eventual loss of the fifth toe in the foot. Because all these features were directly or indirectly involved in the flight apparatus, their modification must have had a significant impact on how pterodactyloids flew, although, surprisingly, this has yet to be investigated in any detail. What recent research has established, however, to which we will return in later chapters, is that these changes also had a profound impact on another aspect of pterosaurs' lives—it opened up a whole new world for them, on the ground.

Returning to the pterosaur tree, we find that pterodactyloids can be separated into four major branches. One of these, the ornithocheiroid clan, is composed of a group of species that are distinctly different from all other pterodactyloids, so they are shown branching off somewhat earlier. It is very difficult to establish exactly how the three remaining clans are related, so I have avoided this problem altogether by having them emerge simultaneously from the main trunk.

Ornithocheiroids: *Built for a Life on the Wing* The ornitho-cheiroids, several of which are shown in *Figure 4.10*, were a diverse, important and highly distinctive group of Cretaceous pterosaurs. Among them were several of the most spectacular animals ever to take to the skies, not least *Pteranodon*, which, with its long, pointed jaws and bizarre crest, is an icon for flying reptiles the world over. Unlike rhamphorhynchoids, but practically without exception among the main pterodactyloid clans, all known orni-thocheiroids were well over 1 meter (3 feet) in wingspan when fully grown and most reached 3 or 4 meters (10 to 12 feet). Several, including *Pteranodon*, grew to remarkably large sizes, with wings that reached 6 or even 7 meters (20 to 22 feet) across.

Ornithocheiroids appear to have been even more highly specialized fli-ers than other pterosaurs and, like modern-day albatrosses and frigate birds, seem to have been particularly well-adapted for soaring. Modifications for this lifestyle are found throughout the body and, in addition to revealing how ornithocheiroids functioned, make it relatively easy to identify fossil remains of these pterosaurs, even, in some cases, just single bones. One obvi-ous adaptation is seen in the hind limbs, which, compared with the forelimbs, are very slender and weak. Presumably, ornithocheiroids spent little time on the ground. Other adaptations are less obvious. Study the shoulder girdle closely, however, and it becomes apparent that the shoulder blade (scapula) is surprisingly short and stout compared with that of other pterosaurs. This means that the wings did not sprout from the body at about half-height, as is usual, but were rooted rather higher, which, according to one recent study,[60] increased ornithocheiroids' stability during flight.

Istiodactylus, represented by several incomplete skeletons and fragmentary skulls from Lower Cretaceous beds that crop out along the southern shores of the Isle of Wight, England,[61] and by an almost complete individual from the Lower Cretaceous Jehol Biota of northeast China, seems to be the least-evolved ornithocheiroid known at present. Uniquely among pterosaurs, the snout was rather wide and shaped a bit like the beak of a duck, except that its edges were rimmed with a set of sharp-edged blade-like teeth. Thus armed,

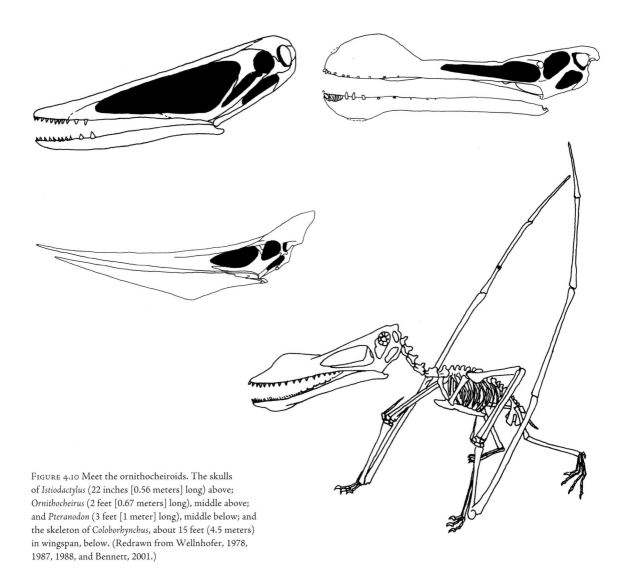

FIGURE 4.10 Meet the ornithocheiroids. The skulls of *Istiodactylus* (22 inches [0.56 meters] long) above; *Ornithocheirus* (2 feet [0.67 meters] long), middle above; and *Pteranodon* (3 feet [1 meter] long), middle below; and the skeleton of *Coloborhynchus*, about 15 feet (4.5 meters) in wingspan, below. (Redrawn from Wellnhofer, 1978, 1987, 1988, and Bennett, 2001.)

Istiodactylus could deliver a powerful "cookie-cutter"-style bite—quite sufficient to snip gobbets of flesh from the carcass of its prey.

The most diverse and important group of ornithocheiroids, found almost everywhere during the Lower Cretaceous, were the ornithocheirids. Initially known only from fragments of jaws from the Lower Cretaceous Cambridge Greensand of England (dismissed by one colleague as "the ugliest pterosaur material I ever saw"), more complete fossils from the Santana Formation of Brazil show that *Ornithocheirus* was a large pterosaur with thick, rounded bony crests on the tips of its jaws.[62] The most distinctive feature of this pterosaur was its well-developed teeth. The first three pairs were very large—all the better to grab their prey with—and serve today as a useful hallmark of the ornithocheirid clan.

This and several other characteristics reveal that *Ornithocheirus* has several close relatives in the Santana Formation, in the Cambridge Greensand, and, most recently, in the Jehol Biota of China, but they have acquired such a plethora of names that it is not clear how they should be referred to correctly.[63] One that we can be sure about is *Coloborhynchus*, first named by Harry Govier Seeley's contemporary and rival, Sir Richard Owen, and represented by several beautifully preserved skulls and skeletons from Brazil and some lumpy-looking bits of jaws from Lower Cretaceous rocks of North America, Europe and Mongolia.[64] One or two individuals of *Coloborhynchus* reached more than 6 meters (20 feet) in wingspan, but not all ornithocheirids were so large. The single complete skeleton of *Haopterus* from the Lower Cretaceous Jehol Group of China measured little more than a meter (3 feet) from one wing tip to the other,[65] and another astounding find from the same rocks, a pterosaur egg, discussed in depth in Chapter 7, contains an embryonic ornithocheirid with wings less than a quarter of a meter (10 inches) across.[66]

The ancestors of ornithocheirids also gave rise to another group, distinguished above all else by the absence of something that most other pterosaurs found indispensable: teeth. This hallmark is reflected in the name of the clan, pteranodontians (meaning the winged toothless ones), and its most important member, *Pteranodon*. First found in the late 1800s in the Upper Cretaceous chalk bluffs of Kansas, well over 1,000 individuals have now been collected from these rocks, making *Pteranodon* one of the best represented and best known of all pterosaurs.[67] Immediately recognizable from its long, scimitar-like jaws and equally spectacular "look at me" crest, this pterosaur sometimes reached wingspans of almost 7 meters (22 feet), although typical adults seem to have been only about half that size.

Nyctosaurus, another toothless pteranodontian, is from the same rock layers as *Pteranodon*, but much rarer and considerably smaller, with a wingspan of only 2 to 3 meters (6 to 9 feet). Although they were long thought to be completely crestless, two recent finds of this pterosaur show that some individuals, perhaps the males, bore an extraordinary, antler-like structure that erupted from the top of the head, rivaling even the best that *Pteranodon* had to offer.[68]

Ctenochasmatoids: *Stressing the Straining*

While ornithocheiroids took to the skies, the ctenochasmatoids, a clan that includes almost all the Upper Jurassic pterodactyloids found so far, set off in another direction—into the water. These pterosaurs, generally only of small or medium size, opted for more teeth, rather than less, as in many other lineages, and became highly adapted to a lifestyle that involved wading in rivers and lakes, using their comb-like dentition to sieve for their supper. The ctenochasmatoids illustrated in *Figure 4.11* demonstrate two key features of the group: a highly modified skull design in which the quadrate bone, upon which the lower jaw hinged, lay in an almost horizontal position, and a neck that was extremely long, achieved not by adding more vertebrae, as birds do, but by stretching several of the existing ones into long tube-like structures.

Pterodactylus, one of the geologically earliest known ctenochasmatoids, was relatively unspecialized. Several hundred specimens of this pterosaur are known from the Solnhofen Limestone and seem to belong to two or three different species, distinguished only by small differences in the shape of their teeth and jaws.[69] Like *Rhamphorhynchus*, there are so many specimens of *Pterodactylus* with evidence of soft tissues that their external appearance at least (shown in *Figure 1.5*) can be restored with some confidence. Most excitingly, *Pterodactylus* has one of the most complete growth series for any pterosaur, ranging from hatchlings only a few days or weeks in age, to adults of half a meter (20 inches) or so in wingspan and even big, old individuals that were half as large again.

The general evolutionary trend within ctenochasmatoids seems to have been toward longer and longer jaws and ever more teeth but, as *Figure 4.5* shows, this was a development that eventually went two separate ways. On the one hand *Gnathosaurus* from the Solnhofen Limestone and its relatives, to be found in Lower Cretaceous rocks of Europe, Asia and South America, opted for fewer, larger, often strongly curved teeth, and seem to have concentrated on larger prey items.

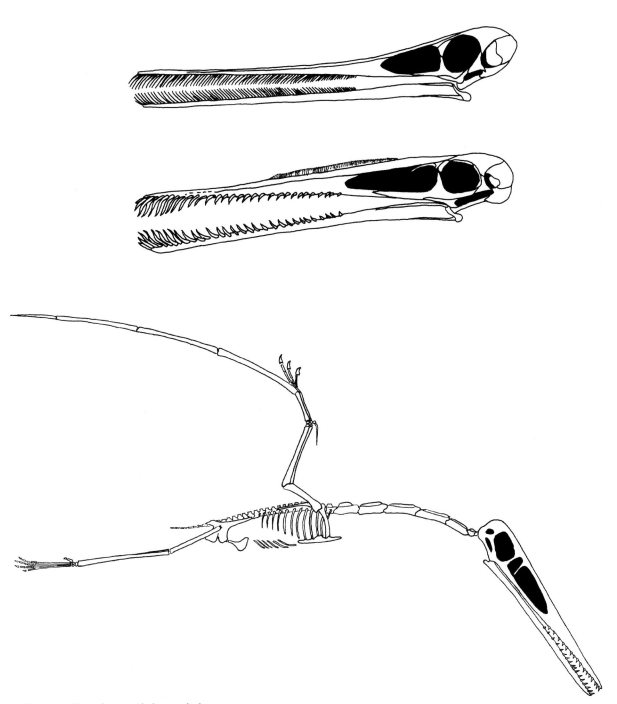

FIGURE 4.11 Ctenochasmatoids, long-necked super-
strainers, illustrated by the skulls of *Ctenochasma*
(4 inches [10 centimeters] long), above; and
Gnathosaurus (11 inches [28 centimeters] long), middle;
and the skeleton of *Pterodactylus*, with a wingspan of
about 20 inches (50 centimeters), below. (Redrawn
from Wellnhofer, 1970.)

On the other hand, *Ctenochasma* and its relatives evolved ever more and finer teeth. This culminated in the so-called flamingo pterosaur, *Pterodaustro*, from the Lower Cretaceous of Argentina. With more than 1,000 long, needle-fine teeth in the lower jaw, this pterosaur was armed with the mother of all filtering apparatuses.[70] Thanks to a series of highly successful expeditions in the 1990s to the Lomo del *Pterodaustro*, the fossil locality in Argentina where this pterosaur was first found, *Pterodaustro*, like *Pterodactylus*, is now also known from a large number of individuals ranging from hatchlings to old adults. Most excitingly of all, an egg with remains of an embryo has just been reported (see Chapter 7) and extends this growth series into the prenatal realm.[71]

Dsungaripterus and the Clam-Cracking Crew Another important pterodactyloid clan, the dsungaripteroids, is principally distinguished by adaptations for cracking open and feeding on shellfish. Naturally, these are most clearly seen in the jaws, illustrated in *Figure 4.12*, the winkle-picking tips of which are long, pointed and toothless, and in the teeth, those at the back being especially massive. In *Dsungaripterus*, they were packed up tight against one another so that they formed small, but doubtless highly effective, anvils. This development is paralleled elsewhere in the robust design of the skull and might even be related to another peculiar dsungaripteroid feature: the remarkably thick and heavy construction of their vertebrae and limb bones.

Early dsungaripteroids were only small or medium-size pterosaurs and represented in the main by *Germanodactylus*, based on just a handful of skeletons from the Solnhofen Limestone, and several similarly sized and shaped species known from fragmentary but distinctive pieces of jaw from Upper

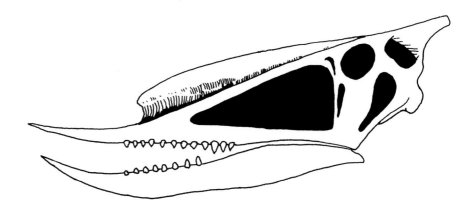

FIGURE 4.12 The quintessential nature of the clam-cracking dsungaripterids is demonstrated by the 16 inch (41 centimeters) long skull of *Dsungaripterus*. (Redrawn from Wellnhofer, 1978.)

Jurassic rocks of Europe and East Africa.[72] More evolved dsungaripteroids, with fully developed shell-cracking teeth, are almost exclusively known from East Asia. Here, extensive remains, including several beautifully preserved uncrushed skulls, have been recovered from Lower Cretaceous lake sediments that crop out in Dsungaria, a remote region of northwest China, and, as detailed in the previous chapter, from similar aged rocks in the region of Tatal, Mongolia.[73] *Dsungaripterus*, which was found and described by the legendary Chinese paleontologist Young Chung-Chien, seems to have been the largest member of the clan, and reached up to 3 or 4 meters (10 to 12 feet) in wingspan.

Quetzalcoatlus *and Other Toothless Terrors*

The azhdarchoids may be the last branch on our tree, but they are most certainly not the least, because among their number they include the largest flying creatures of all time—*Quetzalcoatlus*. Most members of the azhdarchoid clan have only come to light in the last two decades, and the group itself was first recognized just over 20 years ago. As in many other pterosaurs, the most distinctive features of this clan are to be found in the skull, which is depicted in *Figure 4.13*. The complete absence of teeth, a development that occurred quite independently of that in pteranodontians, is striking, but the most extraordinary development concerns the snout, which rises high above the level of the eye and has a huge opening for the nostril.

Early azhdarchoids such as *Tapejara* are best known from the Lower Cretaceous Santana and Crato Formations of Brazil.[74] With its deep, rather parrot-like skull and extraordinary sail-shaped crest, *Tapejara*, a medium-size pterosaur represented by several skulls and skeletons, must have presented a bizarre sight as it flew through the skies. Initially found only in South America, evidence of tapejarids has also turned up in Africa. In a truly spectacular series of discoveries, several complete skeletons, representing a whole flock of these creatures, have recently been found in the Lower Cretaceous Jehol Biota of northeast China.[75] A second group of early azhdarchoids, including *Tupuxuara* and its relatives, grew to much larger sizes, attaining wingspans of 6 meters (20 feet) or so. With a huge crest that ran the length of the skull, they must have looked just as strange as tapejarids.[76]

The most evolved azhdarchoids all belong to a single exclusively Upper Cretaceous group, the azhdarchids. Fossil remains of these pterosaurs, immediately identifiable from the extreme elongation of the neck (achieved as

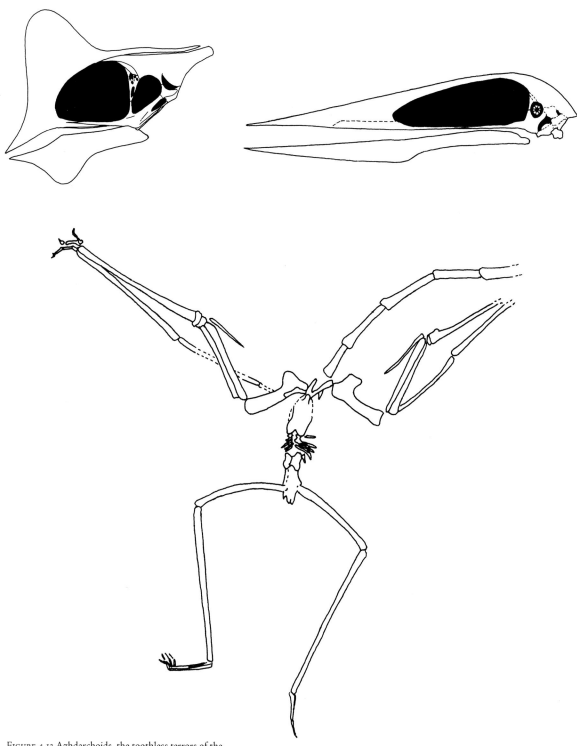

FIGURE 4.13 Azhdarchoids, the toothless terrors of the skies, represented by the 8 inch (20 centimeter) long skull of *Tapejara* (above left), the 12 inch (29 centimeter) long skull of *Zhejiangopterus* (above right), and the skeleton of *Zhejiangopterus*, with a wingspan of about 10 feet (3 meters), below. (Redrawn from Peter Wellnhofer, 1991, and David Unwin and Lü Junchang, 1997.)

in ctenochasmatoids by lengthening individual neck vertebrae), have been found all over the world and indicate the existence of at least five or six different kinds of azhdarchid. Some of these, such as *Zhejiangopterus*, shown in *Figure 4.13* and recovered from volcanic ash beds that are used for building stone in eastern China, were relatively small stork-like forms with wings only 2 meters (6 feet) or so across.[77] Others, most famously *Quetzalcoatlus*, reached huge sizes, with wingspans of 10 meters (33 feet) or more. Discovered in 1973 by Douglas Lawson while he was carrying out field work in Big Bend Park in Texas, *Quetzalcoatlus* is known from the remains of a single forelimb that belonged to a giant individual and several much more complete skeletons, with skulls, of individuals that were only about half this size.[78]

Quetzalcoatlus and its relatives from Spain, France, Jordan and Romania,[79] many of which also reached giant size, existed right at the end of the Cretaceous and were, in one sense, the topmost twigs of the pterosaur tree. Thanks to a steadily increasing array of new discoveries and aided by some intensive computer-based genealogical work, we can now trace our path backward from these terminal pterosaurs, along the twigs and branches and down the main trunk, almost all the way back to the still half-hidden roots of the tree. New fossil finds and new studies will, undoubtedly, bring currently invisible parts of the tree into view and might even redirect some of the branches we have visited here, but the overall shape of the tree probably will not change that much. For now we can turn to other questions about pterosaurs—how they were constructed, grew and flew—but we will return to this tree in the last two chapters to see how answers to these questions may have both shaped its growth and brought it to an end.

5
THE HEAD INSIDE OUT

Sweeping low over the crowded beach, the big male Tupuxuara presented a magnificent sight. It was not just the white-tipped wings, wider than three lanes of traffic, and it was not just the skull, as long as a javelin and with jaw tips to match. It was, above all else, the extraordinary, the magnificent, the show-stopping crest. It towered above the skull, rising up from its roots on the forehead into a huge sail, flaming red at the front, then shading to maroon at the apex and the rear. As he flew by again, broadside on to the flock of pterosaurs, mostly females, he slowed almost to the point of stalling and executed several slow rolls and turns, each one accentuating the majesty of that crest. Several other males came floating by, far fewer now, as the youngsters and the less well-endowed had been winnowed out by indifference. Several females began to stir, then settled back, waiting. The "king" came in again, this time positioning himself so that the rays of the low sun shone through the crest, making it seem as if it were ablaze—a huge flame sailing through the skies. That did it. Three females raised themselves on their stilt-like arms, searched for the breeze, and, with a kick of their hind legs, were soon aloft on outspread wings. Two more joined them as they trailed off behind the triumphant male. Just as always, big crests were in.[1]

FIGURE 5.1 The six inch (15 centimeters) long skull of *Sinopterus*, a toothless azhdarchoid pterosaur from the Lower Cretaceous Jehol Biota of China. (Image courtesy of Wang Xiao-Lin and Zhou Zhonghe.)

Every Body Needs Somebody Ask almost anybody: "What is the most powerful thing in the world?" And they will probably reply: "gravity," or "love," or "money." All these answers and many others, including "Microsoft," are wrong, because the right answer is "evolution." Every single living thing, from the smallest microbe to the biggest blue whale, is a consequence of evolution, even the most complex thing in the known universe—our brains.

Evolution also built pterosaurs. And one of the really clever things about evolution is that it does not start from scratch, with a drawing board, some wobbly pink stuff and lots of sticky-back plastic,[2] it begins with a complete, fully functioning organism such as a lizard-like reptile and turns it, bit by bit, into a beautiful flying creature. What is even more surprising is that evolution does not do this by lopping bits off here and adding bits on there, it just slowly modifies the existing bones, teeth, tissues and organs and remodels them in such a way that they can be used for flight, fishing or climbing trees.

The result, for pterosaurs, was a body whose components (insofar as we know and understand them) can be identified in other nonflying vertebrates, but had become uniquely modified for their mode of life in the skies. As we have previously seen, most fossil evidence of the anatomy of these extinct animals consists of hard tissues—their bones and teeth. In life, however, much of the body was composed of soft tissues: major internal organs such as the heart, liver and lungs, and the blood system, nerves, muscles, skin and so on. Usually, as was explained in Chapter 3, very little evidence of such structures survives the rigors of fossilization. Sometimes, however, a serendipitous concatenation of events presents us with the fossilized remains of pterosaur skin, wing membranes, or even throat sacs, to list just some of the soft structures found so far. When added together (*Figure 3.9*), the the fossil evidence for soft tissues is rather better for pterosaurs than for many other groups of extinct vertebrates, but it still only reveals a small, if tantalizing, part of the whole picture.

Fortunately, there are two other approaches we can use to fill out our knowledge of pterosaurs' soft anatomy. The pterosaur skeleton, like that of other vertebrates, had a highly intimate relationship with the rest of the body, and individual bones frequently bear the subtle marks of this liaison. Crests, ridges, bumps and scars give away the original position and size of muscles, while holes and grooves, usually tiny, but sometimes large, as in the case of the channel for the spinal cord, reveal the courses of nerves and blood vessels as they snaked their way through, around and over the bones. This same principle applies on a larger scale to the braincase and its vital passenger, the brain, whose external shape and general features can be reconstructed from the internal shape of its bony casing.

Yet another potentially useful way of prying into pterosaurs' innards relies not on the fossils themselves, but on comparisons with living relatives which, naturally, come with a full complement of soft parts. The problem here is that irrespective of which of the modern groups of diapsids—lizards, crocodiles or birds—is eventually shown to be pterosaurs' closest living relative (a matter of some debate, as Chapter 4 revealed), any close relationship can be ruled out, so only speculative inferences regarding soft parts are possible. Still, because pterosaurs were certainly diapsids, we can be fairly confident that what is generally true of all living diapsids, for example, that they breathed using lungs, was also true of pterosaurs.[3]

By using a wide variety of research techniques, examining as many different fossils as possible and combining every last bit of available evidence, we can piece together an accurate picture of pterosaurs' skeletal structure and a general view of their soft tissue anatomy. Admittedly, large chunks of this picture are still missing—we have no idea, for example, what the liver looked like, or whether it functioned in quite the same way as in other diapsids— but even in these cases, we can fill in the blanks with best guesses, informed by comparison with living relatives such as crocodiles and lizards, and the knowledge that we have already garnered about pterosaurs. This chapter, and the one that follows, takes pterosaurs apart, from the teeth to the tail and to the toes, to see how they were constructed and what this meant for their major bodily functions such as feeding, breathing and, most vital of all, their metabolism. We begin, however, at the beginning—the head.

Head Start The most complex part of a pterosaur, or indeed any vertebrate, was the head—skull, mandibles and all the associated soft bits—quite simply because so many important components and functions were packed into this single structure. The skull contained the brain, of course, the center of consciousness and neural control, but it also housed key sensory organs for four of the five senses: sight, sound, smell and taste. In addition, the primary passageways involved in eating and breathing also passed through the head and, in the case of pterosaurs, the capture of prey was almost exclusively carried out by the jaws and teeth. On top of all this, quite literally, the heads of many pterosaurs bore crests, some of them extremely large and showy, surely a clue to their main purpose.

As *Figure 5.2* illustrates, pterosaurs had a highly distinctive skull that, superficially at least, looks quite bird-like. The front half was made up of the snout, often bearing teeth, while the rear half was composed of the cranium, consisting of the orbit (the opening that housed the eye) and the braincase, on the base of which hinged the lower jaw. The snout region was stretched forward, sometimes to a remarkable degree, as in *Pteranodon*, and tapered to a sharp point. One or two fossils, most notably a specimen of *Tapejara*, illustrated in *Figure 5.3*, reveal that in pterosaurs, just as in birds, a horny sheath fitted tightly over the front end of the snout and probably helped to protect it from wear and tear.[4]

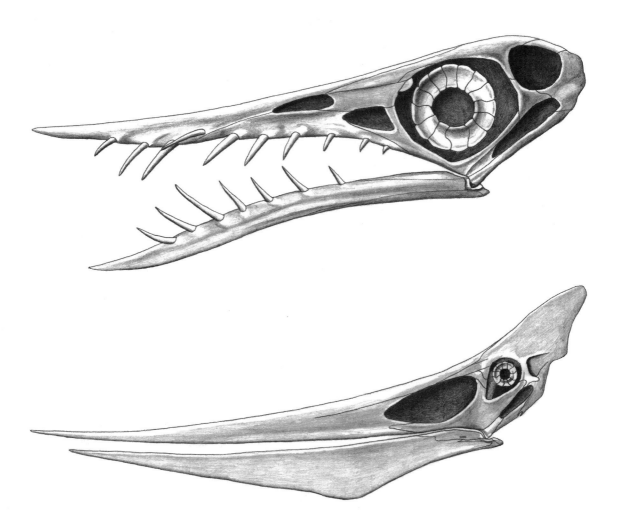

FIGURE 5.2 General anatomy of the pterosaur skull, based on the rhamphorhynchoid *Rhamphorhynchus* (above), with a length of about 4 inches (10 centimeters), and the pterodactyloid *Pteranodon* (below), with a length of about 40 inches (1 meter). (Redrawn from Peter Wellnhofer, 1975, and Chris Bennett, 2001.)

In rhamphorhynchoids, the main part of the snout was pierced on either side by two openings. Those at the front formed the nostrils, the entrance-way to passages that led to the back of the mouth. The function of the rear pair of openings is less clear: They certainly helped to lighten the skull, but they might also have housed muscles or a salt gland through which ptero-saurs, especially the marine forms, were able to dump excess salt.[5] As *Figure 5.2* shows, pterodactyloids dispensed with the bony bar between the front and rear openings and settled for a single large opening, at the front corner of which lay the nostril.

Teeth erupted along the lower edges of the snout, at least in toothed pterosaurs, but the extent of the tooth rows varied considerably. Except in dsungaripteroids, they usually began at the tip of the snout and continued back to below the orbit, but they could terminate much earlier, and in some pterosaurs, such as *Cycnorhamphus*, the teeth were restricted to the jaw tip alone. Like other reptiles, pterosaurs constantly shed their teeth and grew

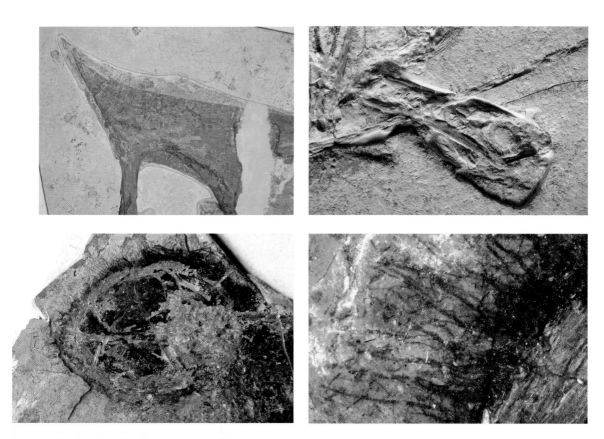

FIGURE 5.3 Fossil evidence for soft-part structures of the pterosaur head. Above left: horny sheath (rhamphotheca) covering the jaws of *Tapejara*. Above right: the throat sac in *Rhamphorhynchus*. Below left: the jaws of the anurognathid *Batrachognathus* fringed by a beard of bristles. Below right: close up of *Batrachognathus'* bristles. (*Tapejara* image courtesy of Dino Frey.)

new ones (often to be seen peeping from the socket of the old tooth) as waves of replacements swept along the tooth row.

Behind the snout lay the orbit. In pterosaurs, this was relatively large and thus able to accommodate a big eyeball, emphasizing the critical importance of sight for these animals. Within the orbit, a ring of small, overlapping, plate-like bones called scleral ossicles helped to support the eyeball. In some really well-preserved fossils, the position and diameter of this ring pinpoints both the exact size and the original location of the eyeball in its socket.

Behind the orbit lay the main part of the cranium, which, in effect, consisted of two bony boxes, one inside the other. The outer box was constructed from bones that roofed the skull and also extended down the sides of the cranium, forming the cheeks. In early reptiles, these "cheeks" were solid,[6] but in pterosaurs, they were pierced by two openings (the upper and lower temporal fenestra), the epitome of the diapsid condition. Deep inside the outer bony box was a second box, the braincase. The roof of the braincase was composed of the same elements as the outer box, but its sides and floor were made up of several distinct bones, often fused together. The lower parts and underside of the braincase were pierced by numerous small openings through which the cranial nerves exited from the brain and ran out into the rest of the head, where they enervated the tongue, eyes, jaw muscles and teeth.[7]

The space between the outer bony box and the braincase was partly filled by blood vessels, nerves and other soft tissues, but mainly occupied by the muscles that attached to the mandibles and operated the jaws.[8] The muscles, shown in *Figure 5.4*, were not confined to this space, but spread out, via the temporal fenestrae, onto the outer surface of the cheek region, thereby attaining a relatively large size and gaining an extensive attachment area. This enabled pterosaurs to power their many different types of feeding behavior, some of which, such as cracking clams, must have required pretty hefty muscles.

The back of each cheek was buttressed by a single, long, pillar-like bone, the quadrate, just behind the top of which lay the ear drum and on the lower end of which articulated the mandible. In early pterosaurs, this articulation seems to have been relatively simple and only allowed the jaws to swing up and down. In some later pterosaurs, such as the ornithocheiroids, the articulation evolved into a more complex screw-like arrangement so that, as the lower jaw opened, the mandibles on each side were pushed outward, increasing the gap between them. Exactly why this was necessary will be explained in the next section.

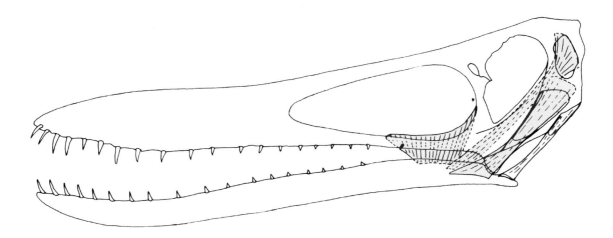

FIGURE 5.4 How pterosaurs operated their jaws.
Muscles that opened the jaws are shown in blue, those
that closed them in pink.

Built from several flat, lath-like bones, the mandibles each consisted of
a long hollow bar that met at the jaw tip. This junction was very short and
sometimes not even fused in early pterosaurs, but became firmly united in
later forms, the fusion spreading farther and farther back along the jaw and
forming an elongated blade-like structure called the mandibular symphysis.
In toothed pterosaurs, the upper edge of each mandible bore a row of teeth,
and in some crested forms, the lower edge supported a decorative bony keel.
In several exceptionally well-preserved fossils, one of which is illustrated in
Figure 5.3, a large patch of wrinkled skin can be seen curving down below
the back part of the lower jaw.[9] This is most likely to have been the fossilized remains of a throat sac, which probably looked rather similar to the
throat pouches of pelicans and within which lodged the leaf lying between
the mandibles of *Ludodactylus*, the "tree biter" that first appeared in Chapter
3. It also seems to explain why the mandibles of some pterosaurs widened as
they were opened: This helped ensure that hard-won prey, such as wet, slippery, wriggling fish, ended up in the sac and not back in the ocean.

Fangs for the Fish   Three aspects of the pterosaur head deserve closer
inspection. The first of these is the teeth. As *Figure 5.5* demonstrates, the
dentition of pterosaurs was remarkably variable: *Dimorphodon* had a long file
of tiny teeth led by several large fangs; *Anurognathus* sported a row of small,
well-spaced, sharp-tipped spikes; while *Dsungaripterus* was equipped with

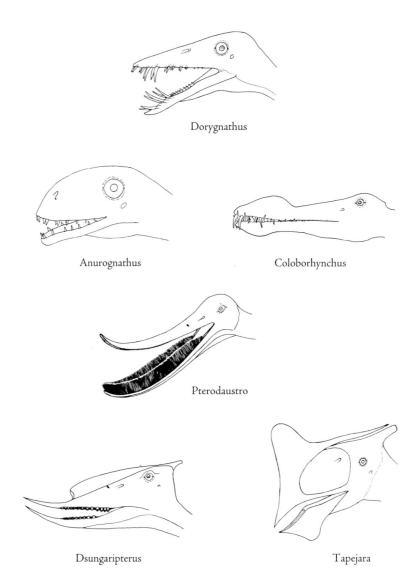

Dorygnathus

Anurognathus

Coloborhynchus

Pterodaustro

Dsungaripterus

Tapejara

FIGURE 5.5 Using only single, simple teeth, but with varied number, size, position and orientation, pterosaurs evolved a remarkable degree of dental diversity. Prey-snagging devices, with a tooth grab at the front and smaller teeth behind, evolved over and over again, appearing independently in many lineages including the rhamphorhynchids, represented here by *Dorygnathus*, and in the ornithocheirids such as *Coloborhynchus*. Small, sharp-pointed, spike-like teeth, well suited for gripping and puncturing the hard covering of insects, are typical of anurognathids such as *Anurognathus*, while large numbers of long, fine filament-like teeth were used by *Pterodaustro* and other ctenochasmatines to strain and sieve for their prey. By contrast, *Dsungaripterus*, and other dsungaripterids, had large, bulbous, clam-cracking teeth at the back of the jaw, while *Tapejara*, its relatives and pteranodontians abandoned teeth altogether and presumably relied on the shape of the jaws to deal with their food.

a battery of clam-crushing dentures. Yet all these different dentitions are composed of a single, rather simple tooth design. Unlike our mammalian teeth with their complex cusps and roots, pterosaur teeth had one large root that anchored them firmly in the jaw and a crown that usually consisted of a single elongated cone, built of dentine and capped with enamel. The striking variation evident in pterosaur dentition was achieved quite simply, just by varying the length and degree of curvature of individual teeth, or by varying their number, size, position and orientation within the jaws.

The simple construction of pterosaur teeth had its drawbacks, though. With a few rare exceptions,[10] the teeth had no cutting edges that could be used to dismember prey or snip off bite-sized chunks, nor did they have cusps and basins, which might have helped to grind or pulp their food. This means that, essentially, pterosaurs were the "fast-food" feeders of the Mesozoic, preferring items that required little or no preparation or processing and could be swallowed as quickly as possible.

Predictably, therefore, the most common type of dentition found in pterosaurs is a prey grab. Typically, this consisted of several pairs of large, slightly curved, sharp-pointed fangs, whose job was to get hold of the prey. This was followed by a row of smaller teeth, whose task was to maintain a tight grip on the victim before it was swallowed. Judging from their construction, and a few fossils in which the contents of the stomach are still preserved,[11] most prey grabs were used to catch fish, although pterosaurs were probably not averse to consuming other delicacies, such as squid. Presumably, most prey was caught on the wing, although some pterodactyloids might have waded in lakes or ponds and hunted in a manner similar to that of herons and egrets today.

Prey grabs evolved on many separate occasions, and although they appear rather similar, each has its own unique features. Dimorphodontids had a long row of tiny, lancet-like teeth behind the prey grab, while the teeth of *Eudimorphodon* and its relatives had several points that may have helped to grip prey items more tightly. Both scaphognathines and rhamphorhynchines had well-developed tooth grabs. In *Rhamphorhynchus*, this was taken to an extreme. In addition to several pairs of murderous-looking fangs, the tips of the mandibles were fused into a narrow, blade-like prow that projected forward from the front of the lower jaw and skimmed through the water surface during flight. As soon as an object was contacted, the tooth-grab-armed jaws snapped shut on what the pterosaur hoped was its prey, but what might on occasion have turned out to be a log.[12]

Tooth grabs evolved in at least two different pterodactyloid groups, reaching their most spectacular development in ornithocheirids. In some species, the main fangs attained 8 centimeters (more than 3 inches) in length, and their highly worn tips suggest that they saw a great deal of use.

Sustenance was to be had not only on the ground or from the waters, it could also be found in the air. The small, sharp-pointed, peg-like teeth of anurognathids seem well-adapted for gripping and puncturing the hard outer covering of insects, which swarmed through the Mesozoic air in the billions. Just like insectivorous birds today, anurognathids had very broad mouths that could gape extremely wide in order to maximize their chances of snapping up a dragonfly that was doing its best to avoid this fate.

The similarity with avian insectivores goes even further. Not only do some of them, such as nightjars, have a very wide gape, but also, like their pterosaurian counterparts, they have short bristles rimming the edges of the mouth[13]—as did anurognathids. A well-preserved example of *Batrachognathus*, illustrated in *Figure 5.3*, shows that the jaws of these pterosaurs were also fringed with short bristles, similar to those of the nightjar, except that in this case, the bristles seem to have been modified from the furry covering of the skin, discussed in the next chapter, rather than from feathers.

Ctenochasmatids opted for a radically different style of feeding—filtration. To do this, they dramatically increased the number of teeth, which reached more than a thousand in the most specialized forms such as *Pterodaustro*, and the teeth themselves became increasingly long and thin. Presumably, ctenochasmatids plunged their open jaws into the prey-filled waters of lakes and pools and, after closing the jaws to form a trap, lifted the prey into the air and allowed the water to drain away. Then, using a long and flexible tongue, they transferred the results of their labors—crustaceans, tiny mollusks and insect larvae—to the back of the throat ready for swallowing.

Heading in an altogether different direction, *Dsungaripterus* and its relatives opted for a life of clam-crushing. The first problem was to apprehend the object of their desire—clams, other shellfish, perhaps even crabs—either by probing for it in sand and mud or, in the case of mussels and oysters, by levering it from its rocky holdfast. Either way, their long, pointed, winkle-picking jaw tips would have served well for these tasks. The second problem was to remove the edible soft parts from their protective wrapping. To do this, dsungaripterids had specially enlarged teeth at the back of the tooth row where the jaw-closing muscles could exert the most effective crunch. In the king of the clam-crushers, *Dsungaripterus*, the hindmost teeth in both the upper and lower jaws were large and chunky, forming a pair of anvils that were firmly embedded in deep sockets and between which oysters were doomed.

Not all pterosaurs relied on teeth as their dinner winners. At least two distinct lineages, pteranodontians and azhdarchoids, made do without any teeth at all. How they fed and exactly what they fed on are not really clear, but in both groups, most species had long, narrow, sharply tipped jaws that were probably used rather like tweezers to pluck up small prey, either while they were flying over water or walking around on land.[14] By contrast, *Tapejara* and its relatives, such as *Sinopterus*, had relatively short, rather powerful-looking beaks that, in some respects, look similar to those of parrots. Perhaps these pterosaurs had opted for a more herbivorous diet that consisted, at least partly, of seeds and fruit.[15]

Gray Matter Matters Tightly enclosed in its bony box, the pterosaur brain left a clear impression of its external shape on the internal surface of the braincase. Casts, or rather "endocasts," as they are referred to, produced from the infilling of the braincase by minerals after the soft tissues had decayed away, can replicate the general shape and external details of the brain with remarkable fidelity. The problem is how to get at the endocast. There are a few fossils, mainly from the Solnhofen Limestone, where fortuitous breaks expose some details, and in one case, in an uncrushed *Dorygnathus* skull from Yorkshire, England, a section of the skull roof was removed to reveal part of the endocast of the brain.[16] But these and other fossils in which internal details of the braincase are visible present only an incomplete picture. Fortunately, a new technique, CAT scanning, in which images from a series of X-rays are reconstructed into a digital (virtual) endocast, means that we can now get the data we need without damaging the fossil at all. Larry Witmer of Ohio State University and his colleagues have applied this technique to two pterosaurs,[17] focusing on the skulls shown in *Figure 5.6*, one of which belongs to the rhamphorhynchoid *Rhamphorhynchus*, the other to the pterodactyloid *Anhanguera*.

The CAT scanning study rapidly confirmed the main conclusion that had been slowly and laboriously arrived at by several older studies[18]—pterosaurs had remarkably bird-like brains. The lobes at the front, concerned with the sense of smell, were very small, suggesting that odors and scents were

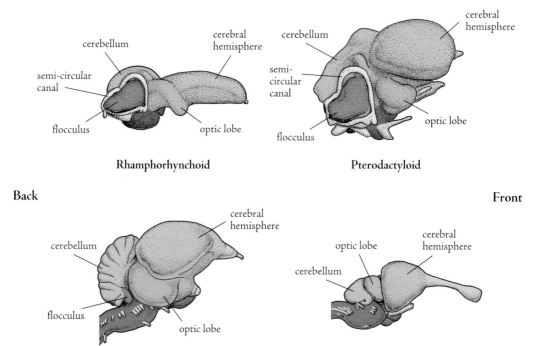

FIGURE 5.6 Schematic drawings of the brain (all drawn to the same size) in the rhamphorhynchoid *Rhamphorhynchus* (above left), the pterodactyloid, *Anhanguera* (above right), a bird (below left) and a crocodile (below right). (Redrawn with permission from Witmer et al., 2003).

relatively unimportant for pterosaurs. By contrast, the cerebral hemispheres, centers for consciousness and cognitive activities that formed the main part of the forebrain, were relatively large and, like birds, but unlike reptiles, even had furrowed surfaces, suggestive of some internal complexity.

Another strikingly bird-like feature of the pterosaur brain was the position and size of the optic lobes, which are part of the mid-brain and connected to each eyeball by an optic nerve.[19] In reptiles, the optic lobes lie on the main axis of the brain and are rather small, whereas in pterosaurs, as in birds, they were situated in a low position, almost beneath the cerebral hemispheres, and were relatively large. This indicates that pterosaurs relied heavily on eyesight and processed considerable amounts of visual information, which is to be expected, because the eyes must have played a key role in critical behaviors such as flight and the hunting and catching of prey.

Where the CAT scanning work really broke into new territory was in showing details of the hind-brain, a region concerned with reflex activities, such as balance and posture. It was already known from older studies that the main component of this region, the cerebellum, exhibited a remarkably bird-like condition in that, although it lacked the intense folding seen in birds, it lapped forward over the midbrain to contact the forebrain. What Witmer and colleagues established, however, was that in pterosaurs the semicircular canals, which form part of the inner ear and act as the main organs of balance,[20] were extraordinarily large. They are large in birds and bats, too, as one might expect in flying animals where a high degree of sensitivity to any changes in orientation is absolutely vital, but they were even larger in pterosaurs.

Taken at face value, this might suggest that, in some respects, the flight ability of pterosaurs was at least as good as, if not even better than, that of birds and bats, but there may be another reason for the large size. The semicircular canals enclose another important part of the brain, a lobe called the flocculus, which receives impulses from several parts of the body, including the neck muscles, the eyes and the skin. Birds have proportionately the largest flocculi of all living animals, but in pterosaurs they were even larger, prompting the question: Why? The answer, detailed in Chapter 8, is rather unexpected and possibly related to pterosaurs' wings.

Apart from their size, the semicircular canals had another surprise, illustrated in *Figure 5.7*. They revealed, for the first time, the likely position in which pterosaurs held their heads during flight and, perhaps even more importantly, on the ground. The theory behind this is quite simple. Studies of mammals and birds have shown that in the posture typically adopted by the head, the semicircular canals are aligned so that the lateral canal is more or less parallel to the ground.[21] Consequently, if we take a virtual pterosaur brain endocast and rotate it so that the lateral canal is also in a horizontal position, we can discover the typical head posture for that particular species. The result for *Rhamphorhynchus*, which probably holds true for other rhamphorhynchoids as well, was unsurprising: The head, together with the neck and body, appears to have lain in an almost straight line, both in flight and when it was moving on the ground.

In pterodactyloids, however, things were quite different. Here, as the results for *Anhanguera* show, the lateral canal is inclined to the long axis of the head, so rotating the brain to bring the canal back into normal alignment results in a head-down posture. From this, we can deduce that in flight, the body and neck were probably near horizontal, while the head slanted

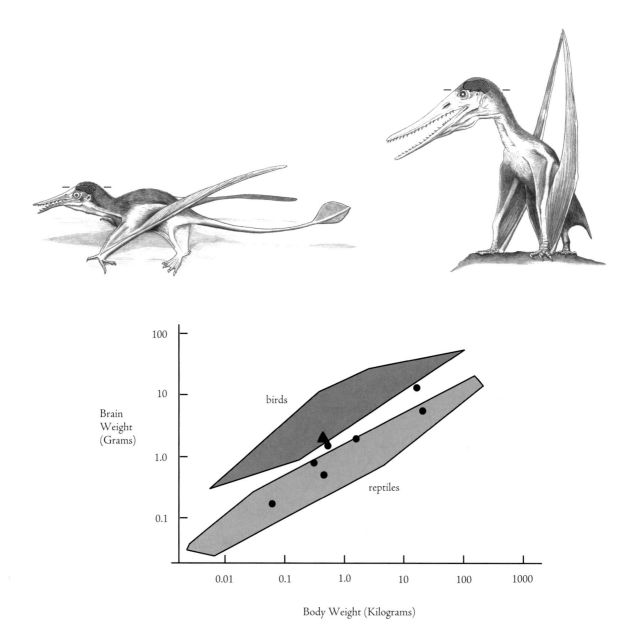

FIGURE 5.7 Ground truth? Pterosaur head orientations interpreted in terms of posture when on the ground. Above left: The horizontal alignment of the lateral semicircular canal, indicated by the red line, is consistent with a crouching posture and forward-directed head in long-tailed pterosaurs, represented by *Rhamphorhynchus*. Above right: In pterodactyloids such as *Anhanguera*, the reorientation of the canal can be interpreted in terms of an upright position and downward-pointing head. Below: graph showing the relative mass of pterosaur brains (circles) compared with their body size, with polygons showing the same relationship for reptiles and birds including *Archaeopteryx* (triangle). (Redrawn from David Unwin, 2003, Larry Witmer et al., 2003, and Jim Hopson, 1980.)

downward, which might have been important during hunting.[22] Down on the ground, though, the combination of this head posture with a horizontal neck and body would have resulted in a rather peculiar position with the eyes facing downward, rather than forward. This problem is easily solved by canting the body and neck steeply upward, bringing the head up into a forward-facing posture. What's more, this position, although quite different from that of rhamphorhynchoids, was easily supported by the relatively long arms of pterodactyloids. It also matches details of the numerous tracks that they left behind—a neat and attractive set of interlocking ideas that will show up again in Chapter 9.

Before we take our leave of pterosaur's brains, we might ask one final question: Do their brains tell us anything about their intelligence? The quickest way to answer this question is to calculate the ratio of brain mass to body mass and compare it with living tetrapods. This comparison gives a very rough guide to degrees of intelligence, in that the highest values, reflecting the largest and most complex brains, are found in men and other apes, while reptiles and amphibians, not known for their intellectual capacities, have the lowest values.

The downside is that calculating this ratio for fossils is difficult, especially for pterosaurs, where most skulls are crushed flat, so the few available data for this group, shown on the graph in *Figure 5.7*, should be treated with caution. Still, the location of pterosaurs right between the clusters for reptiles and birds is suggestive: Pterosaurs seem to have been rather more intelligent than your average living reptile and, as one might expect for a flier, they were able to exert more precise control over their movements. They may also have been a little more sophisticated in terms of their social behavior. On the other hand, although their brains were in many respects quite bird-like, primarily reflecting adaptation to an aerial mode of life rather than any close relationship with birds, pterosaurs might not have been quite as intelligent or endowed with the same degree of behavioral complexity as modern birds.[23]

Does My Head Look Big With This? Visible from a great distance—surely a clue to their function—and quite breathtaking both in their size and variety, the most splendid feature of many pterosaurs was their head crest. Such crests have been known since the mid-1800s, following the discovery of the bony-finned snouts of ornithocheirids in the Cambridge Greensand and the extraordinary weather-vane crest that decorated the head of *Pteranodon*. It is only in the last decade, however, with the finding of numerous new kinds of pterosaurs and of fossilized soft parts that extended crests to new heights, that it has become possible to comprehend the true diversity and function of these extraordinary structures.

Until quite recently, it seemed that bony crests, at least, were confined to the pterodactyloids, but their discovery in four different rhamphorhynchoids demonstrates that they were present in all groups of pterosaurs, as *Figure 5.8* illustrates. Perhaps the most important and stunning find in this respect was the long, low, blade-like crest adorning the head of *Austriadactylus*.[24] This was the first evidence of a crest in a rhamphorhynchoid and, significantly, is from the Triassic, proving that pterosaur head crests were a constant feature of the group's 150-million-year history.

The uneven, incomplete edge of the crest in *Austriadactylus* hints at the possibility that, in life, it was further extended by soft tissues, an idea that has been confirmed by the exceptionally well-preserved remains of another new rhamphorhynchoid, *Pterorhynchus*. This pterosaur, a scaphognathine from the Lower Cretaceous of China, has a low bony ridge on the snout, from which rose a large, keel-like crest that appears to have been constructed from a stiffened and probably rather leathery sheet of skin.[25] Although overlooked by a string of researchers beginning with Richard Owen, *Dimorphodon*, one of the most primitive of pterosaurs, also had a small fin-like crest on the tip of the mandibles. By contrast, other long-tailed forms, most notably *Rhamphorhynchus* and its close relatives, do not seem to have sported any bony head crests at all, although the possibility that they bore flaps or sails of stiffened skin, as yet unrepresented or unrecognized in fossils, cannot be ruled out.[26]

Head crests reached their greatest diversity and exuberance in the pterodactyloids, yet are constructed in two quite different ways, hinting perhaps at separate origins. One type of crest, confined to the ornithocheiroids, had a completely smooth outer edge, suggesting that originally it was covered with a close-fitting, thin layer of skin that added little or nothing to its overall size and shape. All the same, ornithocheiroid head crests were remarkably variable, ranging from pug-like prows on the tips of the jaws and half-moon crests on the top of the snout, to the blade-thin fins that spring from the crown of the head in *Ludodactylus* and *Pteranodon*, and not forgetting the extraordinary forked crest of *Nyctosaurus*. Not all pterosaurs were so gaudily adorned, however. Many crested ornithocheiroids had relatives, in some cases members of their own species, in which the crest was relatively small or even completely absent.

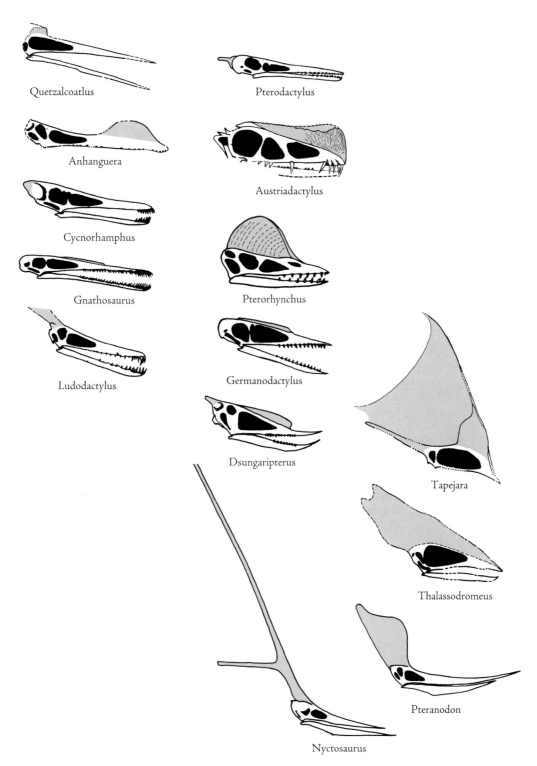

Quetzalcoatlus

Pterodactylus

Anhanguera

Austriadactylus

Cycnorhamphus

Pterorhynchus

Gnathosaurus

Germanodactylus

Ludodactylus

Dsungaripterus

Tapejara

Thalassodromeus

Pteranodon

Nyctosaurus

FIGURE 5.8 Cranial crests show a remarkable degree of diversity in pterosaurs, ranging from the keel-like jaw tip decorations of *Ornithocheirus* to the extraordinarily tall "sail" borne by *Tupuxuara*. In each case the crest is picked out with a pink fill above.

All other pterodactyloid clans, except ornithocheiroids, had a second type of crest, its base composed of bone, but, as shown by several exceptionally preserved skulls of *Tapejara* from the Crato Limestone, its upper part formed from a leathery sheet of skin that was supported internally by a stiff, fiber-like network.[27] These same fossils also show that the narrow, blade-like sail was given additional support through a thickening of the leading edge. Contrasting strongly with the situation in ornithocheiroids, this second crest type is almost always found in the same location, extending from the crown of the head forward along the snout, sometimes, as in *Tupuxuara*, almost reaching its tip. Except for one or two lonchodectids, bony crests on the lower jaw are unknown, although, occasionally, as in the dsungaripterids and at least one species of *Tapejara*, a second bony crest extended upward from the back of the skull. In many cases, the main bony crest has an unfinished outer edge, marking the junction with the base of its leathery continuation, but only in *Tapejara* is the true extent of this development known. Astonishingly, in this pterosaur, the crest reached a height equivalent to five times the height of the skull—if the situation in other pterodactyloids was even remotely similar, then the Mesozoic skies must have been a sight to behold.

Do Ya Think I'm Sexy? Pterosaurs clearly invested substantial resources in growing, maintaining, and coping with their head crests (imagine flying in gusty winds with a weather-vane rooted to your head), which must have been of considerable importance to them. But for what? It's tempting to believe they had some kind of mechanical function, and there are plenty of ideas on offer. Several paleontologists have suggested that perhaps the larger, flap-like crests were employed as a rudder that enabled pterosaurs to steer themselves through the skies. Or, a clever variation, if the softer part of the crest had been able to develop a camber and behave like a sail it might have allowed pterosaurs to tack into the air flow—ingenious, but impossible for *Pteranodon*, or any other bony-crested ornithocheiroid.[28] Other functional explanations for the crests include the following: a device to guide the jaws through the water while fishing, although this must have been restricted to pterosaurs with crested jaw tips;[29] a counterbalance to the jaws, in *Pteranodon* and similar forms, helping to reduce muscle mass that would otherwise have been needed to stabilize the head;[30] or perhaps even a means of dissipating excess heat.[31]

The problem is that all these mechanical explanations, and their variants, suffer from at least two fatal drawbacks. First, with the possible exception of the "radiator theory," they explain only one or two particular kinds of crest.

All the other variations are either unaccounted for or, worse still, would have had exactly the opposite effect. A second problem is that these ideas completely fail to explain a peculiar thing about crests: why they are present in one pterosaur, but not in another. And it doesn't matter if these two fossils are thought to have belonged to the same species, two different species, or even two different genera, because what is the benefit to one pterosaur of investing in a crest when another almost identical individual seems to have managed perfectly well without this expensive adornment?

The alternative is much more attractive: sex. Or, to be more explicit, crests served as display devices to attract members of the opposite gender or discourage members of the same gender. So, what's the evidence? Analyze lots of fossils of *Pteranodon longiceps*, as Chris Bennett did,[32] and one finds that some individuals have a relatively large, well-developed crest, while in others it is much smaller and far less conspicuous. Now, carefully examine the other end of the same animals, and one discovers that big-crested individuals have a relatively small, narrow pelvis, while small-crested forms have a relatively large and wide pelvis, presumably to permit the passage of an egg.

The message seems clear: These were males and females. Interestingly, on average, males of *Pteranodon longiceps* seem to have been somewhat larger than females, a pattern known as size dimorphism and quite common in reptiles,[33] though rare in birds and bats. Apart from reinforcing Bennett's ideas regarding *Pteranodon*, this pattern can also be linked to another rather curious aspect of pterosaurs, discussed in detail in Chapter 7—the remarkable extent to which adult size varies.

Pteranodon longiceps is by no means the only pterosaur where the so-called "dimorphic" pattern of crest size is to be found. It also occurs in species of *Pterodactylus*, *Germanodactylus*, *Lonchodectes*, *Anhanguera*, *Coloborhynchus* and *Nyctosaurus* and several more cases where closely related species, or even genera, appear almost identical, apart from the presence or absence of crests. *Ctenochasma elegans* and *Ctenochasma porocristata* provide one instance, *Brasileodactylus* and *Anhanguera* another. All these examples are most easily and most convincingly interpreted as cases of sexual dimorphism—in which one gender, often (but not necessarily always) the male, bears a display device that is smaller or absent in the other. First documented in detail by Charles Darwin more than 100 years ago in *The Descent of Man and Selection in Relation to Sex*, examples of sexual dimorphism in the modern world are legion with spectacular examples, including peacocks' tails, deer antlers, chameleons' horns and the crests of newts and salamanders.[34]

Another advantage of the idea that crests served for display is that it neatly explains why their size, shape and position varied so much. Quite simply, it did not matter where the advertisement was located—on the tip of the jaws or the back of the head—as long as it was clearly visible to everyone else. This leads to another important point. Among living animals, a secondary effect of display devices is that they often act as a means of distinguishing among different species. Consequently, closely related forms living in the same area, ducks for example, tend to have distinctly different display devices, in their particular case, patterns of feather coloration. If the same held true for pterosaurs, then crest size, shape and position would be expected to vary among species found at the same locality. Which indeed they do as, for example, in the Santana Formation pterosaurs: *Ornithocheirus mesembrinus* has a crest right on the tip of the snout; in *Coloborhynchus robustus*, it is set a little farther back and has a different shape; *Tapejara wellnhoferi* has a very tall, sail-like crest on top of the head; while the crest of *Tupuxuara leonardii* extends almost the whole length of the skull.

That crests were used for display seems fairly certain, but how they were used and under what circumstances is not yet clear. That said, it is most likely that their main function was to draw attention to the owner, to make it appear larger and to impress upon the opposite sex his (or her) superior fitness as a mate. This means that sight, which, as already detailed, was well-developed in pterosaurs, is likely to have played a key role in the affair and that the crested sex displayed in some way, either on the ground or perhaps in the air. It might even be supposed that males and females gathered together for these displays, as some species of birds, mammals and insects do today when they take part in leks.[35]

Wild speculation, surely? Well, not necessarily. Some recently discovered pterosaur track sites, detailed in Chapter 8, seem to show lots of individuals milling around together. Is this a record of some long-ago parade in which bizarrely crested pterosaurs strutted their stuff or craned their necks to catch a glimpse of the new kid in the air?

6
THE BODY INSIDE OUT

Exposed on the rocky headland that projected out into the Tethyan Sea, the lone gingko tree swayed gently in a newly sprung breeze. The upper story of this storm-beaten old giant, now well into its third century, was still bearded with greenery, contrasting with the bare lower reaches, interrupted here and there by splintered stubs of long-lost branches. A perfect roost for the clusters of pterosaurs that clung to the patriarch's white trunk, bespattered with the droppings of uncounted generations. A harbinger of the gale to come, the freshening breeze clattered the leaves and buffeted pterosaurs on the windward side of the tree, streaming out their manes or blowing them back over their heads. The short, fine pelt that ran from their shoulders to the base of the long tail had an iridescent, almost oily, sheen and was jet black, fading rapidly to gray on the flanks and becoming pure white underneath. This monochrome effect was mirrored by the brilliant white of the pelt around the eye, offset against the midnight black color of the rest of the head, and magnified by the mane that adorned the neck. Black at the base and pure white at the tips, its myriad thread-like strands, whipped and whirled by the rising wind, were becoming raveled and snagged in tufts and skeins.[1]

FIGURE 6.1 Fine details in the wall (about 1 millimeter thick) of a pterosaur ulna are made visible by slicing up and polishing fossilized bone until it is very thin, then photographing it using a microscope and polarized light. The bone tissue seen here is termed fibro-lamellar and is typical of, although not entirely restricted to, fast-growing tetrapods, such as mammals and birds. (Photograph courtesy of Lorna Steel.)

Body Parts If it were possible to take apart a pterosaur's body, it would be found to have the basic plan shared by all tetrapods. As shown in *Figure 6.2*, the backbone (vertebral column), which surrounded the spinal cord, the nerves' superhighway, was the principal supporting element of the skeleton, and its successive sections: the neck, the back and the tail, had particular roles. The neck essentially supported the head and gave it the mobility and reach that it needed, for example, while fishing. The main part of the body, which housed most of the major internal organs: the heart, lungs, gut, liver, kidneys and reproductive machinery, was enclosed by the rib cage and supported at the back by the pelvis, beyond which, and bringing up the rear, was the tail.

Just as in most other land-living tetrapods, the shoulder girdles and pelvis were well-developed. The limbs followed almost exactly the same arrangement that we can trace out in our own arms and legs: a single bone, followed by a pair, then several rows of wrist or ankle bones, four digits in the hand (pterosaurs lost the little finger) and five in the foot. Some of the major limb bones were elongated, especially the fourth finger of the hand, but, apart from their relative size and fine-scale anatomy, the arms and legs seem to have been controlled, supplied and moved by pretty much the same nerves, blood vessels and muscles found in all tetrapods.

Pterosaurs had just two pairs of bones whose counterparts do not seem to be present in other diapsids: One of these, the pteroid, was part of the wrist, while the other, the prepubis, was found in the pelvis. This means that the "unique" and highly distinctive appearance of pterosaurs was not due to the evolution of lots of new bits of anatomy, but largely arose through the modification of bones, organs and tissues that already existed in their reptilian ancestors. How the pterosaur's body was modified and what this can tell us about other aspects of pterosaur biology, such as the nature of its metabolism, is explained in this chapter.

Before we plunge in, two supporting themes deserve a little consideration. If one examines the skeletons of several different kinds of pterosaur, carefully ignoring the skull in each case, one feature soon becomes apparent. In terms of its composition and arrangement, the pterosaur skeleton was remarkably conservative. Aside from the tail, the spinal column has a similar number of vertebrae in all known species, the shoulder girdle and pelvis are always composed of the same elements and so, to a large degree, are the arms and legs. Variation does occur, of course, but it usually manifests itself as differences in the relative lengths and shapes of bones or the extent to which they fuse with one another, most noticeably in the backbone. Changes in the actual number of bones were rare and almost always involved the loss of one or more elements—the disappearance, in most pterodactyloids, of the fifth toe and several tail vertebrae being the most obvious examples.[2]

The conservatism of the pterosaur skeleton, a key element of the pterosaur story (see Chapters 10 and 11), is also of particular significance here. Recall that in all vertebrates, the skeleton is both intimately associated with the rest of the body and completely integrated with it. Consequently, when

FIGURE 6.2 The skeleton, above, and fleshed-out reconstruction, below, of a typical long-tailed pterosaur, *Rhamphorhynchus*, with a wingspan of about 3 feet (1 meter). (Redrawn, with modifications, from Peter Wellnhofer, 1975.)

the skeleton exhibits little variation (as in pterosaurs), it is reasonable to assume that other major structures and their primary functions, such as the lungs and breathing, probably did not vary much either. This is certainly true for living vertebrates and the same is likely to have applied to their extinct relatives, including pterosaurs, which is of considerable help in trying to understand these enigmatic animals.[3] It means that whatever can be discovered for one particular species, for example, regarding the breathing apparatus and how it worked, probably applied to pterosaurs in general.

There is, however, one important exception to the general rule of conservatism in pterosaur body architecture in rhamphorhynchoids and pterodactyloids. In these groups, the sections of backbone between the neck and tail, the rib cage and the pelvis are generally similar, suggesting that the major organ systems associated with this region—the heart, lungs, liver and kidneys—were fundamentally the same in all pterosaurs. By contrast, the neck, tail, arms and legs underwent some important changes during the evolution of pterodactyloids from rhamphorhynchoids. These modifications were of fundamental significance for the way these animals flew and walked, and, as the last chapter will reveal, they had critical consequences for the ecology, and ultimately the evolutionary fate, of pterosaurs.

Busy Backbone   In long-tailed pterosaurs, the backbone, or spinal column, illustrated in *Figures 6.2, 6.3* and *6.5*, was composed of at least 50 to 60 vertebrae;[4] it reached as many as 70 in *Rhamphorhynchus*. Pterodactyloids, by contrast, had as few as 33 to 34 vertebrae. The marked differences in these "vertebral counts" were almost entirely due to a single structure—the tail—which was highly variable in pterosaurs, perhaps because it had somewhat greater freedom to evolve than the rest of the spine. Exclude the tail, and one finds that rhamphorhynchoids had 26, or sometimes 27, vertebrae, whereas it was almost always 25 in pterodactyloids.

The main part of the spinal column can be divided into three sections: the neck, the back and the sacrum, each composed of distinctive types of vertebrae, shaped according to the specific tasks of each region. The neck consisted of cervical vertebrae whose number, after more than a century of debate, has now been firmly fixed at nine.[5] The back, composed of up to 18 rib-bearing dorsal vertebrae, varied somewhat in composition, depending on how many dorsals from the rear part were "captured" by the sacrum. Finally, the sacrum, whose primary task was to support the pelvis via several pairs of very short, stout ribs, contained at least three to four sacral vertebrae, although this count ranged up to nine in some pterosaurs, such as *Pteranodon*, the increase consisting almost entirely of captured dorsals.

A Stiff-Necked Breed The largest and most robust vertebrae were generally to be found in the neck, which, compared with other reptiles, was relatively long and even exceeded the length of the rest of the body in several pterodactyloids. *Dimorphodon* and other rhamphorhynchoids had neck vertebrae that were rather short and chunky and topped by a prominent bony spine. Generally, the neck vertebrae were similar in size and shape to one another, with two exceptions. At the beginning of the neck, the first vertebra, also known as the "atlas," on which the head articulated, was very short, often little more than a flattened disc crowned by a simple arch. In later pterosaurs, it was usually fused with the second vertebra, the "axis," which was also rather short and stubby. At the other end of the neck, the last two vertebrae tended to be shorter than those preceding them and had relatively large ribs that stuck out sideways rather than outward and backward.

Big changes came with the advent of pterodactyloids. First of all, most of the neck vertebrae, apart from the last two, lost their ribs—or at least that, until recently, was the generally accepted idea. A beautifully preserved young adult specimen of *Coloborhynchus* in the collections of the National Science Museum in Tokyo[6] has revealed that although the ribs were much reduced in size they were not lost, but became so tightly fused to the neck vertebrae that, in most cases, they could not be distinguished from them.

The most striking feature of the pterodactyloid neck was its greater length compared with that of rhamphorhynchoids. Indeed, in two groups, ctenochasmatoids and azhdarchids, it became extraordinarily long and, in some extreme cases, achieved a length greater than the rest of the body and the tail put together. This was achieved not by adding vertebrae, but by stretching out some of those in the middle part of the neck—usually centered on the fifth—so that, as *Figure 6.3* reveals, they ended up as long, low tubes, their length more than eight times their height.

Despite the enormous variation in the length of the neck, the general construction of the articulations between individual vertebrae was rather similar throughout the group. In a few well-preserved, uncrushed fossils, such as the Tokyo specimen of *Coloborhynchus* mentioned above,[7] it is possible to articulate the vertebrae with one another and then manipulate them

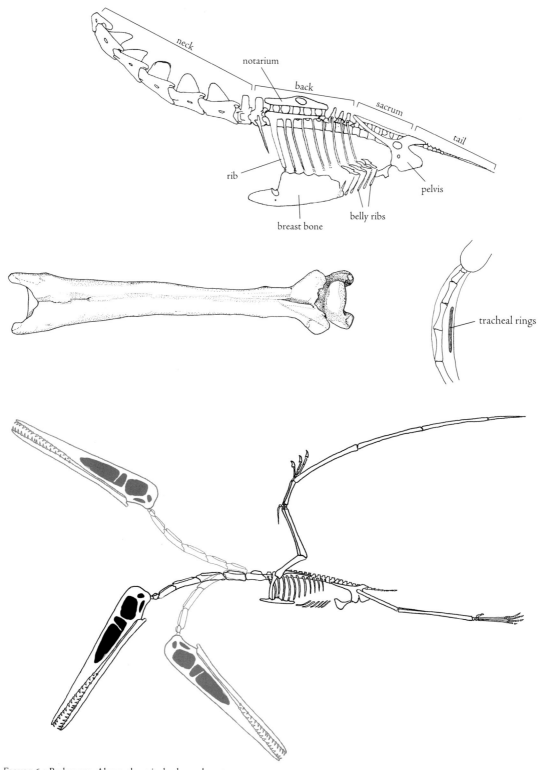

FIGURE 6.3 Body parts. Above: the spinal column, breast bone and ribs of *Pteranodon* are seen in side view. Middle left: one of the extraordinarily elongated neck vertebra of *Quetzalcoatlus*, as seen from above. Middle right: sketch of the neck of *Pterodactylus*, with its supporting tracheal rings. Below: the likely range of neck movements in *Pterodactylus*. (Redrawn from Chris Bennett, 2001, Stafford Howse, 1995, and Peter Wellnhofer, 1970.)

to find out how the neck moved in life. This tells us something rather curious about pterosaurs: They had stiff necks.

The greatest range of movement was at the point of contact between the skull and the atlas, where a ball and cup arrangement allowed the head to be raised and lowered through a wide arc, swiveled from side to side, and even twisted around. Movements between the neck vertebrae themselves were much more restricted: Twisting seems to have been impossible, but pterosaurs could manage a small amount of swiveling to left or right and an even greater degree of nodding. Proceeding backward, these movements became more and more reduced, such that the last few neck vertebrae were practically immobile.

Judging from the best match between articulation surfaces on corresponding vertebrae,[8] the typical "neutral" position of the neck during flight was approximately horizontal. Muscles started from particular vertebrae and continued along the top and beneath the neck, attaching to neighboring vertebrae to the front and eventually becoming inserted into the back of the head. By contracting and relaxing these muscles, a pterosaur could raise its neck and head some way above horizontal and lower them so that the head faced almost vertically downward or even looked a little backward (*Figure 6.3*). The head and neck could also be swung to the left or to the right by muscles that ran forward along the neck on each side, attaching to successive vertebrae and ultimately to the back of the head.

Together, these sets of muscles also helped to stabilize the neck and ensured that the head was held steady during flight, which was especially important for pterosaurs that hunted on the wing. Indeed, the maneuverability of the neck seems to have been quite sufficient for this activity, even though the neck was less mobile than in birds and could not, for example, adopt the "S" shape assumed by ducks and gulls when they swim on water.

Apart from vertebrae, muscles and associated nerves and blood vessels, the neck also contained the esophagus and the main breathing tube, the trachea, which connected the throat to the lungs. Evidence of these soft parts is almost never preserved, the one exception being a large individual of *Pterodactylus*,[9] illustrated in *Figure 6.3*, in which several ring-like objects lie just below the neck vertebrae. Comparisons with birds, which have almost identical structures in their necks, suggest that these are the fossilized remains of cartilaginous rings that helped to support the trachea and prevent it from collapsing or being compressed.

Caged Between the Ribs In pterosaurs, the main part of the body, comprising the chest and abdomen, was relatively short and compact and generally a little wider than it was deep. The principal support for this region was the backbone, which consisted of up to 18 relatively tall, narrow dorsal vertebrae, each tipped with a prominent spine. The first of these vertebrae (the 10th overall in the backbone), was braced on each side by a large double-headed rib that curved out and around to contact a so-called sternal rib, which was short and stubby and articulated with the outer edge of the breast bone (sternum)—a large, flat, shield-shaped plate located in the middle of the chest.

Subsequent vertebrae and ribs followed the same pattern almost all the way to the pelvis, except that by the seventh or eighth pair of ribs (and beyond), contact was no longer made with the breast bone, because it only extended as far as the rear end of the chest. Beyond this point, each pair of dorsal ribs linked up, via a short, intermediate rib on each side, with a set of belly ribs, or gastralia,[10] of which there were typically six to eight sets. The gastralia formed that most useful of items, an in-built corset, which was embedded in the body wall and helped to support it.

Overall, then, the body was completely enclosed in a rib cage that consisted of a series of bony rings. Initially, each ring was formed by a vertebra, ribs and the breast bone in the region of the chest, and then later by a vertebra, ribs and belly ribs in the region of the abdomen. The first ring was the stoutest and had the greatest diameter, while, proceeding toward the rear, each ring became slightly narrower than the last, so that the body tapered toward the pelvis. The last one or two dorsal vertebrae before the pelvis usually had very small ribs or no ribs at all.[11]

In the hip region, the vertebrae usually fused into one another, forming a structure called the sacrum, which was firmly united on each side by short, stout sacral ribs to the bony plates that formed the pelvis. Usually, in adults, this whole arrangement—sacral vertebrae, sacral ribs and pelvic plates—became completely fused into a solid bony frame. In rhamphorhynchoids, this involved at least three, often four and sometimes five sacrals. By contrast, in pterodactyloids, at least four or five sacrals were usually involved, and in large and giant species, the incorporation of an increasing number of dorsals could push this figure up to 9 or 10.

A similar type of bony unification also affected vertebrae in the early part of the back. Again, it only seems to occur in the larger pterodactyloids and in its simplest development involved the fusion of the first three or four dorsal vertebrae to form a structure called the notarium.[12] In some larger pterosaurs, as many as six or seven vertebrae became solidly united. This included the spines on top of the vertebrae, which fused into a long, narrow bar.

The notarium was associated with another development unique to pterosaurs. The far end of the shoulder blade, which always lay close to the first few dorsal vertebrae, came into contact with the bar, and in some cases, this eventually evolved into a true articulation. Because the lower end of each shoulder girdle already articulated with the breast bone, this resulted in what must have been a remarkably strong ring of bone composed of the notarium, shoulder girdles and sternum—exactly the kind of platform needed for a really big pair of wings.

Heavy Breathing In pterosaurs, many of the most important organs—heart, lungs, guts, liver, kidneys and reproductive structures—together with their nerve network and blood supply, were packed into the rib-bound body cavity. We can be fairly certain that these organs were present, because all other tetrapods have them and because it's hard to see how pterosaurs might have managed without them, but that's about as far as it goes. In most cases, there is no direct, or even indirect, fossil evidence from the shape of the bones that might give us more information on what these organs were like and how they functioned. There are, however, two exceptions: the guts and the lungs.

One highly intriguing example of *Rhamphorhynchus*, the "greedy guts" that was featured in the last chapter and is preserved on its side, rather than on its back or belly as is usually the case,[13] has several small banana-shaped objects preserved in the region of the abdomen (*Figure 6.4*). Bearing a peculiar zigzag type of ribbing, they have been interpreted as fossilized organic remains, but do not correspond to any of the plants or animals, or their parts, currently known from the Solnhofen Limestone. Another possibility, proposed by Gunther Viohl of the Jura Museum in Eichstätt, Germany, is that these objects are internal casts of the gut that might perhaps have formed in the intestines. That's not as unlikely as it might seem, because this is one of the few pterosaurs in which the stomach contents—a fish—are also preserved, and they lie immediately in front of the supposed gut casts.

Turning to the lungs, it would seem that because pterosaurs were active, flapping fliers, there can be little doubt that they would have required a fast and efficient respiration, if they were not to plunge gasping from the skies after just a few wing beats. The mystery is that the breathing apparatus of living reptiles, such as lizards and crocodiles, certainly would not have been up to the task, so pterosaurs must have done something else. But what?

Details of the skeleton supply a couple of clues, and, although neither has yet been fully explored or is properly understood, they give some hints as to how pterosaurs breathed. First of all, recall that between the neck and the tail, the backbone is relatively stiff, and in some pterosaurs, including *Pteranodon* and other ornithocheiroids, it consisted largely of fused vertebrae that

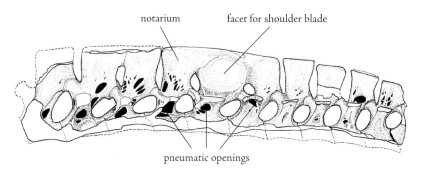

FIGURE 6.4 Guts and lungs leave their marks. Above: the stomach region of Mr. Greedy Guts, a specimen of *Rhamphorhynchus* from the Upper Jurassic Solnhofen Limestone, contains the remains of a fish (seen in the upper middle part of the picture), and several banana-shaped objects (a prominent one is seen at mid right) that might be internal casts of part of the gut. Below: ten dorsal vertebrae of a large ornithocheirid pterosaur from the Lower Cretaceous Santana Formation of Brazil are riddled with pneumatic openings (black fill) left by outgrowths from the lungs. Note that vertebrae four, five and six are fused together to form a notarium, and the fifth vertebra bears a large facet against which the shoulder blade articulated. (Photograph courtesy of Peter Wellnhofer; vertebrae redrawn from Wellnhofer et al., 1983.)

formed the notarium and sacrum. Remember, also, that the body cavity was completely enclosed by the rib cage, the construction of which seems to have allowed little movement in the forward half, but rather more in the region to the rear. Consequently, it would seem that the only effective means of changing the volume of the body cavity and ventilating the lungs was by raising and lowering the gastralia, which was probably brought about by contracting and relaxing muscles in the sides of the abdomen.

The second clue as to lung function consists, literally, of holes in the bones and is referred to as pneumatization. In many pterosaurs, not only did the dorsal vertebrae have a large channel for the passage of the spinal cord, they were also pierced by additional sets of openings, often set in deep recesses located on each side of the vertebrae, or, as shown in *Figure 6.4*, flanking the channel for the spinal cord. The significance of these "holes" is to be found in birds, where openings almost identical in size, shape and position to those of pterosaurs are produced by outgrowths from the lungs. Initially, these outgrowths develop into air sacs, several pairs of which are found in the body cavity. They play a vital role in breathing, helping to move air back and forth through the lungs—a key component of birds' highly effective flow-through breathing system. The important point here is that in many birds, extensions from the air sacs pneumatize adjacent vertebrae, replacing some of their internal substance with air spaces, lightening them but without resulting in any significant structural weakening—very useful for a flier.[14] In many (perhaps all) pterosaurs, several of the dorsal vertebrae were pneumatized, and this often extended into other parts of the spinal column. The shoulder girdles and limb bones of many pterodactyloids are pneumatized, and in extreme cases, such as *Pteranodon*, this system seems to have penetrated almost the entire skeleton.[15]

Link pneumatization, or rather what it implies, air sacs, with the limited mobility of the rib cage, and we can gain some idea of how pterosaurs' lungs may have worked. As the gastralia were depressed, the volume of the abdomen increased and the air sacs inflated, drawing air through the lungs. Then, as the gastralia were raised again by muscle contraction, the volume of the abdomen decreased, compressing the air sacs and forcing air back out through the lungs. If this is correct,[16] it would mean that pterosaurs evolved a flow-through breathing system similar to that found in birds. Studies of this system in the latter group show that it is extremely efficient at gas exchange,[17] which is exactly what birds need to help power the highly demanding activity of flight. Because pterosaurs also appear to have been active fliers, the evolution of a breathing system like that of birds is not unexpected. What's more, if it can be fully substantiated, this discovery will surely prove to be a key step forward in understanding how these animals worked.

A Tale of the Tails The long tail, emblematic of most, but not all, rhamphorhynchoids, was composed of at least 25 "caudal" vertebrae and reached a maximum of about 40 caudals in *Rhamphorhynchus*, where the tail was almost twice the length of the rest of the backbone (*Figure 6.2*). The construction of the rhamphorhynchoid tail, illustrated in *Figure 6.5*, was quite peculiar. The first few caudals were short and similar in size to typical dorsals, but farther back, they became more and more stretched out, such that by the mid-region of the tail, they had become long, low, tube-like structures up to eight times longer than they were broad. Beyond this, the vertebrae shortened again, and the last few were hardly any longer than the first few. This was not a good design for a curly tail, but fine if it were meant to be kept straight, which apparently it was, as other features clearly demonstrate.

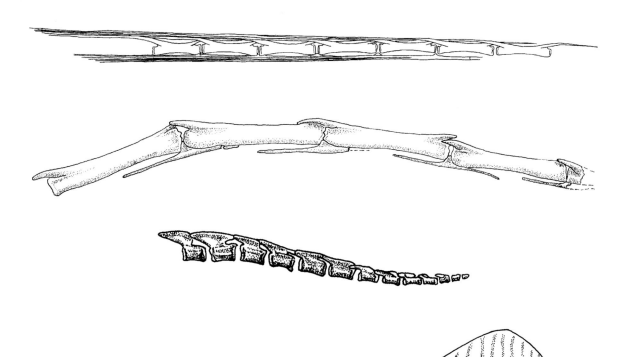

FIGURE 6.5 A tale of the tail. Above: a short section of a typical rhamphorhynchoid tail (*Rhamphorhynchus*), complete with sheathing bundle of bony spars. Upper middle: another long-tailed form (*Eudimorphodon*) in which, unusually, the tail lacks any sheath of bony spars. Lower middle: a short-tailed form (*Pterodactylus*). Beneath: a well-preserved tail flap in *Rhamphorhynchus*. (Redrawn from Wellnhofer, 1975, 1991 and 2003.)

The joints between the first few caudals were rather flexible and allowed the tail to be swung quite freely from one side to the other. The tail could also be waved up and down, although movement in this direction was somewhat limited. Beyond about the sixth caudal, however, the vertebrae became more tightly interlocked, and there was practically no mobility at all, so the rest of the tail behaved like a rather stiff rod. This stiffness was further supported by a most unusual development: numerous extremely long, straight, thin, bony spars that overlapped the caudal vertebrae, both from above and from below. Totaling more than 40 in number in midsections of the tail of *Rhamphorhynchus*, these spars consisted of slender spars of bone that arose both from the front and back of the caudals and from the front and back of small bones called chevrons that lay below the tail vertebrae. Each bony spar ran forward, overlapping as many as five or six vertebrae, and the entire construction—elongate caudals encased within bundles of thin bony spars— endowed pterosaurs with a stiff tail, but ensured that it was whippy, rather than rigid.[18]

The reason for all this skeletal engineering of the rhamphorhynchoid tail was to be found at its far end: a leaf-thin vertical vane. Composed of skin (or so it would seem), and supported internally by a series of uniformly spaced, vertical, rod-like thickenings, fossilized remains of tail vanes have been found in three different pterosaurs and are best known in *Rhamphorhynchus*. In this pterosaur, the vane has a rhombus shape and is slightly asymmetric in outline, showing that, in life, it was oriented vertically, just like the tail-fin of an airplane. Longer and lower in other pterosaurs,[19] the tail vane undoubtedly played a role during flight and probably assisted with both stability and steering.

Interestingly, the pterodactyloids and one group of rhamphorhynchoids, the anurognathids, managed quite happily without this vane. These pterosaurs usually had no more than 15 or 16 simple caudals that gradually became smaller and shorter toward the tip of what was a relatively short tail.[20] Not all pterodactyloids conformed to this pattern, though. *Pterodaustro* had a relatively long tail for a pterodactyloid, with at least 22 vertebrae, and in *Pteranodon* the first six or so caudals had unusual joints that allowed the tail to be moved up and down, but not from side to side, while the last few caudals were fused into a long rod-like structure.[21] Why these features evolved and what they were used for is still a mystery.

Shoulders Pterosaurs had a well-developed shoulder girdle, similar in shape to that of birds (*Figure 6.6*). Its primary function was to serve as a platform on which the wings articulated, via the shoulder joint, and from which arose many of the muscles that powered and operated the wings.[22] Each side of the shoulder girdle was constructed from two long bars of bone that, in mature pterosaurs, fused into a single unit that looked like a "V" turned on its side.

The shoulder blade, or scapula, formed the upper leg of the < and curved up and over the rib cage to lie almost parallel to the backbone. In many pterosaurs, it was held in place by ligaments and possibly by muscles and tendons, but, as mentioned earlier, in larger pterosaurs, its far end often butted up against the notarium. The scapula, and to some extent the vertebrae and ribs in its vicinity, supported the muscles that were responsible for raising the wing during the flight stroke.[23]

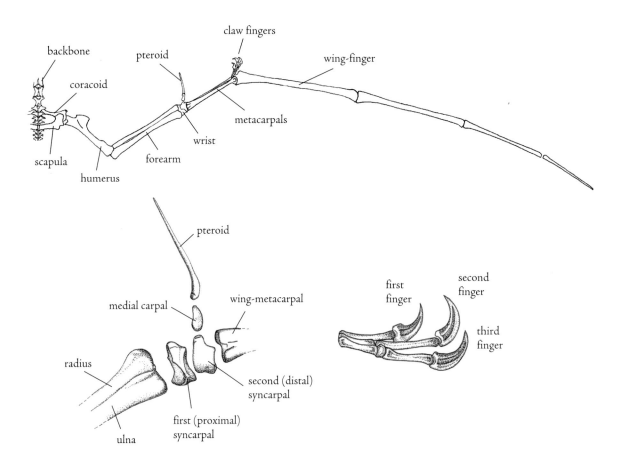

FIGURE 6.6 The shoulders and arms of Anhanguera. Above: the shoulder girdle and forelimb as seen from above. Bottom left: details of the wrist region also seen from above. Bottom right: the claw-bearing fingers of the hand. (Redrawn from Peter Wellnhofer, 1985 and 1991.)

The coracoid formed the lower leg of the < and angled downward and inward, its lower end articulated against the breast bone. Together, the coracoid and breast bone provided the main attachment site for the major muscles that powered the flight stroke, among them the pectoralis, which was far and away the largest muscle in the entire pterosaur body.

The shoulder joint itself developed at the point where the scapula and coracoid met and had a rather complicated saddle-shaped surface. It was quite deeply dished from top to bottom, yet bulged out slightly from front to back. Well-preserved, uncrushed remains of several pterosaurs show that, when engaged with this joint, the upper arm bone, the humerus, was free to move in several directions and even twist backward or forward.[24] As detailed in later chapters, this possibility allowed the arms to be used both for flying and walking.

Lying in the center of the chest, all pterosaurs had a large, shield-shaped breast bone that was flat or gently dished and supported the coracoids and at least six or seven pairs of ribs. Although it looks like a single bone, studies of an immature pterosaur show that it was made up of several elements, including the clavicles, which were thought to have been lost, but now appear to have been hiding in the breast bone all along.

Arms and Fingers The humerus was short, stocky and straight, and its most prominent feature was an enormous wing-like flange that arose from just beyond the head of the bone. Called the deltopectoral process, this flange was the attachment point of the pectoralis muscle—the main powerhouse for the downward part of the flight stroke in pterosaurs. Other muscles that helped lift the wing back up again at the end of the flight stroke also attached to the humerus, and the lower end of this bone also served as the starting point for many muscles that operated the wrist and hand.

The forearm, as in other tetrapods, comprised two bones, the radius in front and the ulna behind. Long and tubular in pterosaurs, this pair of bones articulated with the humerus at the elbow, while the outer ends pressed up tight against the wrist. The construction of the elbow joint meant that, like birds but unlike most other tetrapods, pterosaurs could not rotate their forearms to any real degree, probably an adaptation that helped to resist the tendency of the wing to twist during flight.

So far, the pterosaur arm has hardly differed from that of most other tetrapods, but as *Figure 6.6* shows, when it came to the wrist, things changed—a lot. Somewhere on the evolutionary road to pterosaurs, several wrist bones were lost and, in mature individuals, almost all of those that remained were fused together into two large composite bones called syncarpals.[25]

The first syncarpal was tightly sandwiched between the forearm bones on one side and its neighboring (second) syncarpal on the other. The latter was even more sandwiched by the first syncarpal on one side, the base of the hand on the other and another "medial" carpal[26] to the front. The medial carpal was rather small and chunky, and its front end bore a deep notch into which fitted a special and uniquely pterosaurian bone—the pteroid.

The pteroid—short, flat and spatula-like in rhamphorhynchoids, but long, thin and not unlike a conductor's baton in pterodactyloids—has been and remains the most contentious bone in the entire pterosaur skeleton. Nineteenth century paleontologists argued fiercely over its identity: Was it the equivalent of our thumb (i.e., the first finger of the hand) or was it an entirely new bone unique to pterosaurs? The issue appeared to have been settled in the early 1900s in favor of the latter idea, only for the debate to break out again in the 1990s.[27] The problem is still not entirely resolved, but at present, most pterosaurologists doubt that the pteroid had anything to do with the fingers.

Identity aside, the pteroid has generously provided pterosaurologists with a second issue to fight over: Which way did it point? Inward, toward the body, has long been the almost universally accepted opinion. This is partly because that is how it is usually preserved in fossils and partly because it was widely thought that the pteroid was the attachment point for a tendon that arose from the shoulder region and ran along the edge of the forewing. In 1981, two German pterosaurologists, Dino Frey and Jurgen Riess, proposed a radically new idea: that the pteroid faced forward and served to move the forewing up and down in the manner of a "leading edge flap."

Initially, this idea was sharply criticized and even today, it continues to draw heavy fire whenever it pokes its head above the intellectual parapet. Nevertheless, there is growing evidence to suggest that it might be correct. Complete, superbly preserved wrist bones from several Santana Formation pterosaurs demonstrate that the pteroid *can* be articulated with the medial carpal to face forward and that it *can* also be moved up and down in precisely the manner needed to operate the forewing. Moreover, because of the peculiarly asymmetric shape of the joint surfaces, the pteroid could also be folded inward to point toward the body when the animal was at rest and had furled its wings, as seen in many "relaxed" (i.e., dead) individuals. Support for this idea has also come from another quarter—aerodynamic analyses that reveal that the wings would have worked much more efficiently if the pteroid had pointed forward.[28] The debate continues, and we will return to it in Chapter 8.

The part of the hand equivalent to the palm contained four elongated bones called metacarpals. The first three consisted just of thin, slender rods of bone that lay against the front face of the fourth metacarpal, which was relatively massive and formed the main wing spar. Rather short in rhamphorhynchoids, its much greater length in pterodactyloids is one of the most characteristic features of this group.

Typically, pterosaurs had a four-finger hand.[29] The first three fingers, although rather slender, had well-developed joints and large muscle scars and terminated in a comparatively long bone, on which articulated a claw.[30] In many pterosaurs, including all rhamphorhynchoids, the finger claws were large, deep, narrow, strongly hooked and sharply pointed. In life, the claw shape was further emphasized by the horny sheath within which each claw was enclosed, evidence of which is preserved in several pterosaurs, including *Rhamphorhynchus*, *Pterodactylus* and *Tapejara*.[31] All together, these features demonstrate that the "clawed" fingers had a powerful grasping ability that was probably used primarily for climbing rather than for grabbing and holding prey. In some of the larger pterodactyloids, which might have been just too big to climb, the "clawed" fingers were relatively small, with rather weak claws, while in *Nyctosaurus*, a frigate, bird-like pterosaur that may have spent very little time on the ground, these fingers seem to have been lost altogether.[32]

The fourth or wing-finger, easily the most distinctive and diagnostic feature of pterosaurs, was tremendously enlarged and elongated, and formed more than half the total length of the forelimb. Typically, it consisted of four straight or slightly curved lath-like bones. The first bore a massive double joint on its near end, which fitted tightly onto a thick pulley-like condyle on the end of the fourth metacarpal. This construction enabled the wing-finger to be swung through approximately a semicircle, so that from its fully extended position during flight, it could be folded back against the rest of the forelimb after landing. The remaining joints do not seem to have allowed any significant movement, so in life, the wing-finger must have behaved rather like a stiff spar. Often, in pterodactyloids, the last bone in the wing-finger was relatively short and slender, and in *Nyctosaurus*, it was lost altogether.[33]

Hips and Toes The pelvis, which consisted of a pair of plate-like elements firmly attached to either side of the backbone just in front of the tail (*Figure 6.7*), had several functions. Apart from helping to support the lower abdomen, the pelvis linked each hind limb to the spine, via the hip joint, and also provided a foundation for the major muscles that operated the legs. Each plate was composed of three bones,[34] usually tightly fused together, and its most prominent feature was the hip joint, a shallow, dish-shaped depression that faced outward and upward, and into which the rounded head of the thigh bone (femur) was inserted.

The uppermost part of each plate consisted of a long, flat blade of bone, its inner side firmly united with the backbone by a row of short, broad sacral ribs. This blade extended well in front of the pelvis, where it served as a platform for muscles that both raised the leg and helped swing it forward. In complementary fashion, a shorter process, extending to the rear of the pelvis, provided a base for muscles that also helped raise the leg but, in this case, swung it backward. Below the hip joint, the pelvis bowed out into a large plate, buttressed to the front by a rather narrow pillar of bone. The large muscles that arose from this region played several roles, among the most important of which were helping to pull the leg downward during flapping flight and holding the leg in toward the body when moving on the ground or climbing.

A novel feature of the pterosaur pelvis, not found in any other reptile, was the presence of an additional pair of bones that articulated with the bottom edge of each plate and extended forward beneath the lower part of the torso toward the gastralia. These bones, called the prepubes, varied in shape, ranging from a long, thin rod in *Rhamphorhynchus* to a short, flat spatula in some pterodactyloids, and were occasionally united across the midline to form a single unit. Their function is uncertain, but, by alternately swinging forward and upward and then backward and downward, they may have assisted the gastralia in changing the volume of the body cavity.[35]

The femur, although often of similar length to the humerus, was much slimmer. It had a highly distinctive mushroom-shaped head at the near end, a long tubular shaft and a gently expanded joint at the far end, which contributed to the knee. The head of the femur was bent at an oblique angle to the shaft[36] and fitted neatly into the hip joint, where it was free to swing backward and forward, up and down, or almost anywhere between, and it could also twist around through at least a semicircle. This remarkable freedom enabled pterosaurs to direct the femur out sideways from the body during flight or bring it in beneath the body when moving on the ground. A large scar on the outer side of the femur just below the head and other marks and lines on the front and back of the shaft indicate attachment sites for some of the muscles that powered the legs during flying and walking.

The principal bone of the lower leg, the tibia, was rather similar to its equivalent in birds, called the drumstick—both have a long, tubular shaft that terminates in a large, rounded knob. A second bone, the fibula, was

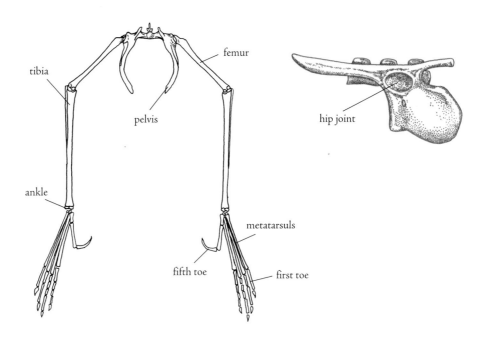

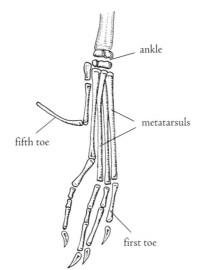

FIGURE 6.7 Hips and legs. Above left: the pelvis and hind limb of *Rhamphorhynchus* seen from the rear. Above right: the pelvis of *Rhamphorhynchus* in side view. Below left: the ankle and foot of *Rhamphorhynchus*. (Redrawn from Wellnhofer, 1975).

splinted against the outer side of the tibia, but was much thinner and, even in early pterosaurs, barely made it to the knob at the far end of the tibia. Indeed, in many pterosaurs, it merely faded into the shaft of the drumstick, and in some ornithocheiroids, it became so completely fused to this bone that the two cannot be distinguished. The construction of the knee joint allowed the lower leg to hinge backward and forward on the end of the thigh bone, but little more.

The pterosaur ankle, composed of two rows of tarsal bones, was similar in some respects to the ankle of birds. The two bones in the first row of tarsals[37] were almost always firmly attached, or fused, to the tibia, and, just as in birds, formed the knob on the end of the drumstick. The second row of tarsals consisted of two, small tablet-like bones whose underside fitted tightly against the foot, while their upper side was free to swivel around the pulley-like knob on the end of the drumstick. The great extent of this pulley meant that from a position extended out in front of the leg, the foot could be swept down and right around the ankle to finish up pointing to the rear, having traversed a complete semicircle.

The first half of the foot consisted largely of four long, slender, matchstick-like bones (metatarsals) that were bunched tightly against one another. Each of these metatarsals supported a long slender toe[38] that ended in a claw. Generally, these claws were smaller than those of the hand and less strongly curved, but, as in the hand, the bone immediately preceding each claw was relatively long, indicating that the foot was primarily adapted for gripping or grasping. Foot webs, consisting of thin sheets of skin stretched between the toes and extending forward as far as the base of the claws, have been found in fossils from several different localities and seem to have been present in most, perhaps all, pterosaurs, because they also show up in lots of pterosaur tracks (see Chapter 9). In some exceptionally well-preserved fossils of *Sordes* and *Pterodactylus* (*Figure 6.7*), there are traces of long, thin, fine fibers that originally lay within the webs and presumably helped to stiffen and toughen them.[39]

The fifth toe was completely different from the other four. The short, hook-shaped metatarsal at its base also contacted the second row of tarsals, but had a much more mobile articulation that allowed considerable freedom of motion. The far end of this metatarsal supported the fifth toe, which, in rhamphorhynchoids, consisted of two long, slender, rod-like bones whose job was to support and control a small flight membrane (the cruropatagium) stretched between the legs. In pterodactyloids, the fifth toe was reduced to just a single, short, stubby bone (concomitantly, the flight membrane was also reduced), and in many species, it completely disappeared, although, curiously, the fifth metatarsal seems to have remained, even in giant forms like *Quetzalcoatlus*.

Hide 'n' "Hair" Pterosaur skin had a tough job. It had to protect its owner from the bumps, jolts and sharp pointy bits of everyday life, stop disease and infection from getting into the body, carefully control the amount of water that was lost so that the animal did not desiccate, and help control the body temperature. Through superficial colors and patterns, it might also have been responsible for advertising pterosaurs to their mates or camouflaging them from predators or their prey. Clearly, it would be very helpful for our understanding of pterosaurs if we could discover what their skin, the single largest organ in the entire body, was like, how it was constructed and how it functioned. Happily, as recounted in Chapter 3, in this case there is at least some fossil evidence, and it has a remarkable story to tell.

Perhaps the most important point to emerge is that pterosaurs, unlike modern reptiles, did not have scales. Or, to qualify this slightly, in general, they did not have scales; as a beautifully preserved fossil of a Brazilian tapejarid shows,[40] the underside of the heel of some (possibly all) pterosaurs was covered in small, diamond-shaped scales, a discovery that, as we will see again in Chapter 9, has been confirmed by the presence of scales in fossilized foot impressions. That the soles of the feet and possibly also the undersides of the "clawed" fingers were scaled is not surprising. These surfaces, more than any others, would have benefited from the reinforcement and protection provided by tough scales, because they came into contact with the ground more frequently than any other part of the body and thus experienced the greatest wear and tear.

Apart from the undersides of the feet, and perhaps the "clawed" fingers, the skin seems to have had a relatively smooth, slightly leathery texture (*Figure 6.8*). Under the microscope, the fossilized skin of a medium-size 2-meter-wingspan pterosaur from Brazil[41] is seen to be crisscrossed by fine lines and grooves that were about the same size and density as those you can find in the patch of skin between your thumb and forefinger. Thin slices of this fossil (shown in *Figure 3.7*) have also exposed some of the skin's deeper secrets, showing that it was composed of the same layers and structures to

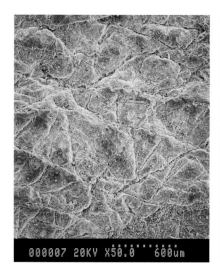

FIGURE 6.8 Pterosaur skin. Left: outer surface of the skin in a Lower Cretaceous pterosaur from Brazil (width of photograph about 1 centimeter). Right: heel region of a tapejarid pterosaur from the Lower Cretaceous of Brazil, showing a patch of fine scales that covered the underside of the foot (width of photograph about 6 centimeters). (Tapejarid image courtesy of Dino Frey and Helmut Tischlinger.)

be found in other tetrapods. Notably, there was a very thin epidermis, while the dermis was much thicker and contained blood vessels, nerves, muscles and other fibers. Overall, pterosaur skin was much thinner and lighter than that of most living reptiles, as is to be expected in a flying organism, where mass had to be reduced to a minimum.

The uncanny resemblance of the pterosaur hide to human skin was further enhanced by another surprising and (for a reptile) most extraordinary feature: hair. It should be said straightaway, however, that pterosaur "hair" was not the same as human hair or even mammal hair in general, and, although superficially it may have looked rather similar, as recounted below, it did not have the same origin or fine structure.

Pterosaur hair is not a new discovery. As early as 1830, Georg Goldfuss, professor of natural history at Bonn University, claimed that he could see evidence of hair in the first fossil find of *Scaphognathus*, and in the early 1900s, the German paleontologists Karl Wanderer and Ferdinand Broili both described what they thought were impressions of hair in Solnhofen pterosaurs.[42] The issue was conclusively resolved by the discovery, in the 1960s, of the aptly named *Sordes pilosus* (hairy devil) in which whole tracts of fossilized hair could be clearly seen in several specimens, one of which is illustrated in *Figure 6.9*.

Fossil evidence of hair has now been reported in at least nine pterosaurs found at several different locations spread across the world and spanning a time interval of more than 100 million years (see Chapter 3). This is a strong hint that most, if not all, of these Mesozoic dragons were hairy. Moreover, by combining information from different fossils, it has been possible to establish some general details of this extraordinary feature. In a typical pterosaur, it would appear that, apart from the beak and wings, much of the head, the neck, the body, the base of the tail and the upper part of the legs were covered by hair. Generally, this pelage appears to have been rather sparse, perhaps no denser than the hair on the forearms of human males, although, as *Figure 6.9* shows, in *Sordes* and *Pterodactylus* at least, and perhaps in many

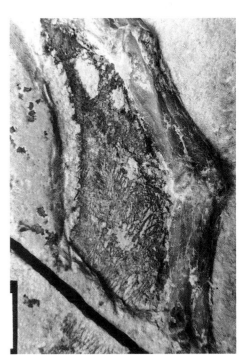

FIGURE 6.9 Pterosaur "hair." Above left: *Sordes pilosus*, with a mane of hair preserved on the crown of the skull. Above right: details of the hair. Bottom left: the hairy body of *Jeholopterus*. Bottom right: the hairy neck of *Pterodactylus*. (*Jeholopterus* image courtesy of Wang Xiao-Lin and Zhou Zhonghe; *Pterodactylus* image courtesy of Helmut Tischlinger.)

other pterosaurs, the crown of the head and nape of the neck seems to have borne a mane of longer and more thickly clustered hairs.

In crow-sized pterosaurs such as *Sordes*, single hairs were about four-hundredths of a millimeter across, which is roughly the same breadth as fine human hair. Typically, pterosaur hairs tended to be short, only reaching about 10 millimeters (less than half an inch) or so in length in *Sordes*, and in the best fossil examples are often curved or sinuous, suggesting that in life, they were quite supple and flexible. Under the microscope, individual hairs are seen to have consisted of single, unbranched strands that were solid, quite unlike the hollow hairs of mammals.

In several Solnhofen fossils, hair impressions are associated with tiny pit-like marks that look as if they have been made by needle pricks. Several paleontologists have suggested that they represent hair follicles,[43] just like those found in mammals, through which hairs emerge into the outside world. Such pits, however, can also be found in areas well away from those contacted by the skin and are probably a geological rather than a biological feature, an idea that is supported by the surprising absence of evidence for these pits in some of the best-preserved pterosaurs, including *Sordes* and *Jeholopterus*. Indeed, in *Sordes*, hairs seem to arise directly from the outer layer of the skin, which, if correct, means that the origin and development of pterosaur hair was quite different from that of mammals. Moreover, it also implies that, unlike mammals, pterosaurs were unable to raise or lower their hairs.

It would seem that the hair on pterosaurs was unique to these animals. Or was it? During the last decade, a series of superbly preserved small dinosaurs, many of them with fossil evidence of the body covering, have been found in the Jehol Group of northeast China, the same rocks that have also produced such pterosaurs as *Jeholopterus*. Some of the Jehol dinosaurs, such as *Microraptor*, sported feathers that are little different from those of birds, while others, such as *Sinosauropteryx*, seem to have borne a simpler covering, consisting of thin, fine filaments—"dino fuzz."[44] In some respects, this fuzz looks remarkably similar to the body covering borne by pterosaurs. Indeed, some scientists have even suggested that they are one and the same thing,[45] implying that something similar was already present in ornithodirans and subsequently evolved into the hair of pterosaurs and dino fuzz, the latter eventually evolving into true feathers.

It's an exciting idea, but there are several difficulties. The way feathers (and by implication, feather forerunners such as dino fuzz) develop from deep in the skin appears to be quite different from the origin of pterosaur hair, which seems to have grown directly from the surface of the skin. More significantly, there is no evidence of dino fuzz in most dinosaurs or any or-nithodiran, and, in any case, this idea only works if pterosaurs *are* ornithodi-rans, which, as already recounted in Chapter 4, is not at all certain. It would seem that, for the present, the case for a common origin of pterosaur hair and dinosaur fuzz is still far from being proven.

Beyond the issue of its origins lies another important question regarding pterosaur hair: What was its purpose? There are several possibilities. By trapping a layer of air against the body, or by acting as a barrier to the sun's rays, pterosaur hair may have played an important role in preventing the body temperature from sinking too low or rising too high. Hair might also have been critical for flight. The friction between a pterosaur and the air through which it flew is termed "drag," and the energy costs of overcoming this can be high, which is the main reason airplanes tend to be very smooth on the outside. There is relatively little friction between air and more air, and pterosaurs may have been able to take advantage of this by using hairs to trap a layer of air close to the body, thereby significantly reducing drag.

Yet another possibility is that if pterosaurs were capable of producing hair with different textures and colors, they may have been able to generate patterns that could be used for display purposes: to attract mates or intimidate competitors. Alternatively, these patterns may have helped to camouflage pterosaurs, hiding them from predators while they were resting or enabling them to get within striking distance of prey.

Revving the Metabolic Engine

Were pterosaurs hot-blooded, like birds and mammals, or cold-blooded, like reptiles and amphibians? Metabolism and physiology touched all aspects of pterosaurs' lives and determined, more strictly than any other factor, how long and how actively they could under-take strenuous activities like flying. Partly because of this, and partly be-cause details of physiology are so difficult to establish for extinct animals, the nature of these processes has been fiercely debated for more than a cen-tury. But, before we review the current state of play, we should scrutinize the question a little more closely.

Birds and mammals are not actually "hot-blooded," of course. Their body temperature, measured in centigrade, typically lies in the mid-30s, which is pleasantly warm—especially compared with a cold winter's day in Berlin.

Of more fundamental importance to their physiology, however, birds and mammals have a body temperature that, by and large, stays the same (homoeothermy) and is often higher than the external temperature and whose "endothermic" heat source is mainly generated internally. This type of physiology is not cheap, though. The basic metabolic rate (determined, essentially, by the speed at which chemical processes take place at the cellular level) is much higher than in reptiles, for example, and demands a much greater input, both of food and oxygen. Among the advantages of this system are that it allows birds and mammals to achieve peak performance very quickly, without the need for any warm-up period and, of particular significance here, to maintain highly energetic activities for hours or even days—a critical requirement for active fliers.

Reptiles and amphibians generally have a lower and more variable "heterothermic" body temperature than birds and mammals. It tends to track the external temperature, so that these animals are warm and active during the day, but cool and sluggish at night. Metabolic rates are lower, hence, reptiles and amphibians require considerably less food than mammals or birds and a substantial portion of the heat energy of these "ectotherms" is acquired from an external source—the sun. Although they need to warm up first, peak rates of performance are not that much different from mammals or birds—try dodging a striking snake—but, critically, reptiles and amphibians are generally incapable of sustaining any vigorous activity for more than a few minutes.

Beyond these characterizations, another important point regarding the question of "hot" or "cold" blood is that these regimes occupy only two regions among an entire spectrum of physiological possibilities. Pterosaurs might, in theory, have had the same, or similar, physiology as amphibians and reptiles or birds and mammals, but, alternatively, they may have utilized a quite different type of physiology not found in any living vertebrate. So, with all these caveats in mind, let's look at the evidence.

Hot or Not? Harry Seeley argued, at a time when good evidence of the outer covering was still unknown, that because pterosaurs were close relatives of birds and therefore must have been warm-blooded, they must also have had some kind of hair-like covering. Later, when "hair" came to light, paleontologists such as Baron von Nopcsa and Bob Bakker argued, quite forcibly on occasion, that this discovery proved that pterosaurs were warm-blooded.[46] Even if, as seems the case, pterosaurs could not raise and lower their hair as mammals do, the hair would still have been able to trap a thin layer of air against the skin. The insulation thereby afforded would have assisted with temperature control, but does this prove that pterosaurs were warm-blooded? No, not really, because such a mechanism would have been useful, regardless of what type of physiology pterosaurs had: hot- or cold-blooded, or anywhere in between.

An alternative approach to answering this physiological question lies with the brain, or, rather, its relative size. As shown in *Figure 5.7*, compared with their body mass, the brains of ectothermic heterotherms (the frog and lizard brigade) are rather small, while those of endothermic homeotherms (eagles 'n' beagles) are relatively large. Moreover, there is a clear, decisive gap between the two groups, into which pterosaurs fall. So, in fact, this line of evidence doesn't really help us very much, except that it raises the possibility that pterosaur physiology was unlike that of any living group.

Another potential physiological indicator is to be found in the microscopic details of bone. Modern amphibians and reptiles tend to grow slowly and cyclically, leaving pronounced growth lines in their bone tissues. By contrast, today's birds and mammals usually reach near-adult size in a single sustained and rapid spurt of growth, often laying down a particular type of tissue, called "fibro-lamellar" bone, that rarely contains any growth lines. Interestingly, as seen in *Figure 6.1*, pterosaur bones also consist largely of fibro-lamellar bone, suggesting rapid growth and an active physiology.[47] We should be cautious, however, about leaping to any firm conclusions from this. Pterosaur bones had very thin walls (see Chapter 8) and, as yet, it has not been possible to find a thick chunk of bone in which a substantial record of growth is preserved. Moreover, fibro-lamellar bone is not confined to "hot-bloods" but also appears in living and extinct reptiles. So, while bone tissue type in pterosaurs suggests a relatively active physiology, it is by no means conclusive.

Happily, one aspect of pterosaurs' biology goes a long way toward resolving the question of their likely physiology—flight. As Chapter 8 details, there is an overwhelming amount of evidence to show that pterosaurs were well-adapted for flying and, presumably, occasionally needed to flap their wings for quite long periods. Such an activity would have required the kind of sustained energetic output that living reptiles, at least, are incapable of delivering and that, today, is only found among mammals, birds and some insects.[48]

The overall picture that emerges from these various lines of inquiry seems to be consistent: Pterosaurs had an active physiology that was more like that of mammals and birds than reptiles or amphibians, although exactly how similar it was to living "hot-bloods" is still uncertain. Critically, powered flapping flight required a physiology that could deliver sustained output, which, in pterosaurs as in birds, seems to have been linked to the evolution of an efficient flow-through breathing mechanism. Other features of pterosaurs, such as the presence of hair, a largish brain and the development of fibro-lamellar bone tissue, are also consistent with an active physiology, although they do not necessarily demand it.

It looks as if the end is in sight for the physiological problem. Or is it? The next chapter covers some of the most exciting fossil finds of recent years—pterosaur eggs and babies—and concludes that, contrary to established opinion, reproduction and growth in these animals may have been little different from that of crocodiles and lizards. Hot-blooded reptiles? Perhaps pterosaurs were even more unconventional than anyone ever imagined.[49]

7
BABES ON THE WING

Bible black crosses against a sun-burnished sky, a small flock of pterosaurs wheeled slowly through the warm air. Far below, in the middle of a large patch of fine sand, the ground briefly quivered, went still, then quivered again. There was a long pause, then suddenly, in a little flurry of sand, a small head thrust its way into the dazzling daylight. Stretching her jaws wide, tiny teeth glittering in the sun, the hatchling pterosaur took her first gulp of sharp sea air, and, exhaling, let her head fall forward on the sand. Seconds later, the activity began again and slowly, spasmodically, the rest of the body, all long gawky limbs wrapped in tissue-thin membranes, emerged from the sand. For a while, she lay motionless, exhausted by the effort of freeing herself from the sandy incubator. A sound, a movement, perhaps just an instinct, but whatever it was, the pterosaur slowly propped herself up on spindly arms and legs and began to wobble in the direction of green. Intent on the safety of the ferns just a few lurches away, she could see tiny dots circling above as they momentarily clipped across the edge of her vision. Involuntarily, her tiny muscles tensed for the first time. Overhead, the sky was waiting.[1]

FIGURE 7.1 Fossil pterosaur egg (about 4 centimeters across) containing the embryo of an ornithocheirid found in Lower Cretaceous lake deposits of the Jehol Group at Jingangshan in Liaoning Province, China. (Photograph courtesy of Wang Xiao-Lin.)

The Generation Game Many of the pterosaur fossils featured in this book belonged to fully grown, mature adults, or nearly so. This is because most of the individual pterosaurs preserved in the fossil record had reached adulthood before they died. The same principle applies to the fossil remains of humans, dinosaurs, turtles and many other groups of backboned animals. Indeed, in the vast majority of extinct species, almost all known fossils appear to have been adults, for the simple reason that when it comes to fossilization, bigger is generally better.

Adult vertebrates have another advantage in preservation: Their bones are usually more mineralized (that is, they contain more of the main mineral component of bone, calcium phosphate) than those of juveniles and thus are more suited to surviving the long journey to becoming a fossil. As a rule, this means that juveniles tend to be uncommon in the vertebrate fossil record, and individuals at very early stages of growth (newborn or even prenatal), are rare or unknown[2]—except, oddly enough, in pterosaurs.

Fortunately for us, the Solnhofen Limestone and many of the other "death traps" in which pterosaurs were preserved (detailed in *Figure 3.9*), took little account of how big or bony individuals were, and, in a commendably egalitarian fashion, preserved any and all that fetched up in their muddy embrace. The end result is that the fossil record of pterosaur growth is surprisingly good. There are many examples of adults and near-adults, plenty of juveniles and several species where hatchlings or slightly older individuals are known. Crowning it all, we now have eggs, each with an unhatched embryo inside, from three different kinds of pterosaurs, one of which is illustrated in *Figure 7.1*.

All stages of growth, from hatchling to adult, are now known for two pterosaurs, *Pterodactylus kochi* and *Pterodaustro guinazui*, and in the latter case, an egg with an embryo even reveals details of a prenatal stage. Filling out this picture, for many other pterosaurs—among them anurognathids, rhamphorhynchids, ornithocheirids, dsungaripterids and azhdarchids—fossil remains of juveniles (some of which were still very young when they died) and adults are known. In combination, these fossils have begun to reveal the secrets of how pterosaurs reproduced and grew, and the story they have to tell is extraordinary.

The process of development in birds and bats is as we might expect. Youngsters invested all their time and energy in growing and only began to fly when they had reached adult (or near-adult) size. Pterosaurs did things the other way around. Recent findings show that these incredible animals probably took to the air soon after birth, and only then did they set about the process of growing up to become adults. In other words, pterosaurs flew first and grew later. And in some species, where the adults had a wingspan more than 10 times that of the hatchlings,[3] they had a lot of growing to do. To use a modern analogy, pterosaurs achieved the equivalent of starting out with a single-seater airplane and slowly enlarging it until it reached the size of a jumbo jet—all while still flying! Quite astonishing, but it also provides one of the keys to understanding another unusual aspect of pterosaurs—how they reached giant size. First, however, we need to go back to where it all started. In the beginning was the egg.

Egg "Amazing," I thought, "absolutely, totally and completely amazing!" There it was on my computer screen: a pin-sharp, full-color picture of an egg, exactly the same size and shape as a hen's egg, but filled with a small, almost perfectly preserved embryo of a pterosaur.

This illustration, shown in *Figure 7.1*, was part of a manuscript written by my colleagues, Wang Xiao-Lin and Zhou Zhonghe, paleontologists at the Institute for Vertebrate Paleontology and Paleoanthropology in Beijing, that had been sent to me for review. A few months later, the finished version of this paper appeared in the journal *Nature* and caused a sensation.[4] Websites, science journals and daily newspapers all raced to cover this astonishing find and, to my delight, at the end of 2004, I even received a New Year's greeting card featuring a large, full-color image of the new egg and its amazing embryonic cargo.[5]

Whether pterosaurs laid eggs, like birds, or gave birth to live young, like bats, had long been an issue. In the mid-1800s, several fossil eggs from the Stonesfield Slate of England were tentatively identified as belonging to pterosaurs, but this idea never caught on.[6] Other reports seemed even less convincing[7] and, compared with the growing fossil record of dinosaur and bird eggs,[8] the complete absence of pterosaur eggs began to seem a little strange, although, as we will see shortly, there is a good reason for it.

The alternative was live birth, which would explain the distinct lack of eggs and does occur in diapsid reptiles, most notably in some lizards and snakes. But to prove this, one would need fossil evidence of an embryo inside, or at least next to, a mother, and such evidence was manifestly not forthcoming (although examples are known for bats[9]). So, the whole question of pterosaur reproduction was a moot point until the Chinese discovery.

Wang and Zhou's paper settled the issue once and for all: Pterosaurs laid eggs. If there was any doubt, it has been dispelled by two further finds, one in Argentina and the other in China. No eggs for 200 years, and then three in six months! The light brown color of the shell contrasting sharply with the creamy white limestone of the two mirror-image slabs on which it is preserved, Wang and Zhou's egg was found at Jingangshan in western Liaoning in rocks of the Jehol Group, a sequence of Lower Cretaceous lake sediments that we first encountered in Chapter 3. Evidently, the exceptional conditions in these lakes that led to the preservation of such soft tissues as the skin and feathers of dinosaurs, the fur on mammals and even a bird's egg complete with embryo,[10] also led to the fossilization of the pterosaur egg, although exactly how it happened is not yet clear.

The second Chinese egg, a rounded capsule adrift in a sea of insect larvae, illustrated in *Figure 7.2*, was recovered from the same fossil site as the first egg and, although of comparable volume, it is somewhat narrower and longer.[11] Otherwise, the two eggs are rather similar but, according to the team of paleontologists who studied them, led by Ji Qiang from the Institute of Geology in Beijing, they had a very important property: Their shells were soft. Bird's eggs always have a hard shell, as do those of crocodiles but,

FIGURE 7.2 Above: a second pterosaur egg (about 3.5 centimeters across) with embryo, from the Jehol Group at Jingangshan in Liaoning Province, China. Below: Pterodaustro egg (about 2.2 centimeters across) from the Lower Cretaceous of Loma del Pterodaustro in Argentina. (Photographs courtesy of Lü Junchang.)

in most cases, the eggshells of snakes, lizards and many turtles are soft and leathery. Ji and his colleagues concluded that pterosaur eggs were soft-shelled too, mainly because they could not find any evidence of calcite, the mineral that forms the hard shell of birds' eggs even though it is present in other fossils, such as mollusks, from the same rocks. The relatively undistorted shape of the eggs and the apparent absence of any shell fragments, or evidence of cracks or breaks, common in fossilized eggs, also points to a soft shell.

An exciting and highly significant discovery, except that in the very same issue of *Nature* that contained the soft-shelled egg from China was another paper, by Luis Chiappe and colleagues, on a *Pterodaustro* egg, complete with embryo, from Argentina.[12] And it had a hard shell. This third egg, from Lower Cretaceous lake sediments of the Loma del *Pterodaustro*, was about the same size as the Chinese specimens but, as *Figure 7.3* shows, even narrower and almost sausage-shaped. Using a scanning electron microscope, Chiappe's team showed that the shell was made up of calcite crystals that formed a single layer, with one very important characteristic—it was extremely thin, only about one-sixth the thickness of the shell of a bird's egg of about the same size. Not, in fact, a hard eggshell after all, because, although by no means always present, such layers are common in eggs with soft shells.[13]

Soft shells explain, in the first instance, why pterosaur eggs are so rare, and secondly, they give us a vital clue as to the reproductive biology of pterosaurs. They are rare, for the simple and obvious reason that in order to become fossilized, soft-shelled eggs required exceptional circumstances—a nesting ground located next door to a fossil "trap" in which soft, as well as hard, tissues had some possibility of being preserved. The chances of this happening were small (although, as recent finds show, not nonexistent), explaining why fossils of soft-shelled eggs belonging to lizards, turtles and, of course, pterosaurs, are hardly ever found, whereas the hard-shelled eggs of dinosaurs, for example, are relatively common.

Apart from improving their chances in the fossil lottery, hard shells also protect the contents of the egg from drying out and, critically, enable the parents to sit on the egg and keep it warm with their own body heat. It is hard to see how pterosaurs could have sat on their soft-shelled eggs without squashing them, so, like many other reptiles and some birds, such as the megapodes,[14] they probably buried their eggs in a mound of vegetation, soft sand or soil. Unlike eggs exposed to the "hot" incubator of body contact, buried eggs develop slowly, even those of megapodes, and are often incubated for two, or even three, months. When they do hatch, however, the young ones are highly precocial, meaning they are able to move, feed and look after themselves almost immediately. Could pterosaurs do this, too? The embryos reveal all.

Embryo With its outsize head tucked down on its chest and arms and legs folded tightly around the body, the embryonic pterosaur described by Wang and Zhou and sketched in *Figure 7.3* looks remarkably similar to the egg-bound embryos of crocodiles and birds, apart, of course, from the wing-fingers that curl in opposite directions, right around the entire animal. Almost all the major elements of the skeleton are present and well-formed, demonstrating that the embryo was almost fully developed and probably only a few days, or even hours, away from hatching when it died.

Usually, in most vertebrate fossils, it would be almost impossible to identify such a young individual,[15] but, happily for us, pterosaurs are one of the few exceptions. The relative lengths of the main limb bones betray their identity, even in the earliest stages of growth. In this particular case, the long wing-metacarpal and remarkably short foot give away the embryo as a member of

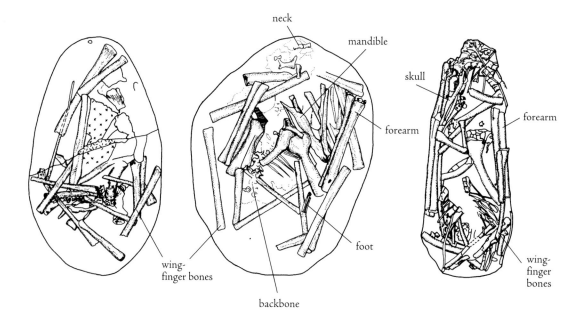

FIGURE 7.3 The three embryos. Left: the second pterosaur embryo from the Yixian Formation (part of the Jehol Group) at Jingangshan in Liaoning Province, China. Middle: ornithocheirid embryo from the same location. Right: *Pterodaustro* embryo from the Lower Cretaceous of Loma del Pterodaustro in Argentina (Redrawn from Ji Qiang et al., 2004, Wang Xiao-Lin and Zhou Zhonghe, 2004, and Luis Chiappe et al., 2004.)

the ornithocheirid clan, although exactly which ornithocheirid—three are known from the Jehol Biota—has yet to be pinpointed. The well-developed skeleton allows a fairly accurate estimate of the wingspan, which is just short of 12 inches (30 centimeters), about the same as that of the smallest known example of *Rhamphorhynchus muensteri*, but significantly larger than that of young individuals of *Pterodactylus kochi* and *Ctenochasma elegans*, some of which were less than 8 inches (20 centimeters) between the wing tips.[16]

The Argentinean embryo is packed into the egg in the same way as the first Chinese embryo (*Figure 7.3*) and also seems to have been close to hatching. It, too, has a well-developed skeleton and is almost exactly the same size as hatched individuals of the same species—more on these later. Although the estimated wingspan of the Argentinean embryo is similar to that of the Chinese specimens, the proportions of its skeleton are quite different. The relatively large feet give it away as a ctenochasmatid, and details of the head and teeth show that it belongs to *Pterodaustro*, which is to be expected, because *Pterodaustro* is the only pterosaur found so far at Loma del *Pterodaustro*.[17]

In the second Chinese egg, the contents seem more disarranged and poorly formed (*Figure 7.3*). Presumably, in this case, the embryo died at an earlier stage of development, when the skeleton was less well-mineralized and became jumbled during the egg's journey to the bottom of the lake. Although from the same fossil site and similar in size to Wang and Zhou's ornithocheirid embryo, according to Ji Qiang and colleagues, this specimen must belong to another kind of pterosaur, possibly the ctenochasmatid *Beipiaopterus*, because its limb bone proportions are quite different.[18]

Discovering one embryo was tremendous, two was quite remarkable, but with three, our cup runneth over (indeed it runneth over a lot, because we can classify them, as well). Now we come to the full significance of these finds: They bear out the prediction made earlier regarding precociality—pterosaurs were go, go, go as soon as they hatched. There are several clues. First, all the major long bones in the arms and legs were already elongated, well-formed and partially mineralized so that they did not bend too much when the animal tried to fly. This condition even extends to that weird wrist bone, the pteroid (spotted in the first Chinese embryo), which, so far as we can tell, was only used for flight—showing that pterosaurs were ready not just for life out of the egg but, as related in Chapter 8, for take-off, too.

One needs more than just a skeleton for flight, though. Wing membranes, the surfaces that generated the lift that kept pterosaurs in the air, are also necessary—and are preserved in the ornithocheirid embryo. Moreover, the presence in these membranes of a key component, wing fibers, show that they, too, were prepared for flight.

So, these embryos had the necessary flight gear, but would it have worked? Two features say yes. First of all, when birds hatch and bats are born, their arms are still relatively short and stumpy. But not pterosaurs. The arms of all three embryos appear to have been fully developed. In the first to be found, each arm is about four times the length of the body (measured between the shoulder and hips), a ratio typical of adult pterosaurs that certainly could fly. Emphasizing the point, the wing-finger of this embryo occupies a little over three-fifths of the total arm length, a proportion that commonly occurs in adult ornithocheiroids and many other pterosaurs that certainly were able to fly.

Perhaps most exciting of all, studies now under way on these embryos are generating aerodynamic data. Wing-loading, calculated by comparing estimates of the weight of the flier with the reconstructed wing area (see Chapter 8 for more details) provides a measure of flight ability that reveals whether an individual could fly. Preliminary results indicate that wing-loading estimates for the embryos are no different from those of other flight-capable pterosaurs. Even though still in their shells, these embryos had the potential to fly.

So, what the embryos have taught us so far is that pterosaurs close to hatching had a full set of flight gear that was the right shape and size, and, had they tried to take off, it seems that it would have worked. But that is still not actually flying. To see if newly hatched pterosaurs really could get airborne, we must move on to the next stage in their development: flaplings.

Flaplings Flaplings, that is, hatchling pterosaurs just a few days or weeks old and yet to experience the growth spurt that would propel them through juvenility to adulthood, are rare, but not as rare you might think. Two tiny individuals of *Pterodaustro*,[19] their skeletons (one of which is shown in *Figure 7.4*) almost identical to those of the pterosaur embryo and from the same fossil site, must have died soon after hatching, possibly even on their maiden flights. The same fate may well have befallen several specimens of the Solnhofen pterosaurs *Pterodactylus kochi* and *Ctenochasma elegans*, illustrated in *Figures 1.6* and *7.4*.

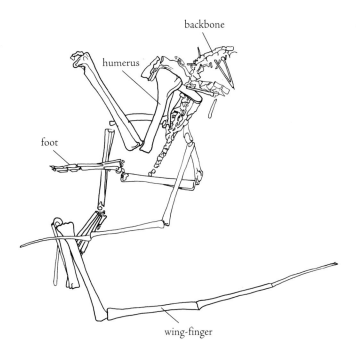

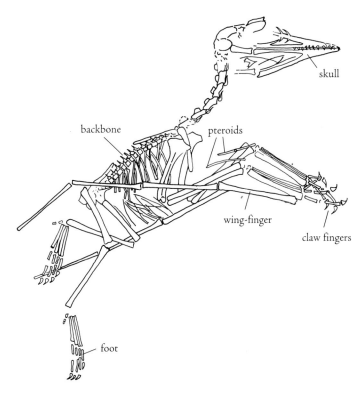

FIGURE 7.4 Above: small "flapling" specimen of *Pterodaustro* guinazui from the Loma del *Pterodaustro* site in Argentina, with an estimated wingspan of 12 inches (30 centimeters). Below: flapling of *Pterodactylus kochi* from the Solnhofen Limestone of Germany, with an estimated wingspan of 8 inches (20 centimeters). (Redrawn from Chiappe et al., 2004, and Wellnhofer, 1970.)

These individuals are tiny enough to have fitted into eggs that, hypothetically, were small enough to have been laid by adults of these species. This suggests, together with such other features as their remarkably large eyes and poorly mineralized bones, that these tots were also flaplings. There is further evidence of such youngsters in the dsungaripterid and azhdarchid clans, although in these cases, it consists only of bone fragments, which say little other than "here were flaplings."

What flaplings, unlike embryos, can reveal is whether pterosaurs were able to fly soon after they hatched. Exactly as you would expect, flaplings had all those features of the flight apparatus that were already present in the embryos. Thus, as *Figure 7.4* illustrates, in flaplings of *Pterodactylus* and *Ctenochasma*, all the long bones of the flight apparatus, the pteroid included, were well-formed and at least partially mineralized. Wing membranes, preserved in a few flaplings, are remarkably complete in the sparrow-sized example of *Pterodactylus kochi*, illustrated in *Figure 7.5*. Using ultraviolet light and a microscope, it is possible to trace their outlines from the tip of the wing-finger down the arm, around the elbow, and in toward the body, then back out around the knee and down to the ankle—exactly as in adults. Increase the magnification and the wing fibers come into view; calculate the wing loading and the result that pops out is little different from the values for adults.[20] These flaplings were ready for the sky.

That they did fly is shown by their final resting place, fossilized alongside juveniles and adults and far from any nesting site. Fossil flaplings have been found in the Solnhofen Limestones of Germany, in the lake beds of the Loma del *Pterodaustro* in Argentina, in the Tatal Formation of Mongolia, and in the case of *Azhdarcho*, in Upper Cretaceous river deposits of Uzbekistan.[21] Could all these be examples of nonflying nest-bound individuals that somehow got washed away and by some fluke finished up in the same burial place as their parents? Once or twice, perhaps, but in every case—hardly credible. They ended up entombed in these rocks because, like their elders, they lived and flew in the vicinity. Besides, the alternative—that they couldn't actually fly—doesn't make any sense. Why would evolution equip flapling pterosaurs with a complete flying apparatus if it didn't intend them to use it?

Juvenile If the fossil record is anything to go by, the Mesozoic skies must have been full of juvenile pterosaurs. They are known for almost all the main clans and are present in more than 20 of the 100 species of pterosaur found so far. Moreover, in two of these species, *Rhamphorhynchus muensteri* and

FIGURE 7.5 Winged baby. Picked out by their yellowy-orange color, the well-preserved remains of the wing membranes of this tiny individual of *Pterodactylus kochi*, about 7.5 inches (18 centimeters) in wingspan and probably only a few days or a few weeks old when it died, match closely to the fossilized wings of adults of the same species.

Pterodactylus kochi, they occur in substantial numbers (at least compared with other pterosaurs), which is a good thing because, when lumped together with hatchlings and adults, they really start to give us some idea of how pterosaurs grew.

Some of these growth patterns, shown in *Figure 7.6*, are much as we might expect because they are common to many backboned animals, including us. So, just like you and me, pterosaurs started off with relatively big heads (because you cannot skimp on space for the neural computer if you want a working organism) that ended up "normal" size in adults, because it grew more slowly than the rest of the body. The shape of the heads themselves changed, too. The snout often grew faster than neighboring regions, so that large-eyed short-faced flaplings finished up with long, low skulls and relatively small eyes.

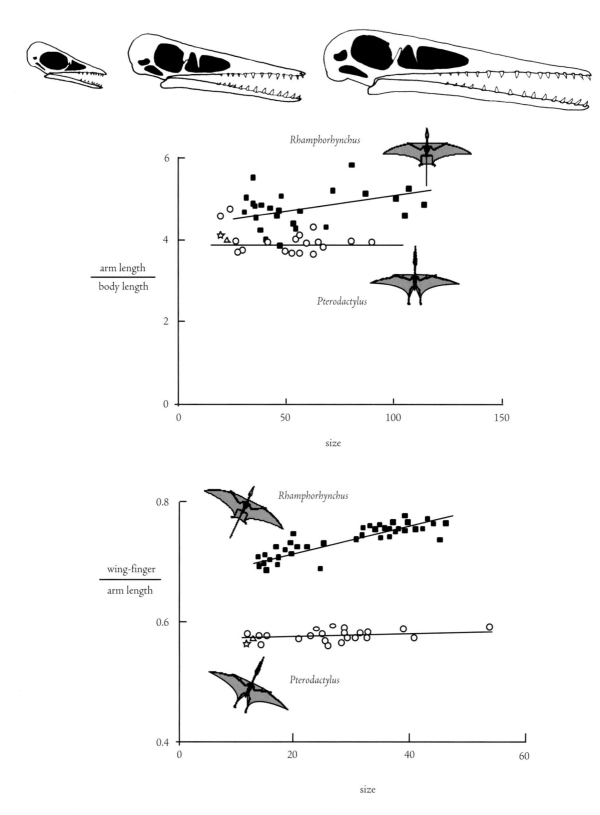

FIGURE 7.6 How pterosaurs grew. Above: three stages in the development of the skull of *Pterodactylus*. Middle: a graph of arm length compared with body length for *Rhamphorhynchus* and *Pterodactylus*. Bottom: graph of wing-finger length compared with the rest of the arm for Rhamphorhynchus and Pterodactylus. The individual illustrated in Figure 7.5 is represented by a triangle; the individual from Figure 7.4 below is represented by a circle. (Skulls redrawn from Peter Wellnhofer, 1970.)

Intriguingly, this parallels some of the patterns we encountered earlier in the pterosaur family tree, such as the evolution of increasingly elongated beaks in ctenochasmatids. Evidently, these changes could have been brought about quite simply by tweaking the timing and speed of growth of different components of the skull—an evolutionary process called heterochrony.

Different growth rates also account for the long necks of some pterodactyloids, as *Pterodactylus kochi* reveals. Although flaplings of this species started out with rather short cervicals, they lengthened more quickly than vertebrae elsewhere in the spinal column. The result, in adult ctenochasmatoids (the clan to which *Pterodactylus* belonged), was a series of long, tube-like vertebrae that formed the fishing-rod neck typical of these pterosaurs.

How the limbs responded to the profound increase in body size experienced, for example, by individuals of *Pterodaustro*, where flaplings of 30 centimeters (12 inches) in wingspan grew to adults of up to 3 meters (about 10 feet),[22] is of great interest because, throughout these changes, they continued to be used for flying and walking. In theory, the length of the arms, at least, should have increased more rapidly than the length of the body; otherwise, wing-loading would have risen dramatically. In practice, it seems that some pterosaurs did follow this pattern. *Rhamphorhynchus* adults, for example, had arms that were relatively longer than those of juveniles, although the difference is smaller than expected. *Pterodactylus*, by contrast, seems to have been completely unaware of any theoretical predictions, and the ratio of arm to body length is the same in flaplings, juveniles and adults of this pterosaur. Why that was so and, even more baffling, how it happened is a mystery that has yet to be solved.

Analyzing growth patterns within the limbs results in even more puzzles. Again, *Rhamphorhynchus* generally followed the rules. Predictably, the wing-finger grew more rapidly than the other long bones and occupied an increasingly greater proportion of the arm. The legs, by contrast, grew relatively slowly and thus were proportionately shorter in adults than in juveniles. Because wing-shape was largely defined by the lengths of the arms and legs, the wings had no choice but to become longer and narrower as individuals grew bigger and older. This must have had some visible impact on the flight behavior of *Rhamphorhynchus*, but exactly what is still unknown.

Pterodactylus, true to form, has some surprises for us. In this case, the wing-finger occupied the same proportion (three-fifths) of the wingspan, from flapling all the way up to big, old adults. The ratio of arm length to leg length hardly changed either, which means that wing shape stayed pretty much the same for all growth stages. It all sounds nice and simple, and yet it is quite the opposite. Bear in mind that most of the long bones of the arm (seven in total) were of different lengths, and all seem to have grown at different rates—yet the cumulative effect of all this was that the wing-finger remained at almost exactly three-fifths the total arm length through-out growth from wingspans of 20 to 80 centimeters (8 to 32 inches). It has not yet been established exactly how *Pterodactylus kochi* did this (although it should be a relatively straightforward task), and finding out why it happened will be a lot harder.

Beyond establishing how pterosaurs changed size and shape, it would be useful to know how fast they grew. So far, however, this problem remains rather intractable. It's easy with living vertebrates—just weigh them regularly—and not impossible with extinct ones either, as some recent successes with dinosaurs have shown.[23] The key to this work lies in annual rings, visible under the microscope in transparently thin slices of bone tissue and formed, so it is claimed, just like tree rings, by changes in growth rates during the year. Paleontologists have linked ring counts, indicative of age, with ring size, indicative of body mass, to estimate the speed at which dinosaurs accrued their bulk (quite fast, apparently). This technique could work for other extinct reptiles—except for pterosaurs. It fails here because, in these fliers, the bones were hollow, with only thin and slender walls, leaving hardly any tissue to record the history of their growth.

A completely different way of aging pterosaurs and estimating how fast they grew has been proposed by Chris Bennett.[24] He pointed out that in *Rhamphorhynchus muensteri* and several other Solnhofen Limestone ptero-saurs, individuals cluster into distinct groups of different sizes (they can vaguely be made out in the graphs in *Figure 7.7*) and suggested that these rep-resented different generations, one per year. Ingenious and, if correct, ptero-saurs (or at least those species included in Bennett's study) must have grown relatively slowly, with *Rhamphorhynchus* taking several years to reach the size a large seagull attains in a few months.

When scrutinized closely, however, neither the clusters nor their annual periodicity is particularly convincing, but Bennett's general conclusion is sound, for another reason. Consider that the lucky young seagull is regularly stuffed with food by its parents and has but one task: to sit on a ledge and grow. Life was not so easy for little pterosaurs. They had to expend at least part of what they ate on powering their flight muscles, catching food and evading predators, so they simply could not have grown as quickly as bird chicks, and they might never have met their parents.

Mums and Dads—Who Needs Them? Imagine if, on the day
you were born, you had to look after yourself, hunt down your supper, watch
out for enemies and start learning to fly. Pterosaurs could do it all, yet at this
point in our lives, we humans are utterly helpless and need years of looking
after before we can even cram a few handfuls of food into the correct orifice.
With one major exception,[25] young birds and bats need looking after, too,
and are fed and protected by their parents, sometimes until long after they
have learned to fly. It has often been assumed that this was also the case for
pterosaurs, an idea long reinforced by illustrations showing a benign parent
delivering fish to its hungrily demanding young.[26] Such images are increasingly at odds with what we know about these animals and have more to do
with the wishful thinking emanating from the analogy trap that we met
back in Chapter 1 than with the likely reality of pterosaurs' lives.

A quick survey of the fossil record of flaplings and juveniles shows that
from Day One these youngsters had well-developed jaws and teeth and fully
functioning wings and legs. Thus equipped, these "tiny adults" did not actually need any assistance from mom and dad and could have fed themselves,
walked around and even flown away, if trouble threatened. It is possible, of
course, that like crocodiles today,[27] parents were on hand to protect their
young from predators, especially just after they had hatched. It probably did
not go much further than that, though, and the idea that pterosaurs must
have looked after their young, because it was assumed they were not able to
fly for weeks or even months after they had hatched, has been swept away by
the recognition that flaplings and juveniles were more than capable of getting airborne.

The image we are left with is of a Mesozoic landscape in which hordes of
young pterosaurs went about their daily business, enthusiastically, if perhaps
somewhat inexpertly, flapping, and occasionally flopping (thereby discovering all the many ways that novice fliers can die while flying—and ensuring
that they left some kind of fossil record[28]). This scene prompts several questions, two of which merit some attention. Were these flaplings feeding on
the same sorts of things as their older, larger "flap off kid, this is my food"
elders? And, in a related vein, what about "small" species of pterosaur, the
adults of which might have competed for food with the flaplings and juveniles of larger species?

Taking the first question first, if we compare a flapling *Pterodactylus kochi*
with an adult, we find that the former is about one-fifth as big as the latter.
As *Figure 7.6* illustrates, the flapling has a row of tiny, spiky teeth set in short,

stubby jaws, quite unlike the large saw teeth set in the long jaws of the adult. Although there is (as yet) no direct evidence in the form of fossilized stomach contents to prove the conjecture, it seems highly likely that these two fed on quite different food items. The flapling would have found it difficult to catch or kill fish the adult could have taken with ease, whereas the adult may not have been as adept at catching small insects or crab larvae as the flapling. The same argument can be made for other species in which young and adults are known, *Ctenochasma elegans*, for instance,[29] and is certainly true of living reptiles such as crocodiles.[30]

It is fascinating to think, then, that as pterosaurs grew from flapling through juveniles to adulthood, their diet may have changed at least two or three times. What's more, this kind of growth ecology might have had some useful advantages. It means that youngsters probably did not have to compete with adults for food (inevitably, they would have lost out most of the time), and it also means that the species was probably not dependent on one food resource. This would have reduced the "load" on particular prey items. It's also possible that in times of environmental crises, such as the Mesozoic equivalents of El Niños, this improved species' chances of survival.

What about "small" species of pterosaur? Might their adults have competed for the same food resource as the young of larger species? No, for the simple reason that there doesn't seem to have been any "small" species, which is even stranger than you might think. Consider that the vast majority of birds and bats are less than one-third of a meter (1 foot) in wingspan—this is certainly the case for living species and also seems to be true for fossil birds and bats.[31] By contrast, adults of the smallest pterosaur species known at present, such as *Anurognathus ammoni*, are at least 40 centimeters (15 inches) in wingspan, and most of them are bigger. A biased fossil record? Hardly. Otherwise, we wouldn't have found all those flaplings and juveniles, nor would we have the thousands of other small vertebrates (fish, amphibians, lizards, birds and mammals), that have also been recovered from the Solnhofen Limestones, the Jehol Group and several other fossil-bearing horizons.

No "small" pterosaurs then, or, to be more precise, no small adults. But lots of small young belonged to pterosaurs that, as adults, were at least 40 centimeters (15 inches) or more in wingspan. This suggests a rather surprising conclusion: Young pterosaurs *were* the small species, or at least occupied some of the living space (niches) in which one might have expected to encounter small adults. If this novel and as yet unexplored notion is correct, it means that the way pterosaurs filled the ecological landscape was completely different from the way it is filled by birds and bats today. It might even explain why there are no really small species of pterosaurs: The spaces they could have evolved into were already occupied by the next generation.

All Grown Up At the start of this chapter, I stated that most pterosaur fossils appear to have reached adulthood before they died. What I did not do, though, was explain what I meant by "adult," as it applies to pterosaurs. How is it possible to tell if a 150-million-year-old squashed-flat skeleton of *Pterodactylus* had actually achieved adulthood? Defining adulthood for pterosaurs is relatively easy: It was the stage at which individuals had completed the main period of growth and had become sexually mature or, put simply, were grown up and able to reproduce. Determining whether a particular individual had actually reached this stage is much harder.

Size, of course, immediately springs to mind, but trying to age a pterosaur on this basis is very difficult because of the way they appear to have grown. Individuals that make up a particular species of bird or mammal tend to have similar growth rates (largely controlled by internal mechanisms and genetics), and they stop growing when they reach adulthood.[32] At this point, they switch all their resources to maintaining the body and reproduction. Consequently, in any given species, most adults are about the same size. Living reptiles, by contrast, have highly variable growth rates (strongly influenced by external factors, such as the availability of food) and although growth slows down when they reach adulthood and start to reproduce, they often continue to grow, albeit slowly.[33] This means that in reptiles, adult size is much more variable, and immature individuals can, on occasion, be much larger than mature adults.

The upshot of all this for paleontologists is that, while size offers a good, if not completely infallible, guide to the maturity of fossil mammals and birds, it is far less reliable when it comes to fossil reptiles, including pterosaurs. The plots in *Figure 7.7* show that at the "adult" end of the growth spectrum, individual size was highly variable, both in rhamphorhynchoids and pterodactyloids, matching the pattern typical of living reptiles. Thus, even when a species such as *Rhamphorhynchus muensteri* or *Pteranodon longiceps* is represented by dozens of fossils, "big" does not necessarily equate with "adult."

Fortunately, there are other ways of aging a pterosaur. Once the major growth spurt was over, growth slowed dramatically and, in some cases, may

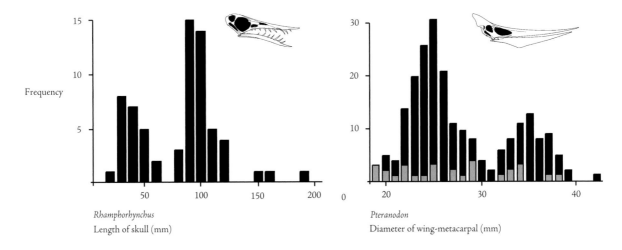

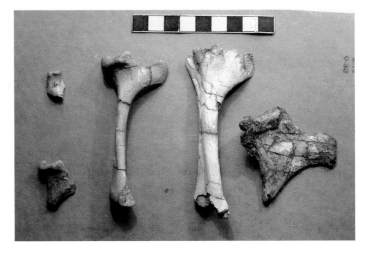

FIGURE 7.7 Size and age in pterosaurs. Above left: size distribution for the skull of *Rhamphorhynchus* showing two main peaks that, according to Chris Bennett (1996), may reflect annual generations. Above right: size distribution for the skull of *Pteranodon*, indicating mature (black) and immature (shaded) individuals, identified using features such as the degree of fusion of bones. Below: a large size range in the Early Cretaceous Tatal dsungaripterid represented by several fragmentary humeri belonging to individuals from about 20 inches (50 centimeters) up to 9 feet (3 meters) in wingspan. (Graphs redrawn from Chris Bennett, 1993, 1996.)

even have stopped altogether. As illustrated in *Figure 7.8*, this shows up in various ways in the skeleton, including on the outer surface of bones. In flaplings and juveniles, bone surfaces often have a dull texture and are rather porous, whereas in adults, they have a shiny, silky appearance and are much denser.[34] These differences reflect changes that were going on inside the bones. In young pterosaurs, new bone tissue was laid down quickly on the outer surface of bone walls and almost as quickly eroded away from their inner surfaces, leaving behind a girder-work of thin, bony struts that supported the hollow bones internally. Later, as adulthood was attained and growth rapidly decelerated, bone was deposited more slowly, forming thin, dense layers on the outer surface of the bone and, occasionally, a narrow back-fill on the inner surface as well: bony fingerprints of adulthood.

FIGURE 7.8 Aging pterosaurs. Compare the smooth, satiny finish, lacking any obvious pores, of the external surface of the wing-metacarpal of a mature individual of *Pteranodon* (above left) with the rough, grainy texture, pierced by numerous pores, of the surface of the same bone of an immature individual of this pterosaur (above right). Similarly, in thin sections, the bone tissue (total thickness about 1 millimeter) of the mature individual (below left) is very dense with relatively few canals, not one of which penetrates through the thin outer layers capping the external surface of the bone (at top). By contrast, the immature individual (below right) has numerous canals, some of which extend right up to the bone surface. (Photographs courtesy of Chris Bennett.)

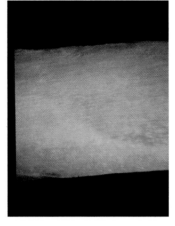 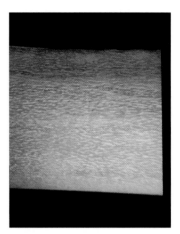

 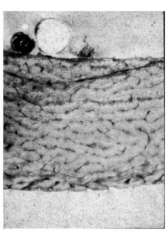

Adulthood also left its imprint on other aspects of the skeleton. Typically, pterosaur long bones, like those of other tetrapods, consist of a shaft, or diaphysis, capped at each end by a bony knobble called the epiphysis. Most of the growth that led to lengthening of the bone took place in the gap between the epiphysis and the diaphysis, but when adulthood was reached and growth slowed, these united and eventually fused. Epiphyses can be found throughout the skeleton and are especially prominent at the elbow, where they cap both the far end of the humerus and the near end of the main forearm bone, the ulna. A similar type of fusion can also unite two or more whole bones to one another. This is common in the skull, but, as detailed in Chapter 6, also occurs in the backbone (notarium and sacrum), shoulder girdle, pelvis, wrist and ankle.

Using these signatures of adulthood, the status of particular fossils (even isolated, fragmentary bones) can often be decided with confidence. The giant 10-meter-wingspan (33-foot) Texan pterosaur *Quetzalcoatlus northropi* is a good example. The epiphyses on the far end of the humerus of the giant individual are firmly fused to the main body of the bone, so this was an adult.[35] But what about the several 5-meter-wingspan (16-foot) individuals that have been thought, on occasion, to be young of the same species? They're adults, too, as the fusion of epiphyses, skull bones, shoulder girdle, wrist and ankle bones show. So, with such a large size gap between these adults and the giant, could they all have belonged to the same species? We will come back to this question in the next section.

Earlier, I referred to the diagram (*Figure 7.7*) illustrating the different sizes of individuals of *Pteranodon longiceps*. It is impossible to tell from these data alone which individuals were mature and which were not, but this can be done using those telltale features of adulthood to be found in the skeleton. The result, shown in *Figure 7.7*, sends us three messages. First, adults show a huge range in size, the largest reaching twice the wingspan of the smallest. This is not unusual for pterosaurs, though. Many species show a similar pattern, even *Quetzalcoatlus northropi*, if one lumps the smaller adults and the giant together. Second, individuals cluster into two groups that, according to Chris Bennett, correspond roughly to males and females—an example of the size dimorphism that we encountered in Chapter 5. Third, and most significant, immature individuals[36] are scattered across the adult size spectrum, emphasizing that, in pterosaurs, size does not an adult make.

Collectively, these messages reinforce the central defining feature of growth in pterosaurs: It was reptilian. Pterosaurs inherited this type of development from their diapsid ancestors; it explains the existence of eggs with flight-ready embryos, flaplings, and adults of all sizes; and it could also explain one other thing—how to build giants.

The Way to a Whopper   There are two kinds of giant pterosaurs: relative giants and absolute giants. Relative giants can be found in many species and are simply individuals that grew much larger than typical adult size (*Figure 7.6*), usually on the order of twice as large (the equivalent for humans would be more than 10 feet tall). While they might also be absolutely very large (the giant airplane-size individual of *Quetzalcoatlus northropi* being an obvious example), they could also be, in absolute terms, rather small. A "giant" example of *Pterodactylus kochi* in the collections of the Museum für Naturkunde in Berlin has a wingspan of less than a meter, but is still twice as big as typical adults of this species.

Absolute giants include those pterosaurs that achieved wingspans of 8 meters (26 feet) or more. The most famous member of this category is the original specimen of *Quetzalcoatlus northropi*, a series of arm bones that belonged to a pterosaur about 10 meters (33 feet) in wingspan. *Quetzalcoatlus* was not the only one, however. *Hatzegopteryx* from Romania seems to have been about as large, and other finds in Spain indicate the existence of yet more giants that might have been even bigger.[37]

Two features of their growth seem to have allowed pterosaurs to become giants. First, they had the unique ability of being able to fly and grow at the same time. As we have already seen, the nature of their flight apparatus meant that it was serviceable as soon as the pterosaurs hatched. Even more importantly, the way they grew allowed pterosaurs' wings to continue to function right up to the usual adult size and beyond—the critical phase in becoming a giant. Birds and bats are excluded from becoming giants, because they are not normally able to take to the air until they are almost full-grown. As soon as they start flying, their growth rate slows, and it stops soon thereafter.

A highly variable growth rate was the second factor critical to gigantism. The adults toward the right end of the plots in *Figures 7.6* and *7.7* could have achieved their relatively large size in two ways: by growing faster than usual, or by continuing to grow for longer than usual (*Figure 7.9*). Both of these departures from the "normal" speed at which members of particular species grew were probably brought about by external environmental factors. The most likely is an unusually abundant supply of food, but exceptionally calm or equable weather and relatively few competitors are among the other possibilities. Sometimes, it must have happened that several factors particularly beneficial for growth coincided and, so long as such conditions persisted, those lucky enough to experience them just kept on growing because they could. This resulted in relatively gigantic individuals that, in big species, where adults typically reached 5 to 6 meters (16 to 20 feet) in wingspan, ended up being absolutely gigantic, too. This prompts another question: Are *Quetzalcoatlus* and its relatives really the largest that pterosaurs could get?

I posed this question to Jim Cunningham, an aerodynamic engineer from Arkansas who is deeply fascinated by pterosaurs. Somewhat to my surprise, he replied that, from his understanding of *Quetzalcoatlus*, a wingspan of 15 or even 20 meters (49 to 66 feet) was not impossible. So, there could be some super-gigantic brutes out there waiting to be discovered. Or maybe they have already been found and are hiding in paleontological collections pretending to be dinosaurs.[38]

Curators everywhere, check your drawers. Something awful from the skies may be in them.

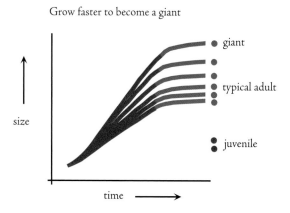

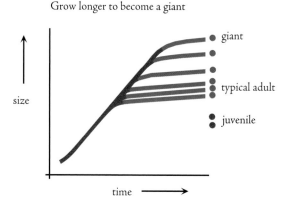

FIGURE 7.9 How to grow a whopper. Giant individuals may have achieved their relatively large size by growing faster (above) or for a longer time (below).

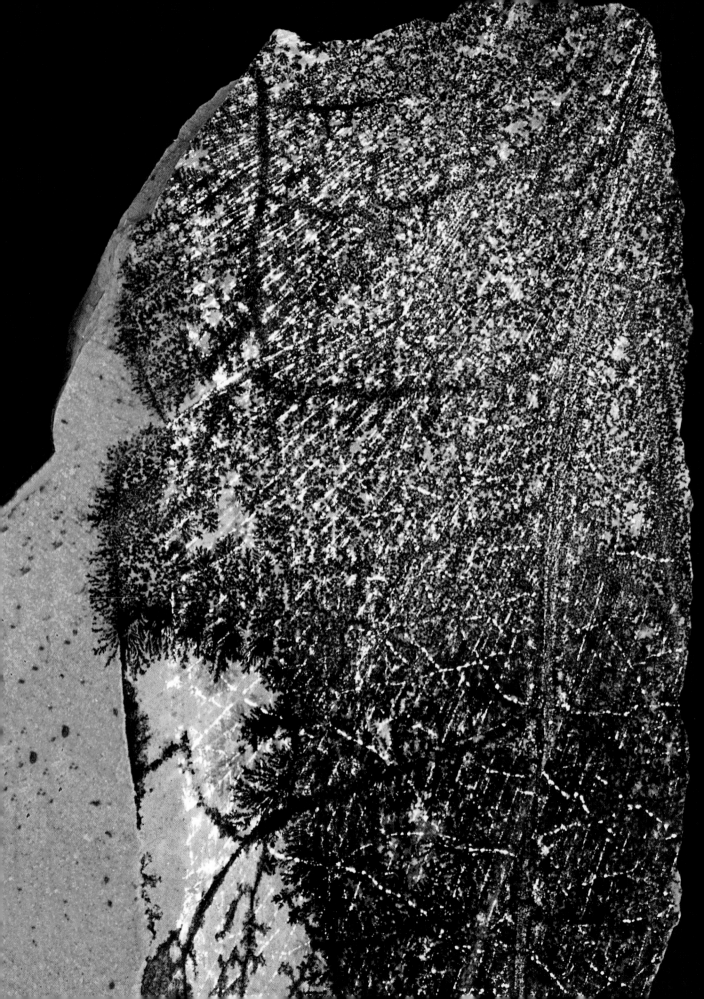

8
HIGH FLIERS

Throwing both feet sharply to the left and dipping his right pteroid, the big male Coloborhynchus banked hard and began to scan the long, narrow beach fringing the inland sea stretched out below. A recent storm had left trees and other debris strewn across the sand or half-submerged in the shallow water. Swinging back his wings, the Coloborhynchus launched into a long, slow glide that took him toward some clearer patches of beach farther out to the west. Half hidden beneath the bough of a large tree fern, a young spinosaur waited, completely still, and watched as the Coloborhynchus swept low over the sand, head swinging from side to side as he checked for danger. Flapping to gain height, the pterosaur turned back into the gentle breeze and began to drop ever closer to the beach. Both pteroids were hard down now, curving the wing deeper and deeper. Trimming his path with flicks of the wrists, the pterosaur allowed his legs to drift downward, and, as the drag began to take effect, his body tilted up and up until suddenly, he dropped, landing on outstretched feet. He stepped forward once, twice, swiftly folded the huge wing-finger back and up, then toppled gracefully onto his hands. The spinosaur didn't wait. Kicking as hard as she could with both legs, she launched herself toward her target, claws swinging and jaws wide, aiming directly at the pterosaur's slender, defenseless neck.[1]

FIGURE 8.1 The main flight membrane (cheiropatagium) of the Dark-Wing *Rhamphorhynchus*, with a wingspan of about 3 feet (1 meter), fluoresces beneath ultraviolet light, revealing the presence of long, thin wing-fibers, a fine network of muscle fibers and the delicate tracery work of the blood system (see *Figure 8.7* for an explanatory diagram of this fossil). (Photograph courtesy of Helmust Tischlinger.)

The True Fliers Club Vigorous fliers that soared, swooped and skimmed over the waves, as gracefully as any bird on the wing? Or clumsy gliders that plummeted from the heights, tail thrashing, eyes glaring, to finish in a tangle of limbs and a plume of dust? Pterosaurs' flight ability has been ascribed to both these extremes and almost every shade between. Indeed, when the first fossils came to light around the turn of the 18th century, some scientists didn't even consider the possibility of flight and thought pterosaurs were sea creatures that had rowed themselves through the water using their long arms.[2] The first person to suggest that pterosaurs could fly, surmising, correctly, in a paper published in 1801 that *"Il n'est guère possible de douter que ce long doigt n'ait servi 'à supporter une membrane qui formoit à l'animal, d'après la longeur de l'extrémité antérieur, une aile bien"* [It is not possible to doubt that the long finger served to support a membrane that, by lengthening the anterior extremity of this animal, formed a good wing] was Georges Cuvier, the great grandpére of pterosaurology.[3]

Cuvier's idea is now universally accepted, in part because of the discovery of such fossilized wing membranes as that illustrated on the opposite page, and the central issue no longer concerns pterosaurs' ability to fly, per se, but their degree of competence. Were they passive gliders or active, powered fliers?

Many different kinds of animals have some gliding ability, among them fish, frogs, lizards, snakes and several groups of mammals, including flying squirrels and the so-called flying lemur or colugo.[4] All that is needed to glide is a surface, be it a fin or a sheet of skin stretched between the toes or limbs, that has (or can adopt) a curved, airfoil shape so that it produces lift as the animal moves through the air.[5] The critical point about gliders is that they depend, essentially, on gravity for their propulsion, but move forward as well as downward because of the lift generated by their "wings." A parachutist, for example, falls downward, slowed only by the drag effect of the chute (which doesn't produce any lift), while a paraglider[6] moves both downward and forward, the latter resulting directly from the lift generated by the aerofoil shape of the chute. No matter how light they are and how effective their flight surfaces might be, gliding animals are passive fliers and constantly lose height, some more rapidly than others, but overall and inevitably, the direction is downward.[7] This means that flights are short and rarely last more than a few seconds. It's useful, it's effective, but it's not true flying.

True fliers differ in one fundamental way from gliders. They have the means to actively thrust themselves through the air at speeds that are fast enough for their wings to generate enough lift for them to be able to maintain height or even climb. Apart from helicopters, where the main rotors do both, in most man-made flying machines, the thrust is produced by propellers or jet engines, while the lift is produced by the wings. In the natural world, thrust is generated by flapping the wings. Indeed, surprising as it might seem, flapping is not what keeps birds up in the sky, it just propels them along. It is the lift generated by their wings as they move through the air that keeps them from falling on our heads. But what this means for birds and other true flying animals is that their wings must do two things at once: generate both lift and thrust. This is difficult, as man's general lack of success with ornithopters[8] well illustrates, and perhaps goes some way toward explaining why true fliers are rare in nature. Those animals that have acquired this ability—insects, birds and bats—have all been tremendously successful, and the reasons are not hard to find. True fliers can cover large distances rapidly and efficiently, they can hunt or find food quickly and effectively, they can easily escape from ground-bound predators,[9] and they even use the sky as a place to display and find mates.

The big gap between the flight ability of gliders and powered fliers and the profound implications this has for the biology and history of the groups concerned means that for pterosaurs, this is the crunch question. Were they card-carrying members of the true fliers club, or not? Opinion has long vacillated between the two sides. Richard Owen,[10] many of his Victorian contemporaries and even one or two modern scientists doubted that pterosaurs were much more than glorified gliders. Othenio Abel, a Viennese paleontologist who published widely on pterosaurs in the 1920s, managed to embrace both views at the same time, envisioning rhamphorhynchoids as gliders and pterodactyloids as flappers.[11]

By contrast, many current researchers favor the view adopted by several earlier pterosaurologists, such as Abel's contemporary, Baron Franz Nopcsa, and one of Richard Owen's many adversaries, Harry Seeley,[12] favored the view that all pterosaurs were true powered fliers. That said, the assumption that pterosaurs were less competent in the air than birds or bats (practically all-pervasive in the 19th century), lingers on even today. But not for much longer. This particular chauvinism has been challenged by some extraordinary new evidence that shows, instead, that pterosaurs were every bit as good as modern fliers and might even have outperformed them.

Engineering With Bone Some of the best evidence for pterosaurs' remarkable flight ability is to be found in their skeletons (*Figure 8.2*), particularly in the limbs. The arms and legs were extremely long, forming spars that supported the wings, and, through muscle action, enabled it to be flapped up and down. All the main limb bones were elongated, especially the fourth finger of the hand. Apart from supporting the entire outer half of the wing, this huge finger, the *sine qua non* of pterosaurs, curved backwards and downwards, imparting a particular twist to the main wing membrane that helped improve stability during flight.

Other modifications of the skeleton for flight are most clearly developed in large and giant pterosaurs. A good example is the fusion of vertebrae in the shoulder region, to form the notarium, and in the hip region, to form the sacrum, which helped stabilize and stiffen the spinal column, enabling it to better withstand the powerful forces generated by flapping flight. The shoulder girdles and hips acted both as anchor points for the big muscles that powered the flight stroke and as buttresses against which the arms and legs could articulate. The shoulder girdle was braced against the breast bone and, in the bigger pterosaurs, also against the notarium, while the hips were firmly butted up against the sacrum or completely fused to it.

Other adaptations of the skeleton for flight were more subtle, but no less important. Manipulate a pterosaur wrist and one finds, as with many other joints, that the bone surfaces fit tightly together and glide over each other with all the precision of highly engineered machinery. This is quite unlike typical reptilian limb joints, which usually tend to be less well-defined and rather looser in operation, but it served pterosaurs in two important ways. First, the tight fit ensured precise control over limb movements—important for fliers, where even small movements can have major effects. Second, this construction enabled the joints to transmit forces effectively, reducing the risk of dislocation and the need for extra muscles to help stabilize the joint or control movements—another small but helpful way of saving weight.

One of pterosaurs' most important flight adaptations—their hollow bones—is best seen in damaged fossils. As *Figure 8.2* illustrates, flat bones from which the skull and such other bulky elements as the vertebrae, shoulder girdle and pelvis were constructed had wafer-thin walls, often only 2 or 3 millimeters in thickness, and were supported on the inside by a fine, honeycomb-like filling also made of bone. Wrist and ankle bones were built in the same way, but in long bones, the honeycomb was restricted to either end, while the main shaft was hollow.

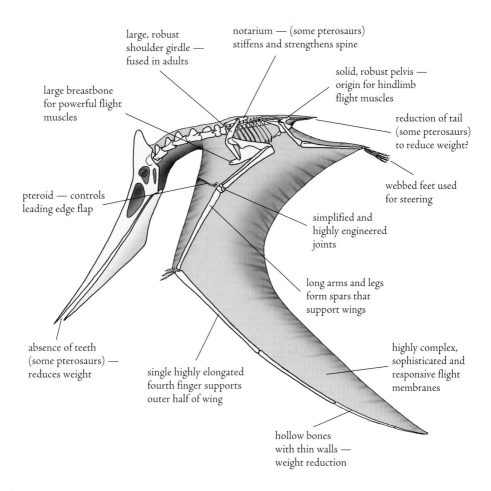

large breastbone
for powerful flight
muscles

large, robust
shoulder girdle —
fused in adults

notarium — (some pterosaurs)
stiffens and strengthens spine

solid, robust pelvis —
origin for hindlimb
flight muscles

reduction of tail
(some pterosaurs)
to reduce weight?

webbed feet used
for steering

pteroid — controls
leading edge flap

simplified and
highly engineered
joints

long arms and legs
form spars that
support wings

highly complex,
sophisticated and
responsive flight
membranes

absence of teeth
(some pterosaurs) —
reduces weight

single highly elongated
fourth finger supports
outer half of wing

hollow bones
with thin walls —
weight reduction

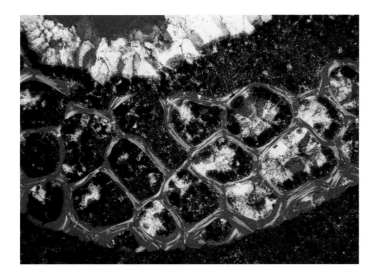

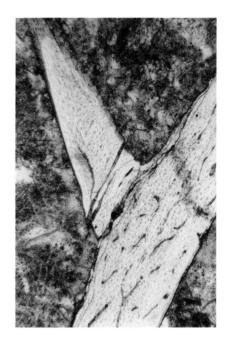

FIGURE 8.2 A design for flight. Some of the many adaptations of the pterosaur skeleton for a life in the air are highlighted in this illustration based on *Pteranodon*. Two special features are seen in the thin sections of pterosaur bones shown below: On the left, a weight-saving meshwork of bone picked out by polarised light, and on the right a thin supporting strut branching off from the bone wall. (Photographs courtesy of Lorna Steel.)

Presumably, the spaces in the honeycomb or in the hollow shafts were usually filled with bone marrow, but in some pterosaurs, the lightening went even further. Outgrowths from the lungs, which we first encountered in Chapter 6, penetrated through the walls of the vertebrae and then expanded, filling them with numerous tiny bags of air. In some species, this gas-filled labyrinth spread further out, tunneling through the shoulder girdles and hips and into the arms and legs, so that in such large pterosaurs as *Pteranodon*, practically every bone in the body was perforated and lightened by this system. This pneumatization had a dramatic impact on the skeleton, which, apart from the teeth, was the heaviest tissue in the body, and might have reduced its total weight by as much as 70 or 80 percent.[13]

But, long, hollow bones can also bring risks. Bump into a tree, or hit the ground too hard, and the slender-walled, tubular bones of large pterosaurs were likely to break or even collapse by buckling.[14] Ever resourceful, though, nature had a few tricks up her sleeve to avoid this catastrophe. Internal angles of the hollow bones were buttressed, walls were thickened in the lightest way possible, by lining their inside surface with a thin layer of honeycomb bone, and hollow shafts were bolstered with thin, perfectly aligned struts (*Figure 8.2*). Each helped to reduce the risk of collapse for the absolute minimum amount of extra bone.

Wing Ding Georges Cuvier was dead before anyone found a pterosaur with recognizable evidence of wing membranes,[15] but the methodology he had devised—comparative anatomy—had not failed. Several examples of *Rhamphorhynchus* with superbly preserved impressions of the wings, one of which is shown in *Figure 3.5*, were found in the late 1800s[16] and confirmed that pterosaurs had indeed flown, using thin membranes of skin, as Cuvier had originally supposed. These and other "star" fossils, such as a beautiful specimen of *Pterodactylus kochi* in the Natural History Museum in Vienna, also illustrated in *Figure 3.5*, belong to a series of more than 30 specimens from the Solnhofen Limestones, all of which preserve details of the wings.[17]

Even after evidence from other fossil localities in Kazakhstan, Brazil and China began to come to light in the mid- to late 1900s, Solnhofen fossils were (and remained) far and away the single most important source of information on pterosaur wings and completely dominated any discussion of their shape, structure or function. Unfortunately, they had a major drawback: While impressions of the flight membranes are often quite distinct in the outer part of the wings, in the region close to the body, they are, with few exceptions, either poorly preserved or completely absent.[18] This lack of clarity about the shape and extent of the flight membranes (patagia), spawned a major controversy that dominated pterosaur research in the 1980s and 1990s.

At the heart of the fight was a simple question: Were the legs connected to the wings (as in bats), or were they completely separate so that the attachment of the flight membranes was restricted to just the forelimbs and body? The answer to this question is perhaps the most important key to understanding pterosaurs. This is, in part, because it has profound consequences for the construction of the wings and the way in which they functioned during flight; and in part because, by analogy with bats, if the legs were fastened to the wings this must have had a huge impact on pterosaurs ability to move around on the ground. This, in turn, has major consequences for their ecology and evolutionary history.

Sordes pilosus gave the first clear answers to the problem. Flight membranes are more completely preserved in this middle-Asian rhamphorhynchid than in any other pterosaur, and, as the example of *Sordes* illustrated in *Figure 3.6* shows, not only were they connected to the legs, but they were also stretched between them. The reality of this rather bat-like construction has been confirmed by new fossil finds[19] from the Jehol Biota of China and the Crato Formation of Brazil and, most recently and spectacularly, by a Solnhofen pterosaur, the so-called "Dark-Wing *Rhamphorhynchus,*" shown in *Figures 1.2* and *8.1.*

Indeed, so much evidence of the flight membranes has come to light in recent years that we now know something of these structures in at least 13 different species of pterosaur, which represent nearly all the main pterosaur clans and span most of the groups history, from the Late Triassic until well into the Late Cretaceous (listed in *Figure 3.9*). In concert, these fossils tell us a great deal about pterosaur wings, including the exact shape and extent of the individual flight membranes, and, of critical significance for the evolutionary history of these animals, how they varied between rhamphorhynchoids and pterodactyloids.

Meet the Patagia Pterosaur wings were composed of three types of flight membranes, also known as patagia. The specific name for each patagium refers to its position within the wing, which is illustrated in *Figure 8.3.* Thus the propatagium (fore-wing) extended along the front margin of each arm, while the cheiropatagium (hand-wing), which formed most of the wing surface, extended between the arm and leg. Finally, the cruropatagium (leg-wing) was stretched between the legs.[20] Since there was a left and a right propatagium, a left and a right cheiropatagium and single cruropatagium, pterosaurs had five flight surfaces in all.

In almost all published restorations of pterosaur wings, the propatagium is shown as just a small triangular piece of membrane located in the angle of the elbow, its front edge curving from the neck out to the wrist, where it is met by the pteroid. Back in the Mesozoic, this might well have been how pterosaurs appeared when they were at rest, the wings partially or fully folded and the propatagium in its "relaxed" position. As mentioned in Chapter 6, however, there is growing evidence to show that, during flight, the pteroid was no longer directed inward, but pointed forwards from the wrist. In this case, as *Figure 8.3* illustrates, the propatagium must have extended much farther out in front of the arm, curving from the neck past the wrist to finish at the base of the fingers, or perhaps even beyond.

Once they have regained sufficient coherence to be understood after their apoplectic fits at the sight of such a reconstruction,[21] most pterosaurologists start listing reasons why it is wrong, usually beginning with the most obvious, and, in their eyes, fatal problem: the absence of any fossil evidence for the propatagium beyond the pteroid.

Two points can be made in reply. First, in those fossils where one or both wings are preserved, they are almost always partly folded, and the pteroid lies in its "relaxed" position, pointing toward the body. In this case, the

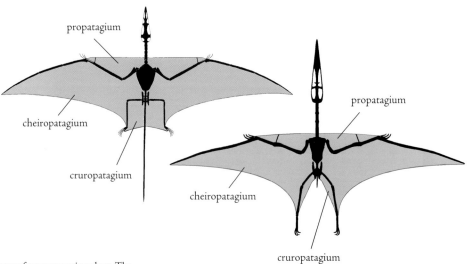

FIGURE 8.3 The two main types of pterosaur wing plans. The typical arrangement of the flight membranes is shown for a rhamphorhynchoid (above left), distinguished by a complete cruropatagium stretched between the hind limbs, and a pterodactyloid (below right), in which the cruropatagium is split up the middle, leaving two small crescent-shaped patagia filling the inner angle of the knee.

propatagium beyond the wrist is likely to have been furled back tight against the wing-metacarpal and, even assuming that it survived long enough to do so, would have left only the narrowest of impressions in front of the arm. Second, and compounding the problem, during the freeing of pterosaur fossils from the enclosing rock, this area is often prepared away to expose the slender metacarpals that articulated with the clawed fingers of the hand. As discussed in Chapter 3, this means that any evidence of the propatagium in this region may be inadvertently destroyed. Despite the odds being heavily stacked against its preservation or survival, traces of a propatagium beyond the pteroid are present in several fossils. Although not easy to see, I have found them in the Zittel wing (*Figure 3.5*), in which, tellingly, the pteroid points forward, as you can see in, and they are clearer in a tapejarid from the Crato Limestone where remains of the propatagium are preserved around the base of the fingers. The best evidence, however, is in a new specimen of the Chinese pterosaur *Jeholopterus*, shown in *Figure 8.4*, where the front (leading) edge of the propatagium can be clearly seen extending beyond the wrist directly to the base of the fingers.

The cheiropatagium, the main wing membrane, originated from the flanks of the body and filled the space between the arm and the leg. The free (or trailing) edge of the cheiropatagium can be clearly traced in the example of *Sordes* shown in *Figure 8.4*, curving inward from the wing tip toward the knee, then, as it approaches the knee, it veers sharply backward to run down to the ankle. The resulting notch, which was probably rather less prominent in life, indicates the region in which the stiffer outer part of the cheiropatagium met the softer, more extensible inner part. We will see, shortly, how and why pterosaurs were able to vary the stiffness of their wings.

Practically the same arrangement of the cheiropatagium is to be found in several other pterosaurs, including the anurognathid *Jeholopterus* (*Figure 8.4*), the rhamphorhynchid *Rhamphorhynchus*, the ctenochasmatids *Pterodactylus* and *Eosipterus*, and a tapejarid from the Crato Limestone. Such a broad distribution suggests that the type of cheiropatagium found in *Sordes* was probably universal for pterosaurs. This conjecture tallies well with the discovery that there is a strong correlation between the distance from the shoulder joint to the wing tip and the distance from the hip to the ankle, both in those species where there is some evidence of the wing and in many others, as well.[22] This close relationship, also found in bats, but not in birds, where leg length is completely independent of arm length, suggests that in pterosaurs the arm and leg were somehow connected, and we now have an excellent explanation for this—the cheiropatagium.

Until quite recently, the existence, or otherwise, of a cruropatagium in pterosaurs was another of those issues guaranteed to raise the blood pressure of most pterosaurologists.[23] Again, *Sordes* came to the rescue. A cruropatagium can be seen in several specimens, including that shown in *Figure 8.4*, where it fills the region between the legs and extends down to the foot and out along the fifth toe. The rear, or trailing, edge of the cruropatagium can then be traced from the tip of one fifth toe across to the other. *Sordes* also demonstrates that the cruropatagium was not attached to the tail, but

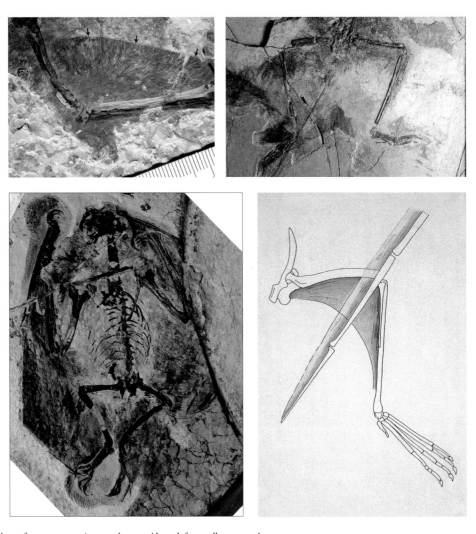

FIGURE 8.4 The critical evidence for pterosaur wing membranes. Above left: a well-preserved propatagium, its front (leading) edge (indicated by arrows) extending along the front of the arm bones to the base of the fingers in a new specimen of the anurognathid *Jeholopterus*, with a wingspan of about 18 inches (45 cm). Above right: the cheiropatagium and cruropatagium of *Sordes pilosus*, with a wingspan of 2 feet (60 cm). Bottom left: superbly preserved wing membranes of the first example of *Jeholopterus* to be found, with a wingspan of 30 inches (80 cm). Bottom right: crescentic cruropatagium preserved in the inner angle of the right knee of *Pterodactylus*, with a wingspan of 20 inches (50 cm). (Photographs courtesy of Lü Junchang, above left, and Wang Xiao-Lin, bottom left).

lay beneath it. This arrangement allowed the tail to be swung from side to side, without interfering with the cruropatagium, and crucially permitted the legs to be lowered without affecting the tail, as the pterosaur came into land, for example.

Evidence of the cruropatagium has also been found in several other pterosaurs, including *Jeholopterus*, *Eudimorphodon* and *Rhamphorhynchus*, and in each case it perfectly matches the details evident in *Sordes*. Because, collectively, these pterosaurs represent four of the five main rhamphorhynchoid clans, it would seem that a large cruropatagium, manipulated by the fifth toes, was typical and quite probably universal of rhamphorhynchoids.

The situation in pterodactyloids was quite different. Several examples of *Pterodactylus*, one of which is illustrated in *Figure 8.4*, show that in this pterosaur, the cruropatagium was split into two membranes, each consisting of a narrow, crescent-shaped structure that filled the inner angle of the knee, the free edge curving from the base of the tail down and out to the fifth toe, or fifth metatarsal, when the toe was lost. Other pterodactyloids, including recently discovered examples of ctenochasmatids, tapejarids and azhdarchids, show an almost identical arrangement,[24] so it seems likely that this paired arrangement was typical for these pterosaurs.

The splitting of the cruropatagium had profound consequences for pterodactyloids. It meant, first of all, that the legs were unshackled from one another, which had a second, and even greater, significance. Because the leg was linked to the arm by the cheiropatagium, it meant that the limbs on the left side of the body could now operate entirely independently of those on the right side. This certainly must have made some difference to pterodactyloid flight (although the details have yet to be established), but it was of even greater importance for movement on the ground. In comparison with rhamphorhynchoids, hobbled by the cruropatagium shackling their legs together, the limbs of pterodactyloids must have been more manoeuvrable (although still hampered somewhat by the cheiropatagia), allowing them far greater and easier mobility on the ground than rhamphorhynchoids ever had.

Patagia—The Inside Story

"Tough and leathery."[25] Until quite recently, this rather unflattering phrase effectively summed up what was known about pterosaur wings. It conjures up images of a simple construction and a correspondingly modest flight ability, but both are quite wrong. Several extraordinary fossils, including a three-dimensional chunk of flight membrane found in a nodule from the Santana Formation of Brazil,[26] the Dark-Wing

Rhamphorhynchus and the hairy devil *Sordes*, have revealed much of the inside story of pterosaur wings, showing them to be as complex and sophisticated as anything to be found in modern fliers, living or man-made.

That each flight patagium consisted of a thin sheet of modified skin has long been suspected and is now generally accepted, but it raises an important question: How were the patagia attached to the arms and legs? The answer is, they weren't—at least not directly. Because the patagia were essentially extensions of the skin that invested the rest of the body, then the limb bones must have been enclosed within them too, and this is just what is seen in the Santana specimen shown in *Figure 3.7*. This extraordinary specimen also illustrates another important feature of the flight patagia: They were remarkably complex internally and made up of several different layers that, seen from the side, looked rather like a generously filled sandwich. It was not a hairy sandwich, because the thin layers of epidermis that bounded it top and bottom seem to have been quite naked and hairless. On the inside were connective tissue, blood vessels, a nerve network, bundles of muscles and a uniquely pterosaurian development: wing fibers.[27]

The largest and most prominent component of the flight membranes, wing fibers have been known about since the late 1800s and were first noticed by the Munich-based paleontologist, Alfred von Zittel, in the isolated wing of *Rhamphorhynchus* that now bears his name. The extremely high quality of the impression left by the cheiropatagium in this fossil shows a fine lineation that Zittel and many later researchers interpreted as an internal system of fibers. Not all scientists were convinced of this, however, and some suggested instead that this lineation might have been left by fine parallel folds in the patagium.[28] Details of Zittel's wing and the presence of discreet, identifiable fibers in the wings of several pterosaurs, including the Brazilian fossil show that Zittel was correct, after all.

Much has been learned about wing fibers, thanks to their preservation in more than 10 different kinds of pterosaur,[29] including *Sordes*, illustrated in *Figure 8.5*. In this, as in other pterosaurs, the wing fibers were embedded within the patagia and typically measured a little less than one-tenth of a millimeter in diameter—about twice the thickness of a human hair. In some spots, unraveled fibers (visible in *Figure 3.6*) reveal that they were composite structures, composed of at least 20 or 30 very fine strands, wound together in a helical fashion.[30] Each strand was only a few hundredths of a millimeter across and probably made of collagen, a material that is common in the skin of vertebrates.

The layout of the fibers within the patagia, compiled from several fossils of *Sordes*, is shown in the somewhat stylized restoration depicted in *Figure 8.5*. The fibers emerged at right angles from the body flanks and ran parallel, or nearly parallel, to the limbs, curving around the inside or outside of major joints, such as the elbow and knee. They did not contact any of the limb bones, however, and clearly were not attached to them, but they did extend to the free edges of the patagia, often meeting them at a steep angle.

In the outer part of the wing, the fibers were relatively long, up to 10 centimeters (4 inches) in *Sordes*, always remarkably straight and usually packed closely together. Moreover, I have never come across any evidence of them unraveling in this region, suggesting that the individual strands from which they were made were either very tightly wound or perhaps even firmly fixed

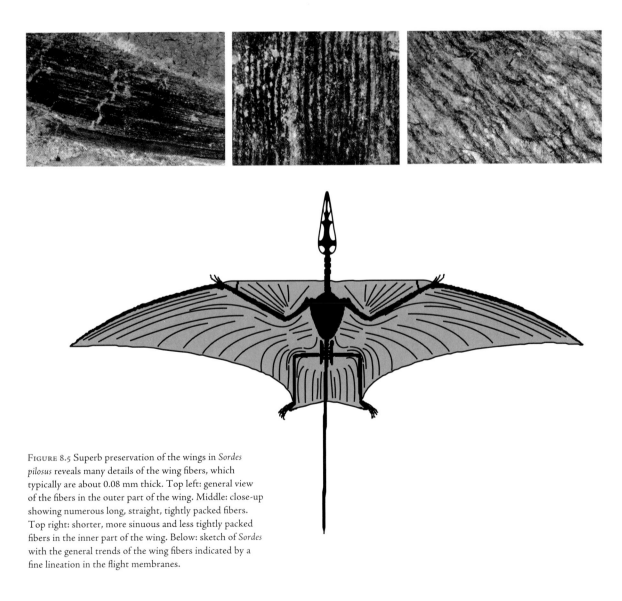

FIGURE 8.5 Superb preservation of the wings in *Sordes pilosus* reveals many details of the wing fibers, which typically are about 0.08 mm thick. Top left: general view of the fibers in the outer part of the wing. Middle: close-up showing numerous long, straight, tightly packed fibers. Top right: shorter, more sinuous and less tightly packed fibers in the inner part of the wing. Below: sketch of *Sordes* with the general trends of the wing fibers indicated by a fine lineation in the flight membranes.

in some way to one another. Regardless of their exact design, it seems certain that the fibers in this region were relatively stiff.

By contrast, fibers in the inner part of the wing are always shorter, usually more widely spaced and often rather sinuous. This, and occasional signs of unraveling, all point to a great deal more flexibility and suppleness of the fibers in this area, compared with their neighbors in the outer part of the wing. It also makes sense from an engineering point of view. Those areas of the patagia near the body are likely to have experienced a lot more stretching, bending and flexing, for example, when the wing was folded, than those out near the wing tip, so we might expect them to be more flexible. But why did pterosaurs need wing fibers in the first place?

Some researchers have suggested that these structures were like the shafts of birds' flight feathers and served to transmit loads experienced during flight to the arms and legs.[31] This is impossible, though, because unlike birds' feathers, which are firmly anchored to the arm, wing fibers had no contact with any of the limb bones and so could not transfer loads to them. Another alternative is that they helped to toughen up the patagia, reducing the risk of dangerous tears or rips. Undoubtedly, this would have been useful, but why were so many fibers needed, especially in the outer part of the wing? Chris Bennett thinks he has the answer.[32]

A problem faced by any animal with a membranous wing is how to tension it so that it doesn't flutter, billow or, even worse, bunch up around the wing spar. Pterosaurs could control tension in the propatagium, cruropatagium and inner part of the cheiropatagium because, effectively, these patagia were stretched between various bones. But what about the outer part of the cheiropatagium? How was this tensioned? As Cherrie Bramwell and George Whitfield discovered in their ground-breaking study of the aerodynamics of *Pteranodon* in the 1970s,[33] stretching this region of the cheiropatagium between the arm and a tendon running from the ankle to the tip of the wing-finger would have been impossible. This was partly because it would have meant keeping the limbs in the same position all the time, but mainly because, according to Bramwell and Whitfield's calculations, the arm bones were just not strong enough to tension the tendon to the degree needed for it to work properly.

This, according to Chris Bennett, is where the wing fibers come in. Although they may have been able to bend to some degree, the fibers in the outer part of the wing, like a strand of uncooked spaghetti, probably could not be stretched or compressed. This would have neatly solved the tension problem because, as can be seen in the restoration of *Sordes'* wings in *Figure 8.5*, the arrangement of the fibers in the wing would have kept the cheiropatagium stretched out from front to back, while still allowing it to be folded, concertina-like, parallel to the grain of the fibers. Fossils of *Sordes* and *Rhamphorhynchus* also show that there was a tendon running along the rear edge of the cheiropatagium which would have helped stabilize the whole structure, resulting in a flutter-free, foldable, flight surface.

Smart Wings The story is not over yet, though, because the patagia contained far more than just wing fibers. Illuminating the Dark-Wing *Rhamphorhynchus* with ultraviolet light (see *Figure 8.1*) reveals the presence of what appears to be a blood vessel network that radiated outward from the shoulder region, right across the cheiropatagium to the outer tip of the wing.[34] The discovery of the same system in other fossils, such as the Zittel wing, confirms that it was not an artifact of preservation, so it seems reasonable to assume that it was present in all pterosaurs, where, as in *Rhamphorhynchus*, it served to nourish the living tissues of the wing.

What appears to be another development of the blood system, found in the chunk of membrane from Brazil and visible in *Figure 8.6*, takes the form of numerous, small, interconnected chambers that formed a thin, mattress-like structure, immediately beneath the surface of the patagium in the region near the body.[35] If this system was connected to the blood supply, then

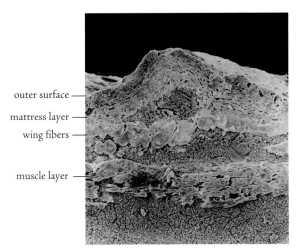
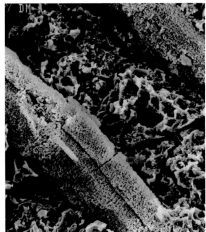

outer surface —
mattress layer —
wing fibers —

muscle layer —

FIGURE 8.6 The structure of the wing membrane in a 6 foot (2 meter) wingspan pterosaur from the Santana Formation of Brazil. The main picture shows a cross-section, about 1 millimeter in thickness, through the membrane. The thin layer at the top is the epidermis and beneath it lies the "mattress" layer. The row of bundles beneath the mattress are wing-fibers, seen in cross-section, and underneath them lies a layer of muscle, a single fiber of which is shown at high magnification in the inset. (Photographs courtesy of David Martill.)

one possibility is that pterosaurs used it to keep cool during vigorous activity, such as active flight. Experimental work on birds and bats has found that if they flap their wings hard, or for long periods, the muscles involved generate large amounts of heat, and doubtless the same was true for flapping pterosaurs. Pterosaurs might have solved this problem by blushing with their wings—gorging the mattress with hot blood and allowing heat to radiate away through the thin epidermis. Another completely different possibility, however, is that the mattress layer might have been part of the air-filled labyrinth connected to the lungs, so that not just the wing bones, but even the patagia themselves, were partially pneumatized.

Investigation of the Brazilian wing membrane using a scanning electron microscope reveals another important component of the pterosaur flight membranes: muscle. As *Figure 8.6* shows, at low magnification, masses of closely packed muscle fibers are clearly visible beneath the much larger wing fibers. By cranking up the magnification, it is possible to distinguish the muscle type. Fine bands running across individual fibers show that it was striated (voluntary) muscle, and thus under conscious control like the muscles that operate our arms and legs.[36] An incredible amount of detail, but that's not all. Bathing the Dark-Wing *Rhamphorhynchus* in ultra-violet light seems to reveal that muscle fibres were to be found there as well, and the completeness of this fossil shows that they radiated throughout the entire wing, forming a network that ran perpendicular to the wing-fibers (*Figure 8.7*). An astounding discovery and one with some profound implications for pterosaur wings.

At the very least, pterosaurs would have been able to use the muscles within their wings to exert control over tension in the patagia. Contraction would have produced a tightening of the wing membrane with a concomitant flattening effect, while relaxation would have resulted in slackening and a greater degree of upward bowing. Such shape changes are important, because they directly affect the amount of lift generated by the flight surface—generally speaking, flatter wings mean less lift, more curved wings greater lift. So, just by using the muscles in their patagia and without even moving their arms or legs, pterosaurs may have been able to control the lift generated by the wing as a whole, or even within particular regions of the wing—extremely useful tricks for highly sophisticated fliers.

This is already far more than anyone might have expected of pterosaurs, and yet these amazing animals might have gone even further. In order for it to function, the muscle fiber network in the wings must have been supplied by a nervous system that operated the muscles and sensors (proprioceptors—nerve endings embedded in muscle), which would have given feedback about the degree to which each fiber had contracted. This meant that pterosaurs were equipped with a system that would have enabled them to monitor the local degree of tension or slackness right across the wings and to update this picture many times per second. Because of their large surface area, however,

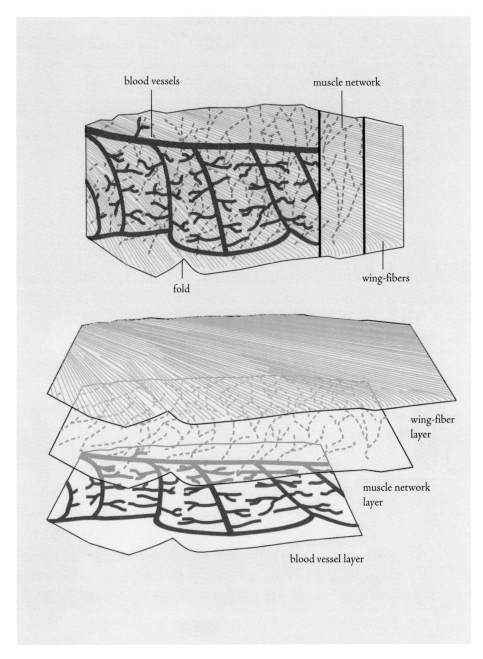

FIGURE 8.7 A diagram, based on the Dark-Wing *Rhamphorhynchus* fossil, illustrating current ideas regarding the construction of the pterosaur wing membrane. The sketch at the top shows the wing as it is preserved, while the lower sketch shows the three main layers from which the membrane was constructed, with wing-fibers at the top, then the muscle network and, at the bottom, the blood vessel system. (Redrawn from Dino Frey, et al., 2003.)

the flight membranes would have generated a formidable quantity of information that, in turn, would have required an equally formidable processing capability of the brain. This is interesting because, as you will recall from our visit to this mental center in Chapter 5, the floccular lobes, the regions of the brain that dealt with sensory input from this part of the body were relatively huge in pterosaurs. This, it would seem, is where all that processing was carried out.[37]

Equipped with such a system, pterosaurs would have been able to perceive exactly how the wing was performing during flight. And, by altering wing shape through localized contraction and relaxation of muscles fibers within the membrane, they could respond extremely rapidly to changes brought about, for example, by snatching up a large food item such as a fish, or flying into turbulent air. With such "smart-wings," pterosaurs had a flight apparatus that was probably even more effective than that of birds or bats. Very clever stuff indeed.

Muscle Engines

In flight, birds and bats propel themselves through the air by flapping their wings, with most of the thrust being generated in the downstroke part of the wing beat, although some sophisticated fliers are also able to do this during the upstroke, when the wings are being returned to their starting point.[38] Flapping is hard work, though, and requires a lot of energy and muscle. Birds and bats have plenty of both. Their "hot-blooded" physiology delivers a great deal of energy, much of which goes to fuel large, well-developed muscles like the pectorals that are the main power source for the downstroke. If pterosaurs were members of the true fliers club, it follows that they, too, must have had large muscle "engines."

In fact, telltale marks of these "engines"—muscle scars, bony processes, even the size and shape of whole bones—can be found all over those parts of the skeleton (mainly the shoulder girdles and hips), to which the muscles that powered pterosaur flight were anchored (*Figure 8.8*). The presence of some big bulky pectorals, the main muscles that powered the downstroke, is demonstrated by the broad extent of the breast bone, from which they began, and the huge bony "deltopectoral" process of the humerus, to which they attached. They did not work alone, though. Prominent scars on the lower part of the shoulder girdle give away the position of other muscles that also attached to the head of the humerus and added their strength to the downstroke.[39]

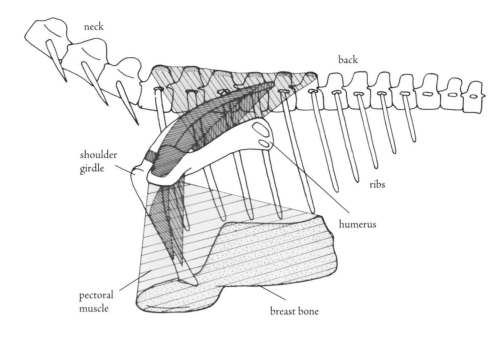

FIGURE 8.8 Sketch of the principal muscles that powered the flight stroke in the Lower Jurassic pterosaur *Campylognathoides*, with a wingspan of about 3 feet (1 meter). Muscles that flexed the wing down are shown in pink; those that lifted it back up again in blue. (Based on Bennett, 2003.)

So, pterosaurs had plenty of muscle to power the wing beat—or did they? Dissect the breast region of a duck or a pigeon, or any bird that flies, and you will find that the breast bone has a large, deep keel running down the center line, attached to either side of which are big, thick slabs of muscle. Such a keel has not yet been found in any pterosaur, which has led some researchers to conclude that pterosaurs were less well-endowed with muscles than birds and that their flight must have been correspondingly weak and ineffective. At the same time, this seems to be quite at odds with all the other evidence that suggests that pterosaurs were highly competent fliers. So what's going on? One possibility is that the pterosaur breast bone did have a keel, but it was made of cartilage and did not fossilize. Perhaps, but the absence of any scars or marks on the breast bone that might have been left by such a keel does not encourage this idea.

Another more likely explanation is that pterosaurs didn't have a keel because they didn't need one. In birds, the massive development of the breast bone is partly related to the presence not just of the pectorals but of a second pair of muscles called the supracoracoideus muscles. Like the pectorals, the supracoracoids slant upward toward the humerus, but, instead of inserting directly into this bone, they loop through a channel between the shoulder girdle bones, which acts as a pulley, and then come back down to attach to the humerus from above. This peculiar development, unique to birds, means that their supracoracoideus does not pull the wing down, but, quite the contrary, helps to raise it back up to its starting position.[40] It has sometimes been suggested that pterosaurs may have evolved the same system, but a recent investigation of the shoulder girdle and its associated muscles by Chris Bennett has cast severe doubt on this idea.[41] Rather, it seems that in pterosaurs, the supracoracoideus was relatively small, as in other diapsids, and assisted to some degree with the downstroke.

So, perhaps this is why the pterosaur breast bone had no keel, but that raises a different problem: How did pterosaurs raise their wings? Again, Bennett's study provides the answer. Lines and scars on the shoulder blade (scapula) and the neighboring parts of the spinal column and rib cage reveal the anchor points for several muscles that slanted forward and downward to attach to the upper side of the humerus. It seems that their combined effort would have been more than sufficient to raise the wing, but their line of action might have caused problems, especially for large pterosaurs. As these muscles contracted, the load they imposed on the skeleton could have pulled the shoulder girdle out of position and distorted the spine and rib cage—painful and inefficient. Recall, though, that in many, if not all, big pterosaurs, the vertebrae in this region fused to form the notarium, which also supported the shoulder girdle by bracing the scapula. This explains why these unusual features developed: They were adaptations of the skeleton to the loads that pterosaurs experienced during flight.

In addition to the muscle engines that worked the arms, pterosaurs had a second set of engines at the back—the muscles that powered the legs. Several large, powerful muscles were anchored to the broad, gently curved surfaces of the pelvis below the hip joint and attached at the other end to the thigh bone. Presumably, their contractions were synchronized with those of the pectorals, ensuring that the rear part of the wing followed the same path as the front half and significantly increasing the overall power that pterosaurs could deliver during the downstroke. (This might be yet another reason pterosaurs did not need a keeled breastbone.) Raising the leg back up again was brought about by muscles fastened to the upper part of the pelvis, above the hip joint, and possibly also to the sacral vertebrae, which perhaps explains why these vertebrae and the hips were so firmly united in pterosaurs.

Airborne So far, we have focused mainly on the construction of the pterosaur flight apparatus: its bony girder work, the shape and structure of the wing membranes and the muscles that powered the system. Now, we can move on to consider how it all worked when pterosaurs did what they did best—flew. Flight is a complicated matter, though, so let's begin with the simplest possible situation, a pterosaur soaring gracefully in the rising air of a thermal, its outstretched wings seemingly quite still (*Figure 8.9*).

Observing this pterosaur from a distance, we would have been able to make out an almost straight head and neck, a slightly humped body and a long, straight tail, if it were a rhamphorhynchoid, or something rather similar, but with a short tail and a head that slanted downward, if it were a

FIGURE 8.9 A generalized pterodactyloid in flight, as seen from the front (top left), from the side (middle right) and from above (bottom left).

pterodactyloid. As the animal drifted more nearly overhead, we would have seen that the arms were flexed at the elbow and the base of the wing-finger formed a zigzag outline that was interrupted by a spike-like pteroid projecting forward from each wrist. Sticking straight out on either side of the hips were the thigh bones, hinged at the knee with the lower leg, which was directed backward, parallel to the tail. Looking closely, we might have been able to spot that the feet were vertically aligned, their soles facing each other, first toe at the bottom, fifth toe (in those pterosaurs that had them), on the top and directed inward along the rear edge of the cruropatagium.

Climbing up on a hill in order to view the pterosaur from the side, we would have been able to observe how the wings curved from front to back, forming the lift-generating cambered profile that was keeping the object of our study in the air. Using a powerful pair of binoculars, we might also have been able to detect how the pterosaur was controlling the camber, by contracting or relaxing the muscle fibers in the wing membranes, or by moving the arms and legs toward or away from one another. Then, as the pterosaur began to turn toward us, we might have been able to make out yet another camber control mechanism, the pteroid-operated propatagium, deepening the curve of the wing as it dipped down, flattening it again as it was raised back up. Once the pterosaur had turned fully head-on to us, we would also have seen that, apart from curving from front to back, the wings also curved from their root to the tip, following the gently arched profile of the arm right out to the end of the wing-finger.

As the pterosaur passed overhead again, we would have had an excellent opportunity to see how it was able to change the profile of the wings. By flexing the hinge joints at the elbow, wrist, wing-finger and knee, the wing became shorter from base to tip, and relatively deeper from front to back. Opening these joints out again returned the wing to a longer, narrower profile. These were not the only options, though. By varying the extent to which each of the joints was flexed or extended, pterosaurs could generate many intermediate profiles. Each of these could be further varied by changing the position of the arms and legs with regard to one another, through movements at the shoulders and hips.

Occasionally, even a soaring pterosaur would have needed to flap, and we would have been alerted to this by the raising of the wings in preparation for the main wing beat. Later, examining high-speed film shown at a considerably slowed rate (the whole process was probably too fast for the naked eye), we would have been able to see, as illustrated in *Figure 8.10*, how the wings swept downward and, because of the direction of pull exerted by the pectorals and its assistants, were also swung forward a little and gently rotated so that the front edge turned to face slightly downward. By the end of the downstroke, the wings of most pterosaurs would have been nearly beneath the body, but not in the case of *Pteranodon* and other ornithocheiroids, specialist fliers where the shape of the shoulder joint meant that the wing beat could go no farther than just below horizontal.[42]

During the upstroke, the wing was lifted, swung backward and counter-rotated so that it returned to a raised position, ready to begin the next flap. The operation was probably similar to that in birds and bats, except that

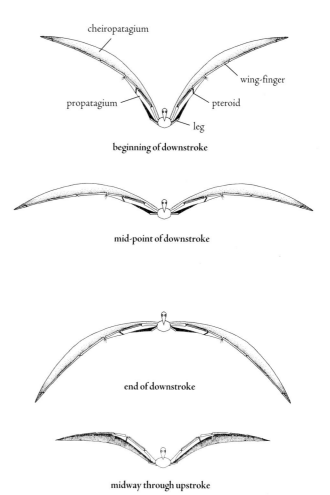

beginning of downstroke

mid-point of downstroke

end of downstroke

midway through upstroke

FIGURE 8.10 How pterosaurs flapped their wings. The sequence begins with the wings raised, runs through the middle position to reach the bottom of the downstroke and then shows the wing half-way through the upstroke, during which it is partly folded (Based on Kevin Padian, 1983, Kevin Padian and Jeremy Rayner, 1993, and Peter Wellnhofer, 1991.)

often these living fliers partially fold the wing, because this reduces the size of any forces generated in the upstroke that might counter the forward thrust produced during the downstroke. If necessary, pterosaurs could also have done this by flexing the main limb joints.

Rock and Roll and How to Avoid It Any flight, no matter how short, involves more than just flapping and soaring. Like birds and bats, pterosaurs needed to be able to exercise precise control over their orientation and position, not only to be able to change direction, climb or dive, but also so that they could carry out difficult and complex maneuvers, such as taking off, landing, skimming for fish or hunting insects. Apart from being able to steer, a flying pterosaur also needed to be able to maintain stability, for example, if it flew into turbulent air or as a response to the often destabilizing effects of maneuvering. It seemed to earlier researchers that pterosaurs, especially pterodactyloids, were poorly equipped to carry out these tasks, but our much improved understanding of their anatomy, especially of their wings, shows that pterosaurs had a wide variety of options for steering and maintaining their balance (*Figure 8.11*).

The vertical flap on the end of the rhamphorhynchoid tail, like the tail fin on an airplane, would have helped to counteract any tendency to yaw (spin) around to the left or to the right. Alternatively, by swinging the tail to one

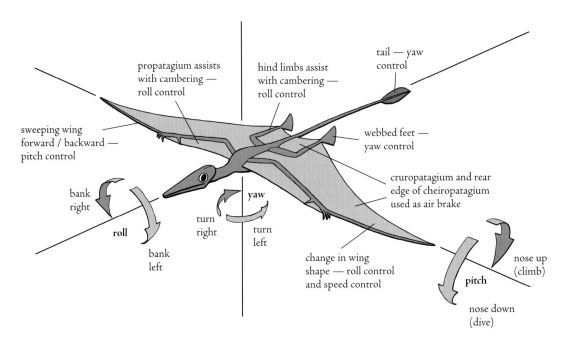

FIGURE 8.11 Diagrammatic restoration of *Sordes pilosus* illustrating how the wings, feet and tail may have been used for flight control. (Based partly on Helmut Tischlinger and Dino Frey, 2002.)

side, rhamphorhynchoids could initiate a turn, counteracting this effect by swinging the tail the other way. Exactly the same effects could also be produced by the webbed feet, which, vertically aligned and hinged at the ankle, could be swung from one side to the other, and thus effectively acted as twin tail fins. Logically, we might expect the foot flaps to have been particularly important in pterodactyloids, because they lacked other vertical stabilizing surfaces (head crests were destabilizing), and this is borne out by attempts to fly models of pterosaurs. Steven Winkworth, a London-based inventor, found that his radio-controlled model of *Pteranodon* was very difficult to control until he added twin tail fins, which took the form of large vertically aligned feet.[43]

Yet another flying model, this one of *Quetzalcoatlus*, designed and built by Paul MacCready, revealed the potential danger of roll to pterosaurs.[44] Although fairly stable once airborne, it was found that if the model began to roll over into a sideslip it was very difficult to prevent this continuing to the point where it stalled and fell out of the sky. Live pterosaurs almost certainly didn't do this because, unlike the model, they could counteract this tendency quite simply by increasing the lift generated by the wing on the downside of the roll. This extra lift was produced by deepening the camber of the wing membranes, which pterosaurs could do in several ways—most effectively by lowering the pteroid, but also by reducing tension in the cheiropatagium, either by relaxing the muscle fibers in the membrane or simply by moving the arm and leg slightly closer together.

By swiveling the arm at the shoulder and the leg at the hip, pterosaurs could swing the wings forward to pitch the animal upward, in order to climb, or backward, pitching the animal downward to enter a dive. Alternatively, they could counteract unwanted movements in these directions, for example a downward pitch caused by the apprehension of a heavy food item, by swinging the wings in the opposite direction—forward, in this particular case. Fine control of pitch might have been achieved by flexing and extending the wing-finger.

Another possibility is that pterosaurs controlled pitch by using the pteroid or the fifth toe to change the shape or position of the propatagium and cruropatagium, respectively, although the relatively small size of these membranes means they were probably mainly used for fine tuning. The primary function of the cruropatagium was probably for braking—lowering the fifth toes; even more effectively, dropping the legs would have slowed flight speed dramatically, but in a stable and controlled fashion.

Up and Down When I was learning to fly a hang glider, I was instructed that the least dangerous thing to do was to fly high and fast, whereas the most dangerous was to fly slowly and near the ground. This poses a small, but important, problem: In order to get into the air, or vice versa, one needs to be on or near the ground and moving relatively slowly. Consequently, taking off and landing are not only among the most difficult of flight maneuvers, they can also be the most dangerous (the best time to recall this is when the 400 tons of jumbo jet that you are traveling in is 10 feet from the runway). The same, albeit on a smaller scale, applies to birds and bats and also to pterosaurs, albeit on a smaller scale. Irrespective of how safe or dangerous it was, we can be certain that pterosaurs did take off and land, in some cases perhaps tens or even hundreds of times a day. So, how, exactly, did they do it?

Landing is, in some respects, easier to explain. As can be seen in *Figure 8.12*, when pterosaurs came into land they probably depressed the pteroid and propatagium as steeply as possible, so that the deeply cambered wings continued to generate lift, even when the animal was flying relatively slowly.[45] Then, when they were just above ground level, they lowered their legs. This brought the cruropatagium and the rear part of the cheiropatagium straight down into the air flow so that they acted as air brakes, rapidly slowing the pterosaur and probably ending in a complete stall that dropped them feet first onto the ground. From this position, they could quickly fold the fore-limbs and topple forward onto their hands.

It seems highly likely that rhamphorhynchoids and smaller pterodactyloids also had the ability to land on steep or vertical surfaces. The landing sequence was probably initiated by a shallow dive toward the target, followed by an increasingly steep climb up to the point where, as the wing stalled, the pterosaur intercepted the tree trunk or cliff face upon which it planned to land, gripping it simultaneously with both pairs of hands and feet. Such an

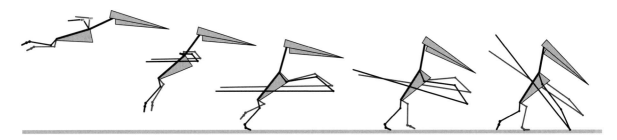

FIGURE 8.12 Sketch reconstruction showing how pterodactyloids may have landed. From the normal flight position the legs were dropped, slowing the pterosaur down, eventually coming into contact with the ground. After two or three steps the pterosaur came to a halt and then fell forward onto its arms, folding the wings as it did so. (Based partly on Sankar Chatterjee and Jack Templin, 2004.)

athletic exercise was no doubt possible for small or even medium-size pterosaurs, but seems less feasible for large or giant species.

Getting airborne may have been relatively easy for small pterosaurs, especially if they habitually used a high starting place, such as a tree or a cliff. In these circumstances, the main requirement was to push themselves sufficiently far from the starting surface to avoid striking it with the wings, once they were in flight. This could be achieved by a powerful leap, using both fore and hind limbs.[46] Presumably, pterosaurs spread their wings as soon as they were clear of the tree or cliff and used the initial fall to gain sufficient airspeed and lift so that, as a rule, they were able to pull out of this starting dive before they splattered themselves all over the ground below.

Taking off from the ground must have been more difficult because, as we will see in the next chapter, earthbound pterosaurs moved on all fours. As *Figure 8.12* illustrates, they probably began by rearing up on their legs and simultaneously extending their arms out into the flight position. Then they began to run forward, using the lift generated by the wings to support the front half of the body. As soon as the wings were producing enough lift to support the entire body weight, the hind limbs kicked off and the animal became airborne. Most, if not all, pterodactyloids were probably capable of such a takeoff, especially if helped by a facing wind, or the possibility of running down an incline. Whether rhamphorhynchoids could manage this is far less certain. With their hind limbs connected by the cruropatagium, they could not move quickly or easily on the ground and probably preferred to land and take off from trees or cliffs.

What a Performance At the start of this chapter I made two claims: that pterosaurs were true flapping fliers and that they were at least as competent in the air as birds and bats are today. Only a real live pterosaur could settle the issue for certain, but the evidence we have seen so far is highly suggestive. Pterosaurs had extremely light, but strong, skeletons, highly complex wing membranes, a well-developed flapping ability with enough muscle and more to sustain powered flight, and plenty of options when it came to maneuvering, or just ensuring that they stayed right side up. These and other features all point to superb flight ability. But, when it came down to it, how good were they really? Estimates of pterosaurs' aerodynamic properties and inferences that can be made from their ecology and locations where they were fossilized give us some measure of their likely performance.

Numerous attempts, some dating back to the early 20th century, have been made to calculate pterosaurs' likely aerodynamic abilities.[47] The biggest challenge for pteronautical engineers is to come up with accurate estimates for the two values fundamental to this type of work: body mass and wing area. It's not easy to find those, especially the mass estimate, though, because both depend on a fairly complete, but as yet far from attainable, understanding of pterosaur anatomy.

There are several ways of estimating mass. The most popular involves the parceling up of the body into geometric shapes, the combined volume of which can be converted into a mass estimate simply by multiplying it by a typical density value for vertebrates.[48] Except, of course, that pterosaurs were not typical vertebrates: They had hollow bones and air sacs (in some, if not all, species), both of which would have reduced the density value. Which goes some way toward explaining the origin of some of the wildly differing mass estimates that have been proposed, for example, for *Quetzalcoatlus*—anywhere between 50 and 250 kilos (110 and 550 pounds), although the lower value is surely nearer the mark.[49] Estimating wing area is easier, but much depends on exactly which restoration of the flight membranes is chosen. Bird-like reconstructions of pterosaurs, for instance, with long, narrow wings, free of the hind limbs and with no cruropatagium,[50] have a wing area that is at least one-fifth to one-quarter less than the now generally accepted arrangement described earlier in this chapter.

Despite these difficulties, almost all analyses of the aerodynamic performance of pterosaurs have come to roughly the same conclusion: They were highly competent fliers. In terms of their basic flight parameters, such as wing loading (essentially, a pterosaur's body mass divided by its wing area), glide angle (the rate at which a pterosaur descended in still air if it didn't flap its wings) and stall speed (the air speed below which the wing ceased to produce enough lift to support the body mass and the pterosaur began to fall out of the sky), some pterosaurs fall inside the range of values for birds and bats,[51] as *Figure 8.13* illustrates. Moreover, most pterosaurs seem to have been highly efficient and effective in the air, with low wing-loading, low glide angles and a slow stall speed, comparable to some of today's most proficient fliers, such as the albatross.[52] Indeed, in some respects, highly evolved forms such as *Pteranodon* and its near-relative *Anhanguera* actually seem to have performed better than modern fliers.

FIGURE 8.13. The flight performance of pterosaurs compared to birds and bats. The horizontal axis represents size while the vertical axis represents wing loading, which is calculated by dividing the weight of the animal by its wing area. Birds and bats all fall within the dotted line bounding the main cloud. Swans and quails, for example, have high wing loadings, whereas in kites and frigate birds it is relatively low. Pterosaurs, which are typified by remarkably low wing loading values, in some cases lower than in any living bird or bat, are indicated by the following letters: a, *Eudimorphodon ranzii*; b, *Dorygnathus banthensis*; c, *Campylognathoides zitteli*; d, *Scaphognathus crassirostris*; e, juvenile *Rhamphorhynchus muensteri*; f, g, near adult *Rhamphorhynchus muensteri*; h, large, old adult *Rhamphorhynchus muensteri*; i, *Pterodactylus antiquus*; j, *Pterodactylus micronyx*; k, juvenile *Ctenochasma elegans*; l, *Pterodactylus kochi*; m, *Pteranodon*; n, *Dsungaripterus weii*; o, *Nyctosaurus gracilis*. (Redrawn from Grant Hazlehurst and Jeremy Rayner, 1992.)

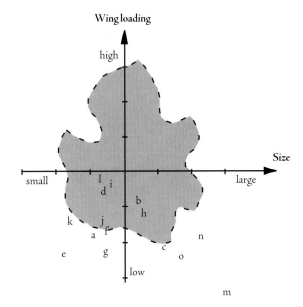

These results are not surprising, though, especially if we consider other lines of evidence, such as pterosaur feeding ecology. If you have ever watched seagulls, terns or other birds snatching up objects from the surface of rivers, lakes or the sea, you will appreciate how much skill and effort it takes. The flier must be highly maneuverable and able to make rapid changes in flight speed, and it helps if it can take off from the surface of the water (although, apparently frigate birds can't, which seems odd). And don't forget that the many different pterosaurs that lived this way were not going after a dead target, but live prey that didn't wish to end up sliding down anyone's gullet, so pterosaurs had to be sharp-sighted and quick-witted, too.

The same must have been true for anurognathids, the lineage of pterosaurs that seems to have gone in for hunting insects. In all probability, dragonflies, beetles, flies and moths, like their aquatic neighbors, were unenthusiastic about taking a trip through a pterosaur and doubtless exerted themselves quite considerably to avoid this fate. If they were not to go hungry, anurognathids must have exerted themselves a little bit harder, which, if swallows, swifts, nightjars and fly catchers are anything to go by, required an outstanding aerial performance.

Yet another giveaway of pterosaurs' excellent powers of flight is hinted at by the nature of some of the sites where they were fossilized. Several azhdarchids and many ornithocheiroids have been found in sediments that were deposited at locations hundreds of miles from what was then land. One or two might have drifted out there, but not hundreds, or, as in the case of *Pteranodon*, more than a thousand individuals.[53] The more likely alternative is that *Pteranodon*, like other pterosaurs, was a strong and highly competent flier that regularly flew hundreds of miles out to sea.

Flappers All? No matter which way we look at the issue, there can be little doubt that what pterosaurs did best was to fly, and some species were really good at it. Were all pterosaurs competent, active fliers though, or was Abel right when he suspected early forms of being relatively incompetent? In fact, every single species of pterosaur that we know of has features indicative of good flight ability. Even the earliest forms, such as *Dimorphodon* and *Eudimorphodon*, have all the main flight adaptations: a well-developed shoulder girdle, a breast bone, a humerus with a large deltopectoral crest indicating powerful pectoral muscles, a well-developed wing-finger, flight membranes with wing fibers, and so on, and the same goes for all other pterosaurs, even the giant forms.[54]

Indeed, if we compare pterosaurs with birds, a rather strange fact emerges. While many birds are good fliers, some are rather feeble, and numerous species have given up flying altogether. So far, however, not one, single, flightless pterosaur has turned up, even though flightless forms are common in the avian fossil record.[55] On the other hand, there is a clear parallel with bats, because there are no flightless bats, either. We will return to this intriguing situation in the next chapter.

Because all known pterosaurs seem to have had the same basic equipment for flight, does this mean they all had the same flight style? Almost certainly not. With their long tails tipped with a flap, most rhamphorhynchoids probably had a noticeably different type of movement in the air from, for example, pterodactyloids with their short tails, although exactly how they might have differed has yet to be established. What is more certain is that, with their relatively short wings, *Anurognathus* and its relatives flew rather differently, for example, from long-winged forms, such as *Rhamphorhynchus*. This makes sense, because anurognathids are likely to have hunted their insect prey in or around vegetated areas where relatively short wings were an advantage, whereas *Rhamphorhynchus* skimmed for fish over large bodies of water where its long wings were unlikely to encounter any obstacles.

Different-sized pterosaurs are also likely to have varied quite dramatically in their modes of flight. Species of small to medium-size pterosaur, perhaps up to 2 or 3 meters (6 to nearly 10 feet) in wingspan, are likely to have depended predominantly on flapping flight, although they undoubtedly were capable of gliding and soaring, if necessary. By contrast, as their size increased, pterosaurs are likely to have found the energy cost of flapping becoming increasingly expensive.[56] Consequently, larger fliers like *Pteranodon* and *Coloborhynchus* and giants like *Quetzalcoatlus* are likely to have relied heavily on soaring flight, only flapping their wings to land or take off and then circling around for hours and hours, riding from one thermal to the next, sailing the skies for days, or perhaps even weeks, at a time.

Not even the very best fliers can stay up indefinitely, though, and, sooner or later, every pterosaur had to come back down to Earth. So, in the next chapter, we take a stroll through one of the fiercest debates of all: How did pterosaurs move around on the ground?

9
GROUNDED

Dawn spread across the eastern sky, lighting a still, calm sea. Glistening mud flats, still shiny wet from the ebbing tide, stretched to the horizon and beyond. Oblivious on their scissoring legs, millions of tiny crustaceans scuttled, searching for damp safety in puddles and channels. Two small theropods, their downy plumage a lustrous reddish gold in the first rays of the sun, stalked snickety snack along the strand line, clawed fingers among the flotsam, alert for anything, dead or alive, that might have washed up overnight. A speck in the sky drew near, membrane wings fluttered gently in the early morning breeze, and a pterosaur, almost pure white, apart from its dark beak and yellow eyes, floated across the beach. Delicately, it touched down on feet first and gracefully bowed forward onto its hands, folding wing-fingers and stowing the wings in tight folds against the body as it did. Mud oozing from either side of its webbed feet, the pterosaur flip-flopped to the edge of a shallow pool, waded into the water and began sieving for shrimp, using its long, tooth-festooned beak. Slowly, drifting down like giant snowflakes, other pterosaurs arrived at the beach. Some gravely inspected muddy pools, while others skittered to and fro, probing for worms, or used their beak tips to pick off small insects mired on the mud. Soon, as on thousands of other days, the beach was crisscrossed with footprints and trackways, wending here, stopping there, starting from nowhere as another pterosaur touched down. When the tide came back, the flats would be erased, unless, by some chance, a storm, a flood, perhaps even an Earth tremor, sent a blanket of sediment to record the day's events on its undersurface.[1]

FIGURE 9.1 Tramping across a beach 150 million years ago, near what is now Villaviciosa in northern Spain, a pterosaur left this beautifully preserved footprint (heel at the top of the picture) complete with pads (preserved as bulges) that cushioned the toe joints and a scaly texture on the sole of the foot, which matches almost perfectly with the scaly skin on the heel of a pterosaur illustrated in Figure 6.8. (Image courtesy of José Carlos Martínez García-Ramos and Laura Piñuela.)

Walk, Waddle, or Worse?   The product of no little mental effort, "Walk, Waddle, or Worse" was my original title for a short article published in *Nature* in May 1987[2] that plunged me headlong into the fiercest of all debates about pterosaurs—how they moved around on the ground. This was no new dispute—quite the opposite. Scientists have argued about how pterosaurs stood and walked for most of the last two centuries. Choose almost any period in this interval and just a few minutes' sleuthing of the literature usually uncovers some startlingly different ideas, from pterosaurs gaily skipping along on their hind limbs to hapless-looking creatures sprawled out on all fours. Works like Harry Seeley's *Dragons of the Air*, in which *Dimorphodon* was reconstructed in these and other poses,[3] show that some authors were even ready to embrace several possibilities at once.

By the 1970s, however, the debate seemed to be on its last legs. The main scientific papers from this period paint pretty much the same picture: pterosaurs walking on all fours, limbs sprawled out to the side like those of living reptiles such as lizards and crocodiles, with the hands and feet flat on the ground (*Figure 9.2*). Moreover, with their limbs encumbered by flight membranes, it was presumed that progression must have been slow and ungainly, and probably not much better than a waddle. Such poor locomotory ability on land would have made pterosaurs an easy meal for predators, especially small theropods, and so the idea became firmly established that pterosaurs spent little time on the ground and preferred to rest on trees and cliffs when they were not airborne.[4] Even such large pterosaurs as *Pteranodon* were thought to have behaved this way and, reflecting a longstanding comparison with bats that goes back to restorations published in the early 1900s by Othenio Abel and beyond, were depicted hanging upside down from cliffs.[5]

Then, in 1983, like Lazarus coming back from the dead, the debate suddenly sprang to its feet again. In that year, Kevin Padian, who had just finished a Ph.D. on the pterosaur *Dimorphodon* at Yale University under the tutelage of a legendary figure, John Ostrom, published two seminal papers in which he revived the idea of bird-like pterosaurs.[6] Critically, Padian argued that pterosaurs did not use their forelimbs for walking, but were bipedal, and stood on their hind limbs alone, which, as in birds, did not stick out to the sides, but were tucked in beneath the body, as shown in *Figure 9.2*. Furthermore, Padian suggested that, rather than being flat-footed, pterosaurs stood on the tips of their toes, as most birds do today, and walked with a rapid, efficient and upright gait, completely unhindered by the wing membranes, which, according to Padian's new reconstruction, were entirely confined to the arms and body.

By the early 1990s, opinion was sharply divided between the four-legged, tree-hanging sprawler and the two-legged sprinter. Other alternatives had also come into view, not least the possibility of a four-legged pose with the body supported on more or less upright limbs.[7] But which of these, if any, was correct? And did all pterosaurs walk the same way? It was vital to find answers to these questions, not just because they could tell us about the ecology of pterosaurs, their origins and their history, but also because discovering how pterosaurs moved on the ground was one of the keys to understanding the basic nature of these animals.

By the end of the 1990s, the problem was solved, and this chapter shows how. The evidence involves fossil tracks, a virtual pterosaur called Robodactylus, and a raincoat, and the answers turn out to be rather surprising. Neither of the two main theories was correct, nor, so it would seem, did all pterosaurs walk the same way. The story begins, as always, with pterosaurs themselves, and what their anatomy—bones and soft tissues—tells us about how they walked.

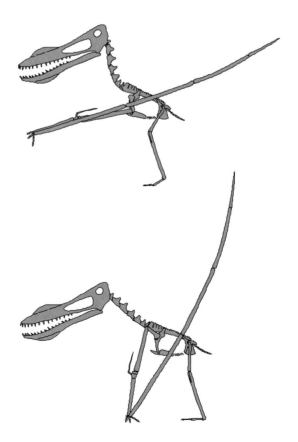

FIGURE 9.2 How pterosaurs may have moved on the ground. *Anhanguera*, reconstructed as a sprawling quadruped (above), and a bird-like biped (below).

Four Legs Good, Two Legs Bad?   The picture-postcard two-dimensionality of most pterosaur fossils is the main reason it has taken ptero-saurologists more than 200 years to work out how these animals walked. Fossils of *Pterodactylus* and *Rhamphorhynchus* from the Solnhofen Limestone may look extremely beautiful with complete, naturally articulated skeletons and perhaps even some fossilized soft parts, but if the joints are crushed flat, as they almost always are, it is practically impossible to establish how they operated. Without detailed information on the possible range and direction of movements of the limb joints—shoulder, hip, elbow, knee and all the others—we cannot determine how far a pterosaur might have been able to tuck its legs under its body, or whether it could or couldn't walk using its forelimbs.[8]

Happily, this frustrating situation came to an end in the late 1980s, when several skeletons of pterosaurs with beautifully preserved, uncrushed shoulder and hip girdles, arms and legs were found in concretions from the Santana Formation of Brazil.[9] Investigation of these and other spectacular fossil finds made at about the same time, including the superbly preserved hind limb and foot of a rhamphorhynchoid from Jurassic rocks in Mexico,[10] showed how pterosaurs are likely to have conducted themselves on terra firma.

One of the first big issues to be resolved concerned the forelimb. Did it have a role to play in walking? The key to this question was the shoulder joint, and the new three-dimensional fossils showed how it worked. The saddle-like joint surface on the shoulder girdle faced sideways and also to the rear. When the humerus was fitted into this joint, it was found that, in addition to all the "up and down" movements involved in flight, this bone could also be swung backward and inward so that it lay nearly parallel to the body (*Figure 9.3*). In this position, the forearm, wrist and base of the hand slanted forward and downward, more or less in a straight line from the elbow, so that the claw fingers of the hand contacted the ground slightly in front of and to the side of the body. The wing-finger was folded backward along the outside edge of the arm, ensuring that the wing membranes were kept well clear of the ground.

What the new fossils also revealed was the rather peculiar way pterosaurs deployed the clawed fingers of the hand. As we saw in Chapter 6, during flight, the equivalent of the underside of the claw-fingers (i.e., the direction in which they flexed) faced forward. This means that when the forelimb was swung down and around for use on the ground, the fingers no longer pointed forward, but extended out to the side. In this position, it would have been difficult for pterosaurs to step up onto the tips of their fingers, so, in all likelihood, they probably slapped them flat down on the ground as their weight came onto that particular arm and then simply lifted the hand back up again when the time came to move on for the next step.

By flexing and extending their arms at the elbows and, if need be, amplifying this movement by raising and lowering the humerus, pterodactyloids, at least, could swing their arms backward and forward in a highly effective fashion that would have allowed them to walk or even run on all fours. The most efficient position for the arms was tucked up close to the body where they were best able to support the pterosaur's body weight but, if they wished, pterosaurs may also have been able to walk with their arms partially or even fully extended out to the side. Pterosaurs probably had enough muscle to pull this off, but it would have required a lot more effort and might have brought their beaks uncomfortably close to the ground.

So, the Santana fossils appear to tell us that, if needed, pterosaurs could have used their arms for walking. But could they climb with them as well?

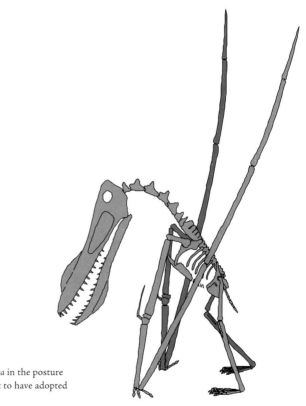

FIGURE 9.3 Restoration of *Anhanguera* in the posture that pterodactyoids are now thought to have adopted when walking.

Almost certainly. With the upper arm folded back against the shoulder (in the normal position for walking) and the elbow joint fully flexed, pterosaurs could bring their hands up fairly close to their heads, in which position they could easily have been used for climbing. That many of them could climb is given away by two features of their clawed fingers. First, the shape of the claws themselves—strongly curved, sharply pointed and powerful, but with a remarkably narrow, blade-like form—is perfectly designed for climbing.[11] Second, the bone on which each claw articulated was relatively long, an adaptation that helped to increase the gripping ability of the claw and is a typical feature of climbers in the animal world.

The key issue with regard to the hind limbs concerns the degree to which they stuck out sideways from the body, or could be swung in below the hips, so that they supported the pterosaur from beneath. In fact, the shape of the hip socket did little to restrict movement at this joint, but its alignment, facing outward, upward and a little backward, meant that the thigh bone could not be swung inward beneath the body without bumping into the hips or the muscles that were attached to them. Instead, it slanted forward, a little downward and somewhat outward so that the knee, lower leg and foot were placed slightly wide of the body. Seen from the front, or back, as in *Figure 9.3*, pterosaurs must have appeared rather like a cowboy who has spent a bit too long in the saddle. Still, by moving the thigh up and down and flexing at the knee and ankle, the legs would have worked well enough for walking, although running at high speed might have been more difficult.

What's afoot with the foot, though? Could pterosaurs stand on their tiptoes or were they flat-footed? The fossil from Mexico, shown in *Figure 9.4*, has answered this question. Ignoring the fifth toe, which was busy with the cruropatagium and probably had no major role in walking or climbing, the shape of the joints between individual bones in the other four toes shows that, while they could be strongly flexed, they could not be lifted above the main horizontal axis that ran lengthways through the foot. (It is the same with the fingers of our hands. We can flex them tightly into a fist, but, normally, they will not go far in the other direction.) In this respect, pterosaur feet were completely different from, for example, those of birds where, typically, the toes can lift well above the axis of the foot, which allows birds to raise their heels up and stand on their toes alone. Pterosaurs could not do this, forcing us to conclude that they must have tramped around in a distinctive flat-footed fashion.

EUDIMORPHODON ranzii

FIGURE 9.4 Right: The beautifully preserved right foot of *"Dimorphodon" weintraubi* seen from beneath, with toes one to four fully extended. From this position the toes can be flexed (coming out of the picture toward the viewer), but cannot be further extended. This means that this pterosaur must have walked in a flat-footed fashion and could not stand up on its toes. The longest complete toe (#3) is 2 inches (5 centimeters) in length. Left: *Eudimorphodon* in the act of climbing. (Photograph courtesy of Mick Ellison and Jim Clark; sketch courtesy of Todd Marshall.)

We are not finished with the foot yet, though. As in the clawed fingers of the hand, the bone immediately preceding each claw was relatively long in most pterosaurs and, in some cases, such as *Sordes*, formed more than three-quarters of the entire length of the toe. As for the fingers, this demonstrates that the toe claws had a really powerful grip, which tallies with the well-developed condition of the claws in many pterosaurs. In anurognathids, for example, they are large, extremely narrow, deeply curved and sharply pointed, and their similarity to the claws of the hand renders it highly likely that they were also used primarily for climbing.

Indeed, the presence of large, sharp-pointed claws on both the hands and the feet of rhamphorhynchoids and many of the smaller pterodactyloids is far more consistent with a climbing ability than with other functions, such as grabbing prey. To deploy the foot for climbing was quite easy. Pterosaurs raised the thigh bone up so that it was level with the body and then flexed the leg tightly at the knee and fully extended the foot at the ankle. This position, which raised the foot up near the body and into roughly the same plane as the hands when they, too, were being used for climbing, enabled pterosaurs to place the sole of the foot flat against a tree trunk or cliff face, so that the claws could be brought into play.

The Raincoat Test The ability of pterosaurs to move around on the ground was not just determined by the shape of their bones and joints; soft tissues must also have played a major part. Details of muscles and nerves are still sketchy and the role that they played is still far from clear, but one set of structures is quite well understood and undoubtedly had a major impact on earth-bound pterosaurs: the wing membranes.

As detailed in the previous chapter, the cheiropatagium, the main flight membrane, attached to the body between the shoulder and hip and ran out along the arm to the tip of the wing finger and down the leg as far as the ankle. Consequently, the cheiropatagium linked together the arm and leg and fastened this whole structure to the body. In rhamphorhynchoids, the legs were also linked to each other by the cruropatagium, which extended from the base of the tail down the side of the thigh and shin as far as the ankle and then out along the fifth toe. Effectively, in these early pterosaurs, all four limbs were linked to one another.

There can be no doubt that this shackling of the limbs must have hindered pterosaurs as they sought to move around on the ground, but it is not so easy to appreciate exactly how this might have worked, which is where the raincoat test comes in. If you have an old raincoat in the closet, you can start right now; if not, then a visit to your local charity shop should remedy the problem. To get the full effect, the raincoat should reach to your ankles and have a continuous hemline without any slits up the back. Now, don the raincoat and use bicycle clips[12] to fasten it to each leg. You should still be able to move around in this "rhamphorhynchoid" design, but, unless the material of the raincoat is highly elasticized, you will find that even walking is rather laborious, and running is almost out of the question. Notice, also, that it's rather hard to negotiate even small obstacles without snagging them on your cruropatagium. If you wish to find out how good rhamphorhynchoids were at wading, then you could try standing in a stream or small river, but only if you are prepared to get very wet, all over.

When you feel that you have fully explored what it is to be a rhamphorhynchoid and are ready to move on to pterodactyloids, you will need a sharp knife. Now, starting from the hem, slit the raincoat up the middle of the back, working upward as far as the top of your legs—unless you have a very steady hand, it might be best to take the raincoat off first. Next, put the raincoat and bicycle clips back on, and, voila—you are a pterodactyloid. Notice that it's quite a bit easier to move around and, with a bit of co-ordination of your arms and legs (oh, did I mention that you should be on all fours?), it is even possible to run.

What the raincoat test reveals is that all pterosaurs were hindered to some degree in their movements on the ground, but to a far greater extent in rhamphorhynchoids than pterodactyloids. Indeed, rhamphorhynchoids must have been so hampered by the cruropatagium that their walking ability was probably not much better than that of bats. By carefully coordinating their limb movements, pterodactyloids, by contrast, could walk quite effectively and may even have been able to run. Perhaps more importantly, they were able to negotiate small obstacles much more successfully and could have waded into quite deep water without being adversely affected by their flight membranes. It also seems unlikely that the flight membranes would have prevented any pterosaurs from climbing. Pterodactyloids, with their split cruropatagium, may have been a little more agile, but, as bats demonstrate, a complete cruropatagium is not that much of a hindrance.

To summarize, pterosaur anatomy seems to tell us that pterosaurs could have been four-legged if they wished (although it does not exclude the possibility that they occasionally stood on their hind limbs alone), that they most likely placed their hands and feet flat on the ground rather than standing on the tips of their fingers and toes, and that pterodactyloids seem to have been more adept at walking and running than rhamphorhynchoids. A good start, but this still leaves several unanswered questions. Were pterosaurs habitually two- or four-legged, and how exactly did their limbs work as they progressed from one step to the next? The second question is particularly difficult to answer, because we need to consider how all the joints worked together to move the limbs, and we need to do this for all four limbs simultaneously. Hard to picture in one's head—but much easier with a computer.

Robodactylus and Roborhamphus: Pterosaurs in Cyberspace   In the summer of 1997, Don Henderson and I had our first baby. We christened it Robodactylus and, although it was a willful child and often misbehaved, we loved it all the same. Robodactylus was a virtual pterosaur, an inhabitant of cyberspace and the offspring of a joint project that Henderson and I had conceived in order to resolve, once and for all, the question of how pterosaurs stood and moved on the ground.

The project began while Henderson and I were based in the Department of Earth Sciences at Bristol University in southwest England.[13] Henderson was using some new software for his doctoral research on dinosaur locomotion, and, after several exploratory discussions, we decided to try to apply the same approach to pterosaurs. The well-preserved skeletons of *Anhanguera* and other Brazilian pterosaurs supplied us with the basic details that we needed on the length and shape of bones and the range and direction of movements at each of the major joints. Using this information, Henderson built a computer model that was run using 36 fiendishly complex equations.[14] These ensured that the limbs behaved in a coordinated fashion, but did not restrain Robodactylus, beyond preventing "impossible" movements.

The first thing we did with our baby, whose portrait you can see in *Figure 9.5*, was to road-test the two main models for a walking pterosaur: the four-legged sprawler and the two-legged skipper. Two legs didn't even get out of the starting gate—well, it did, but as this mainly involved falling on its beak, it did not really count as a start. Just to get this model on its feet meant forcing the limbs into impossible positions, and it was so unstable that it could not take a step without falling over.

The problem with trying to stand a pterosaur on just its hind legs, as several pterosaurologists had already suggested[15] and Robodactylus demonstrated, is that most of the animal's weight lies in its chest region, well in

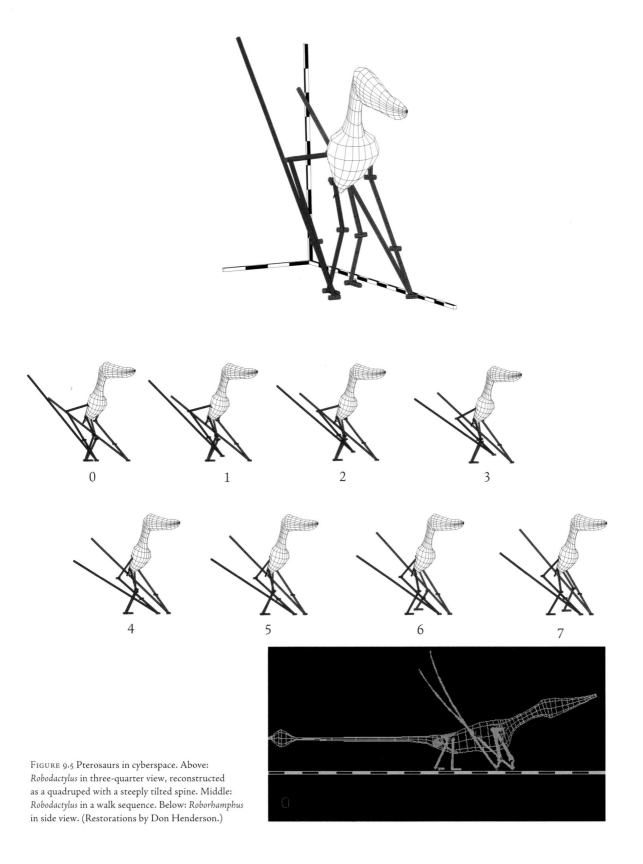

FIGURE 9.5 Pterosaurs in cyberspace. Above: *Robodactylus* in three-quarter view, reconstructed as a quadruped with a steeply tilted spine. Middle: *Robodactylus* in a walk sequence. Below: *Roborhamphus* in side view. (Restorations by Don Henderson.)

front of the feet, rather than above them. To achieve any degree of stability, the feet must lie beneath the center point, through which the weight effectively acts (technically referred to as the center of mass[16]) for at least some of the time when the animal is walking and all the time if it is standing still. For a pterosaur, however, trying to stand on its hind limbs alone, this does not happen for any of the time, so, as mentioned in the previous chapter, probably the only way it could achieve such a pose was by supporting at least some of the body weight using the wings, for example, during take-off and landing. Otherwise, the two-legged skipper was a non-starter.

Don and I turned to the four-legged sprawler, confident that we had now solved the problem of pterosaur locomotion, but it didn't work either! It really rocked, but only from side to side, and its rate of forward progress would have embarrassed a snail. Admittedly, it was stable, most of the time, but this was counterbalanced by the tendency of the model to try to tear itself into pieces if we sought to move it forward at anything like a normal pace.

It took several months of brainstorming before we found a solution: The body was in the wrong position. We had arranged Robodactylus so that its backbone was almost horizontal, but, as Chris Bennett had already suggested,[17] there was another possibility—tilting the backbone up at a fairly steep angle. Suddenly it all fell into place. The joints and limbs worked in harmony, the model was stable and it even moved forward at a reasonable rate: Robodactylus was go!

We had discovered how pterodactyloids stood and walked. The backbone was canted quite steeply, propped up by the long arms, which probably supported most of the body weight and powered pterosaurs along when they were on the ground. This makes sense, because in pterosaurs, the arms are more heavily built and muscled than the legs. The main job of the legs was to support the back end, and, because they are relatively much shorter than the arms, they also determined how long each step could be and thus how fast a pterosaur could walk or run. Seen from the front or back, the arms and legs sloped downward quite steeply, but did not tuck in beneath the body. Think of a cowboy walking on crutches.

Our cowboy Robodactylus also prompted a quite unexpected explanation for another peculiar feature of pterodactyloids—the sharp angle between the head and neck. You might think that tilting the backbone from horizontal to near-vertical would leave the head pointing up at the sky, which it would in rhamphorhynchoids, where the line of the head and neck are the same. In pterodactyloids, however, the knob of bone that articulated with the neck was repositioned under the skull, rather than at the back. So, even with the body in a near-vertical position, the head typically pointed forward, agreeing perfectly with the results of the recent study of the brain, as explained in Chapter 5 and illustrated in *Figure 5.7*, and it would have been easy for Robodactylus to peek down and see where it was putting its feet. Perhaps this explains the evolution of the pterodactyloid skull, but if so, what about rhamphorhynchoids? Would Robodactylus work for them?

In the summer of 2003, Don and I had our second child. It, too, lived in cyberspace, but was quite different from our first-born: It had a long tail and short wrists because it was a rhamphorhynchoid, but we loved it all the same, and we called it Roborhamphus. The general construction of Roborhamphus was based on the skeleton of the Upper Jurassic Solnhofen Limestone pterosaur *Rhamphorhynchus*, with details of the joints from a few relatively uncrushed examples, such as the Copenhagen specimen and material from the Kimmeridge Clay of England.[18] We used this information to build a computer model that Don constructed in the same way as for Robodactylus.

Initially, we thought Roborhamphus was going to be a problem child. If we tilted the backbone upward, then it would have had to walk on its hind limbs alone, which, apart from the balance problem, presented another difficulty—how to stop the tail from bashing into the ground. Clearly, a Robodactylus pose was out of the question. So, we tried the traditional four-legged posture with near-horizontal backbone and, to our surprise, it worked.

The performance was not impressive. Roborhamphus could only manage small steps, because the effective limb lengths to the hands and feet were relatively short, and this also left the body and tail uncomfortably close to the ground. Anything other than a modest walking pace seemed unlikely. On the other hand, this pose was consistent with at least one important rhamphorhynchoid feature—the orientation of the head. Remember that the body, neck and head lay in an almost straight line in these animals (as also, for example, in lizards and crocodiles), and this fitted neatly with the horizontal walking posture of Roborhamphus.

Computer modeling had shown the ways, or at least two possible ways, that pterosaurs could have moved on the ground. But were these the only ways that it had happened? This was something that Robodactylus and Roborhamphus could not tell us. What pterosaurologists needed was direct evidence of how pterosaurs had deported themselves, and that is exactly what they got. In fact, they had had that information since 1957, but it was nearly 40 years before it was widely accepted for what it really was—pterosaur tracks.

Pterosaurs—Guilty of Pteraichnus? Fossil tracks are wonderful. Each one is like a tiny movie with its own little story of movement, behavior, meetings and meals. If we can work out who made the track, we can open a new window onto the past, and through it, we can "see" how animals, extinct for millions of years, moved and behaved. In the case of pterosaurs, this is especially important, because, at a stroke, tracks could show us not only how pterosaurs walked, on two legs or four, on the tips of their toes or flat-footed, they could also reveal their exact gait and stance, how fast they could move, perhaps how they landed and took off. Tracks might even be able to take us much further and reveal the kinds of places pterosaurs visited when they were on the ground and what they did there. But to do any of these things, we first need some tracks.

In fact, we now have lots of tracks, thousands and thousands of them, but we have only known this for sure since 1995. The first pterosaur tracks were found in the early 1950s in the Carrizo mountains of Arizona by William Stokes, a pioneer of paleoichnology.[19] The trackway, seen in *Figure 9.6*, consists of nine sets of hand- and footprints made by a flat-footed animal walking on all fours, and was given the name *Pteraichnus* (literally "pterosaur track"), because, as Stokes recognized from its peculiarities such as a three-fingered hand and four-toed foot, it was the spoor of a pterosaur.

Paleontologists happily accepted Stoke's interpretation of *Pteraichnus*, not least because it fit with traditional notions of pterosaurs, but only up until 1980. Padian's proposal that pterosaurs had stood on their hind limbs alone was clearly incompatible with *Pteraichnus*, and one of them had to go. The problem was investigated by Padian and Paul Olsen, an American paleoichnologist based in Columbia University. After carrying out track-making experiments with a caiman, they concluded that *Pteraichnus* had been made by an Upper Jurassic crocodile.[20] One or two other traces found in the 1980s, including a supposed pterosaur take-off run reported from Clayton Lake State Park in New Mexico were also reinterpreted as crocodile tracks, so that by the early 1990s, it seemed that pterosaur tracks were quite unknown.[21]

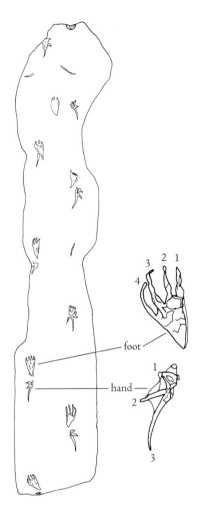

FIGURE 9.6 Pterosaur tracks from the Upper Jurassic of Utah were first found by William Stokes, who gave them a particularly apt name: *Pteraichnus*. Part of a trackway is shown on the far left, while details of the four-toed foot (about 3 inches [7 centimeters] long) and three-fingered hand are illustrated in the middle. A single footprint, with claw marks toward the top of the photograph, is shown on the right. (Track redrawn from William Stokes, 1957, and prints from Kevin Padian and Paul Olsen, 1984. Photograph courtesy of Jo Wright.)

Then suddenly, in 1995, two publications changed everything. In one paper, by Jean-Michel Mazin and his team from the Université de Poitiers, a stunning new track site in Upper Jurassic limestones at Crayssac in France was described. In a second, Martin Lockley and his "dinotrackers" group from Denver, Colorado, reported on new discoveries of copious *Pteraichnus* tracks from the Upper Jurassic of North America.[22] The critical breakthrough, made by both teams, was the discovery of details in the new *Pteraichnus* tracks that had been produced by anatomical features found only in pterosaurs, and not in any of their backboned contemporaries, including crocodiles.

Indeed, as illustrated in *Figure 9.7*, when inspected closely, the match between tracks and anatomy turns out to be extremely good. In a typical pterosaur hand such as, for example, that of *Pterodactylus kochi*, the three claw-bearing fingers increase in length from the first to the third and splay apart so that the third lies almost at right angles to the first. This is exactly what we see in the prints. Still more convincing is the presence, in a few handprints, of the impression of a very long fourth finger that seems to have been directed away from the other fingers toward the body. Only one structure, found in pterosaurs and no other animal, could have made such a feature—the wing-finger. Its rarity in tracks is not surprising though, because, in all likelihood, pterosaurs probably tried to avoid scuffing the wing on the ground when walking.

The pterosaur foot skeleton also shows a good match to the footprints, which have a characteristic elongated triangle shape, a narrow, often deeply impressed heel and four toes. As we saw earlier in Chapter 6, webbing between the toes seems to be widespread in pterosaurs, and the same feature is commonly found in their footprints. Occasionally, foot impressions are so well-preserved, as, for example, that in *Figure 9.1*, that it is possible to establish how many bones there were in each toe, and even how long they were. The correspondence between print and skeleton is perfect. Both footprints and pterosaurs feet have two, three, four and five bones, respectively, in toes one through four, and they also match in another important way. Remember that in each of the first four toes, the bone preceding the claw is always relatively long. The same is true of the footprints.

The evidence in favor of pterosaurs as generators of *Pteraichnus* is very good, but, just in case you are still not completely convinced, here is one final discovery that sets the issue beyond doubt. Pterosaurs have very long arms, much longer, in fact, than any other tetrapod, living or extinct. If they chose to walk with their arms fully extended, which, as discussed earlier, they probably could if they wished, then their hands would leave prints far away to the left and right of the main footprint track. Satisfyingly, such a track, discovered by Mazin and colleagues at Crayssac, has been found. It consists of left and right handprints that lie far apart from one another (*Figure 9.7*) and confirms that pterosaurs did, on occasion, walk on outstretched arms. More importantly, it proves that pterosaurs were guilty of *Pteraichnus*.

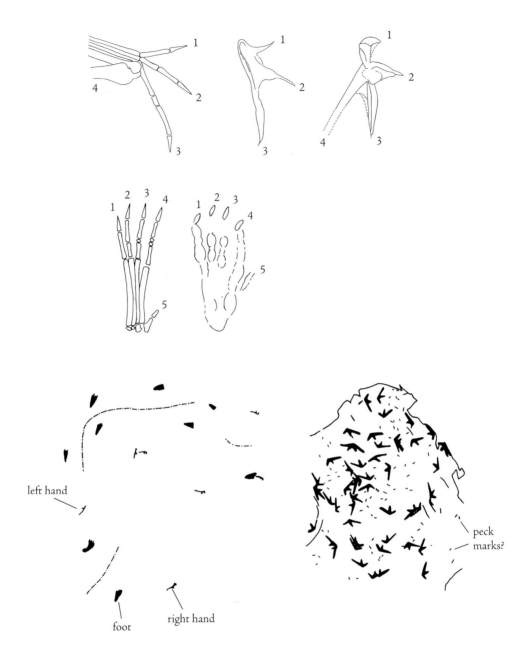

FIGURE 9.7 The hand (above left) and foot (middle left) of *Pterodactylus kochi* show an almost perfect match to prints of the hand (above middle and above right) found at the Pterosaur Beach near Crayssac, France, and the foot (middle right) from the Alcova Lake track site in Wyoming. Below left: Pterosaurs' special trick—widely separated handprints, made by walking with the wings outstretched, found at the Pterosaur Beach, Crayssac. Below right: These "hand-only" tracks, impressed into mudstones of the Upper Cretaceous Blackhawk Formation, were found in the roof of a coal mine near Helper, Utah. The tracks seem to have been made by a pterosaur shuffling around in a small area on mud flats bordering a lake, perhaps picking up food items and leaving what seem to be "peck" marks. (Redrawn from Mazin et al., 1996, and Unwin, 1997.)

Building a Track Record Once paleoichnologists knew what to look for, they started finding "pteraichnites" (pterosaur tracks) almost everywhere (*Figure 3.9*). In the last 10 years, more than 30 track sites, some yielding thousands of prints, have been found in North and South America, Europe and Asia, with particular concentrations in the United States and Spain. The track record also spans a time interval of more than 100 million years, the oldest tracks dating back to the Upper Jurassic, the newest from near the end of the Upper Cretaceous, and they are especially abundant in the mid- to late Upper Jurassic and in the early Lower Cretaceous.

The aptly named "Pterosaur Beach," exposed in a quarry near the village of Crayssac in southern France and illustrated in *Figure 9.8*, is the most important pterosaur track site in the world. Here, numerous millimeter-thin layers of rock, each perhaps the result of a single, sediment-laden tide sweeping across a large mud flat, are covered in thousands of prints. They were discovered by Mazin and his team, who spent eight summers painstakingly uncovering them, mostly working through the night, when they were better

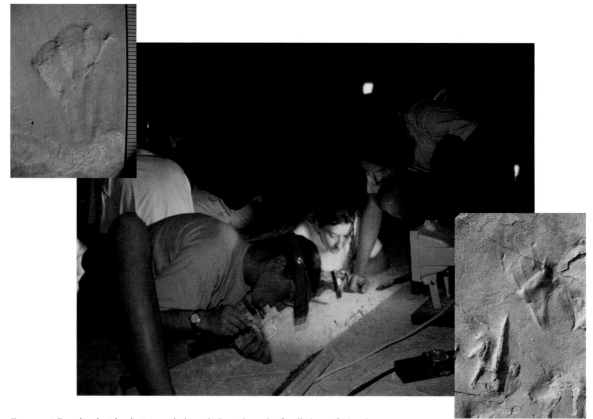

FIGURE 9.8 French paleoichnologists work through the night under floodlights to find and expose tracks on the 145-million-year-old "Pterosaur Beach" near Crayssac in southern France. Insets: superbly preserved prints found at this site include a footprint (left) that shows lobes of mud that were forced out by the webs between the toes as the foot pressed into the mud, and a classic hand/ footprint pair (right) demonstrating typical features of the hand, placed slightly outside and behind the foot and with three finger impressions directed outwards and backwards. (Photographs courtesy of Jean-Michel Mazin [main and inset left] and Helmut Tischlinger [inset right].)

able to see and excavate the prints, using low-angled flood lighting. More than 22 distinct trackways, the longest reaching 3.6 meters (11 feet) and with more than 40 steps, have been found and are affectionately referred to by the Mazin group with names such as "Emile" and "Lucien." And the Crayssac pterosaurs were not alone. Other tracks show that they shared the mud flat with a wide variety of animals, including crustaceans, turtles, lizards, crocodiles and small theropod dinosaurs.

Elsewhere, more than 10 track sites have been found in the Upper Jurassic of the western United States. These include the Sah Tah Wash locality, where Stokes discovered the original *Pteraichnus* tracks, and Alcova Lake, Wyoming, which produced the first evidence of rather peculiar "hand-only" trackways. Back in Europe, a wealth of pterosaur tracks, some associated with bird footprints, have been found in the Lower Cretaceous of northern Spain.[23]

Across the channel in England, pterosaur prints found on three slabs of rock in Dorset, seemingly made by a large pterosaur, possibly an ornithocheiroid, are so distinctive that they have their own name, *Purbeckopus*.[24] One of these slabs spent 40 years in a garden path, where it was regularly trodden upon, while another was used as a headstone for a dog's grave!

Yet another unusual pterosaur track, named *Haenamichnus*, is known from Uhangri on the southeast coast of South Korea. Among the hundreds of prints that have been excavated, some associated with bird tracks, there is an enormous foot impression, over a third of a meter (1 foot) long, shown in *Figure 9.9*, that can only have been produced by a giant pterosaur, most probably an azhdarchid.[25] There is a good match between the hands and feet of these pterosaurs and details of the Korean prints and, tellingly, azhdarchids are the only Upper Cretaceous pterosaurs that reached sizes large enough to produce such a monster print.

A striking feature of pterosaur tracks is that in almost all cases they have four-toed footprints and are most likely to have been made by four-toed pterosaurs—pterodactyloids. By contrast, five-toed rhamphorhynchoid tracks are exceptionally rare. A few possible examples have been reported from Alcova Lake and Crayssac, but, as yet, not one of them has been certainly confirmed as of rhamphorhynchoid origin.

This absence is even more striking if we compare the fossil record of pterosaurs to the geological distribution of their tracks, as shown in *Figure 10.2*. During the long time interval from the Upper Triassic to the end of

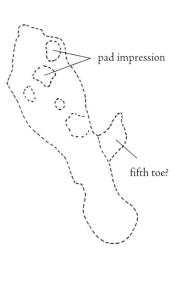

FIGURE 9.9 The mark of a giant. This footprint, more than 1 foot (0.35 meters) long, was made by a huge pterosaur walking along the edge of a lake in the Upper Cretaceous in what is now southwest Korea. A photograph of the footprint is shown on the left and a sketch of the foot is seen on the right. (Photograph courtesy of Koo-Geun Hwang.)

the Middle Jurassic, when only rhamphorhynchoids have been found, pterosaur tracks are completely unknown. Might this be due to a gap in the fossil record, a lack of preservation, rather than a true absence? Probably not, because tracks of other land-living animals, including several kinds of small, lightly built reptiles, are well-known from this period and particularly common in the Upper Triassic and Lower Jurassic.

So, rhamphorhynchoids do not appear to have left a single track for more than 60 million years, and even when both they and tracks are abundant in the Upper Jurassic, they seem to have made a pitifully small contribution to the print record. Pterodactyloids, on the other hand, seem to have gone out of their way to march up and down on beaches or lake shores, ensuring that they left paleoichnologists with plenty of work. I refer to this as the mystery of the missing rhamphorhynchoid tracks, but all will be explained before we reach the end of this chapter.

Walk Like a Pterosaur The main value of pterosaur tracks is that they tell us exactly how these animals moved on the ground (*Figure 9.10*). Even isolated prints are highly revealing. They show that the entire foot, from the heel to the tips of the toes, was plonked down in a flat-footed fashion, much as you or I normally walk. A major difference between humans and pterosaurs is that, with some encouragement, we can stand on our toes; tracks tell us that pterosaurs never did, in complete agreement with studies of the foot, which, as we learned earlier, show that the toe joints do not permit such capers. Handprints reveal that the forelimb was supported by the clawed fingers alone, perhaps partly supported by the large knuckle-joint at the base of the wing-finger, with the hand directed out and backward and the wing-finger folded up over the body.

The presence of both hand- and footprints in tracks show, undeniably, that pterosaurs were fully quadrupedal. Strangely, many track sites are dominated by handprints, and some tracks are made up solely of such traces (*Figure 9.7*). Were pterosaurs able to walk on their hands alone, waving their feet in the air? Probably not. In side view, it is obvious that much of the bulk of a pterosaur was concentrated in the front half of the body and most of its weight was supported by the arms. Consequently, the hands, which had a smaller surface area than the feet and must therefore have exerted higher loads on the ground, are much more likely to have sunk into the sand, or

FIGURE 9.10 *Pterodactylus* goes for a stroll along the beach, leaving a typical pterosaur track. (Redrawn from Martin Lockley et al., 1995.)

mud, leaving prints. At the back end, the lightly loaded feet might not have impressed at all and—presto!—hand-only trackways. The reverse situation, footprint-only trackways, have not been found, so it may well be that, apart from brief moments during landing and taking off, grounded pterosaurs preferred to stay on all fours.

Tracks also provide information about the likely position and movements of the limbs of a walking pterosaur. Almost without exception, trackways are relatively wide, the left and right footprints separated by a gap equivalent to at least twice the width of the foot, so the arms and legs must have been directed somewhat out to the sides in a rather straddle-legged stance. Linking this with the orientation of the footprints, which usually point forward or a little outward, it seems that pterosaurs must have walked by swinging their limbs backward and forward in a plane more or less parallel to their line of progression, with most of the movement taking place at the shoulder in the forelimb, and at the hip and knee in the hind limb. This nicely corresponds with Robodactylus, but is quite different from the way birds and mammals, including humans, walk. As we progress, we swing our feet and (in quadrupeds) hands forward *and inward*, beneath our bodies, producing a narrow trackway with prints that lie approximately in a straight line.

To really see the difference, try doing the pterosaur walk the next time you go to the beach. You need two sticks to represent the forelimbs, and you should begin by leaning forward and supporting yourself on the sticks. Make sure your "fore" and hind limbs are at least two foot-widths apart. Don't forget, no swinging your arms or legs inward, only backward and forward, and remember, when you look back at your trackway, your handprints should lie just behind the footprints. Sounds easy, but I bet you can't produce a perfect pterosaur track on your first try.

Hanging Out With the Rhamphorhynchoids   Now that we are familiar with the main lines of evidence—anatomy, computer models and tracks—we can begin weaving together these different threads into a single tapestry in which different pterosaurs are depicted in various poses on terra firma. The first group to which we should divert our gaze is the rhamphorhynchoids. We can be fairly certain that rhamphorhynchoids did at least visit the ground on occasion. Roborhamphus shows that they were capable of getting around, albeit slowly, and virtual tracks generated by this computer model look similar to the one or two supposed rhamphorhynchoid tracks that have been found so far, which lends encouragement to this idea. When it comes to stance and gait, there is a highly convincing degree of consistency among the main lines of evidence, all of which point to a low-slung, four-legged pose, with the limbs held fairly close in to the body and the hands and feet slapped down flat on the ground.

It cannot be denied, however, that with their relatively short arms and all four limbs shackled together by flight membranes, rhamphorhynchoids must have been somewhat hampered in their walking movements and probably struggled to reach running speed anywhere other than on completely clear terrain, such as one might encounter on a beach. Couple this restriction with the problem, for long-tailed species, of rearing up on the hind limbs, and it begins to seem as if rhamphorhynchoids might have had real difficulty in taking off from the ground.[26]

The problem for rhamphorhynchoids, or any other relatively small, slow-moving animals that had no quick means of escape, be it into water, holes in the ground, or the air, was the severe danger posed by theropods. Common throughout the period during which pterosaurs existed, theropods would undoubtedly have availed themselves of the opportunity of a light snack, should the chance have arisen, as the discovery of their teeth embedded in pterosaur bones so graphically demonstrates.[27] Safety lay in the heights, on tree trunks or cliff faces, where resting rhamphorhynchoids were relatively secure from claw-wielding horrors such as *Coelophysis* and *Allosaurus*.

Because rhamphorhynchoids probably spent most (or all) of their "down" time hanging around, we might expect to see adaptations for climbing. Indeed, there are several. The body was flattened, enabling them to hang close to the surface to which they were clinging, and the arms and legs folded up in such a fashion that the wing membranes could be neatly stowed away and the claw-bearing fingers and toes brought to bear (*Figure 9.4*). As detailed earlier, the claws are exactly like those of squirrels and birds that habitually climb on trees, and the bone on which they pivoted was relatively long, a construction that served to increase their gripping ability. Indeed, the fingers and toes seem well-adapted to a "cling and hang" behavior, analogous in some respects to the way mountaineers use ice axes and crampons, except that in this case there were 14 points of contact rather than six.[28] This also provides an important clue as to how rhamphorhynchoids climbed—predominantly, if not exclusively, head upward, rather than head downward as has occasionally been suggested.[29] This enabled them to deploy both hands and feet and to clamber up or down steep or vertical surfaces, perhaps using the tail as an additional support or prop.

This climbing lifestyle also neatly explains the mystery of the missing rhamphorhynchoid tracks. These early pterosaurs probably spent very little time on the ground: They don't seem to have fed there, they probably didn't rest there, and they probably avoided going there, if at all possible. Consequently, the few tracks that they may have left stood little chance of surviving the great geological lottery that is called fossilization.

Pterodactyloids Conquer the Ground Having established what it was that rhamphorhynchoids got up to when they were not winging their way through the skies, we can now get down to the pterodactyloids. Some species, mainly of small to medium size, show adaptations like those found in rhamphorhynchoids, which suggests that they also had a competent and capable climbing ability. Their bodies were also flattened, and they, too, could fold the arms and legs such that the wing membranes were stowed away and the claw-bearing fingers and toes were deployed for action. Above all, the claws had a similar shape to those of rhamphorhynchoids, indicating that they, too, served as climbing hooks.

Unlike rhamphorhynchoids, however, not every pterodactyloid was equipped in the same way. In some small- to medium-size forms, the foot claws are relatively small and only weakly curved; this seems to be generally true of large and giant forms, although these details are known for relatively few of the bigger pterodactyloids. In addition, some pterodactyloids, such as dsungaripterids, have relatively small finger claws, while the ornithocheiroid *Nyctosaurus* seems to have none at all. Presumably, many of these pterodactyloids only climbed when they really had to, and were not necessarily particularly good at it, while *Nyctosaurus*, at least, gave up climbing altogether. It's hard to imagine giants like *Quetzalcoatlus* clinging to a tree or cliff either, although we should not exclude the possibility that newly hatched and very young individuals availed themselves of the safety of the heights.

The big difference between rhamphorhynchoids and pterodactyloids is that the latter seem to have spent far more time on the ground than the former ever did. We know this primarily because of their tracks, which, as *Figure 10.2* demonstrates, have almost exactly the same distribution in time and space as do the fossil remains of pterodactyloids. Moreover, the different sizes and forms of the tracks show that they were made by several different groups of pterodactyloids and not just one "track-happy" clan.

Anatomical studies and Robodactylus also reveal some of the key secrets of pterodactyloids' successful conquest of the ground. The elongation of the wrist, jacking the body up into a steep position, and the reorientation of the head and neck were clearly important, but the critical development seems to have been the splitting of the cruropatagium, which freed the limbs, allowing them to move independently. This, in turn, probably allowed pterodactyloids to move faster and rather more adroitly than rhamphorhynchoids, and, because they were already steeply canted, it was relatively easy for them to lean back and balance briefly on the hind limbs. Equipped with these abilities, take-off from ground level may have been substantially easier for pterodactyloids, making it simpler for them to "colonize" the ground in the first place.

As for rhamphorhynchoids, all three lines of evidence paint a clear and consistent picture of how pterodactyloids stood and walked. Almost invariably, they moved on all fours, hands and feet flat on the ground, limbs close in near the body, but not quite beneath it, resulting in a highly characteristic "cowboy on crutches" style. The limbs swung forward and outward, backward and inward, but, as the Crayssac tracks show, pterodactyloids could vary the degree to which the limbs were extended outward.

Not only do tracks show how pterodactyloids stood and walked, they are also beginning to throw some light on what these pterosaurs actually did when they were on the ground. In most of the cases where prints form recognizable trackways, they seem to record pterosaurs plodding along rather slowly at estimated speeds of just three to four kilometers an hour (around two miles per hour), but one trackway from Crayssac, "Lucien," is an exception. Here, the spacing between prints is relatively large, and, irrespective of which Upper Jurassic pterosaur is matched to the track, it has to run in order to fit its feet into successive steps.[30]

Many track sites are dominated by what seem to be "random" prints, often concentrated in patches and giving the impression of one or more individuals shuffling around in a limited area (*Figure 9.7*). While this could, conceivably, indicate the beginnings of insanity, it probably reflects feeding behavior, the track-making pterosaurs moving slowly over a small region, searching for and picking up food items from the mud or sand, using the tips of their jaws and leaving "peck" marks, which have been discovered at several sites. The discovery of prints preserved on many different layers, for example at Crayssac, may also be related to this feeding behavior. It shows, at the very least, that pterosaurs made repeated visits to the same beach site and it may even mean that they lived there for part of the year, or perhaps even the whole year round. Why? Because that's where the food was.

At some localities, such as "Pterosaur Beach," it appears that pterosaurs were happy to get their feet wet, leaving prints as they waded in shallow water, perhaps in search of food. They might even have deliberately gotten out of their depth. Peculiar scratch marks at several sites in North America and at Villaviciosa in Spain may have been made by pterodactyloids floating in shallow water, their feet scuffing the bottom as they paddled along.[31] Some researchers[32] have suggested that pterosaurs might have used a duck-like posture, floating on the surface of the water, but this seems out of the question, because, aside from the difficulty of deploying legs that were part of the flight apparatus, the pterodactyloid neck was too stiff to fold into the "S" shape adopted by swimming birds. Presumably, therefore, they must have floated, or swam, in some other as yet unknown but uniquely pterosaurian fashion. Quite what they did is not clear, but it must have been an extraordinary sight to behold.

10
THE PTEROSAUR STORY

The last pterosaur in the world was old. Her joints creaked when she flew, her wings were ragged, and she was almost blind in one eye. Most days she didn't even take to the air, but hobbled around picking off small morsels—a crab, a crayfish, the rotting remains of a dinosaur half-buried on a sand bank. Occasionally, she tried to preen, using her long, slender beak to groom her pelt, patches of which had come away and much of the rest of which was infested with various parasites—some burrowed under her skin, while others clung tenaciously to their hairy home. It wasn't the parasites that had done her in, although they certainly didn't help. One heavy landing too many had left her kidneys barely functioning and a liver on the edge of a nervous breakdown. Soon, she would no longer be able to muster enough energy to take off, and then it would be over in a matter days. Perhaps vaguely aware of the time trickling away, she began flexing her joints. Then, turning to face into the freshening breeze, she swept out her wings, waited for the lift to build and kicked off. She was airborne for the last time. Powered by the occasional flap of her huge wings, she slid into a thermal and slowly began to rise, wings canted, curving round and round, carving giant circles in the sky. Up and up she went, until the landscape became a featureless mass of greens and browns and grays, veined with the shining threads of rivers slowly turning from silver to gold in the evening light. Finally, high in the thin, cold air, the thermal gave out, and she banked around to face the setting sun. The night was behind her now, skimming fast over the surface of the world, but it would be a while before it caught her.[1]

FIGURE 10.1 Stormy Weather. The "Hairy Devil" *Sordes pilosus*, a crow-sized pterosaur, out hunting on a blustery day in the Late Jurassic, in the region of what is now Middle Asia. (Painting by Todd Marshall.)

All the Mesozoic for a Stage Curious, eventful and littered with strange characters—one might expect the pterosaur story to have been told on half a hundred occasions. And so it has been, but often only as part of the backdrop to those hoggers of the limelight: the dinosaurs. When pterosaurs have occasionally made it to center stage the performance barely lasts one scene. *Dimorphodon* briefly lurches into view; those trusty troupers, *Rhamphorhynchus* and *Pterodactylus*, go through their fish-juggling routine; *Pteranodon* sweeps past, all beak and crest; and *Quetzalcoatlus* is winched on for the grand finale. Cue volcanoes and meteorites, and there is still time enough to retire to the bar for a quick drink while the King of the Tyrant Lizards eats the final curtain.

A caricature, certainly, but not so far from the truth. Accounts of pterosaur history are usually quite generalized and, even where they do entertain some details, tend to stick to a fairly simple successive sequence. These narratives are not really histories in the true sense: They do not trace lineages through time and, beyond recognizing that pterosaurs had an origin and eventually became extinct, have rarely attempted to define or describe major events, let alone explain them.[2] This situation is quite understandable, however, when one considers the circumstances that have prevailed until now. A notoriously patchy fossil record riddled with gaps, some dismayingly long; only a very generalized understanding of pterosaur relationships, founded on a handful of the better-known species; and completely divergent opinions regarding the basic nature of these animals.

Happily, this sad situation no longer applies. The sheer quantity of pterosaur fossils in paleontological collections, the lifeblood of pterosaurology, has dramatically increased in the last couple of decades. And it is not just that we have more material. Many important new finds, such as that shown in *Figure 10.1*, have been made in regions where these animals were almost unknown. South America and China, both of which have yielded a plethora of pterosaurs in recent years, are obvious examples, but other discoveries from Greenland to New Zealand and even Antarctica are no less important.[3] Not only have they filled in some large geographical gaps, they also show that pterosaurs were found all over the world.

Perhaps even more importantly, some of the recent finds, such as the new pterosaurs from the Middle Jurassic of Argentina[4] and the Lower Cretaceous of China,[5] lie slap-bang in the middle of a big gap in the geological sequence of the pterosaur fossil record (*Figure 10.2*). These discoveries are especially useful because they shine much needed illumination into some of the long Dark Ages of pterosaurian history.[6] Nor should we forget the multitude of tracks and traces left by pterosaurs happily wandering beside long-forgotten river banks and sea shores, now being discovered in the thousands by paleontologists alerted to their identity.

These exciting finds have added a whole new dimension to the pterosaur story, and yet, even when lumped with all the other new evidence, it has to be admitted that the fossil record of these Mesozoic dragons is, like my mother's Yorkshire pudding, rather more hole than filling. It will never be as rich or dense as that of ammonites, or even those lumbering hulks, the dinosaurs, but at least enough is now known for us to be able to discern the basic outline of the story.

More fossils are only part of the solution, though. The clutch of genealogical studies that have pieced together the relationships of pterosaurs to one another has been just as important to solving this problem. These investigations are modest compared with some of the thumping great studies that have recently been carried out on dinosaurs and birds, for instance. They have yet to include all known species of pterosaur, but, as we learned in Chapter 4, many of the major branches of the pterosaur tree and their approximate position to one another are now widely agreed upon. This has given pterosaurologists a solid frame on which to hang their pterosaur story and has helped in two further, less obvious, ways.

First of all, it has equipped pterosaurologists with the ability to pinpoint and make use of anatomical features that are unique to particular clans. Here is an example: While rooting through the pterosaur collection in the Natural History Museum in London, I once came across a small piece of bone that was recognizable as part of a peculiarly twisted crest, originally found on the near end of the upper arm bone of one particular group of pterosaurs. Previously classified only as "pterosaur," thanks to our genealogical data, I was able to identify this fragment, which had been collected in the early 19th century by men quarrying chalk in Kent, southern England, as evidence of an ornithocheiroid pterodactyloid that had lived (or at least died) in that region about 90 million years ago, in the early Upper Cretaceous. Not something to get terribly excited about, perhaps, except that there are thousands of odd pterosaur bones and bone fragments sitting quietly in museum drawers from Buenos Aires to Beijing. Often dismissed as "junk," our genealogical data can put this material to work by linking it to major pterosaur groups or particular clans, sometimes even to a particular genus or even

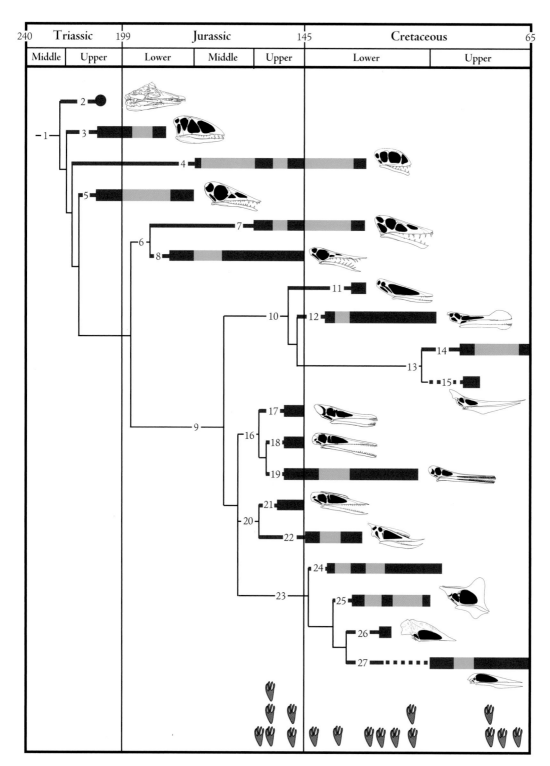

FIGURE 10.2 Pterosaur history. The known extent, in time, of each major clan is shown by the thick bars. Each bar consists of

fossils have yet to be found, although the clan must have existed during this period. The thick line leading up to each bar denotes a minimum length of time during which the clan (or its ancestors) must have already existed but is not yet known from fossils (also known as the ghost range). Each foot symbol indicates one of the principal pterosaur track sites. 1, *Pterosauria*; 2, *Preondactylus*; 3, *Dimorphodontidae*; 4, *Anurognathidae*; 5, *Campylognathoididae*; 6, *Rhamphorhynchidae*; 7, *Scaphognathinae*; 8, *Rhamphorhynchinae*; 9, *Pterodactyloidea*; 10, *Ornithocheroidea*; 11, *Istiodactylus*; 12, *Ornithocheiridae*; 13, *Pteranodontia*; 14, *Nyctosauridae*; 15, *Pteranodontidae*; 16, *Ctenochasmatoidea*; 17, *Cycnorhamphus*; 18, *Pterodactylus*; 19, *Ctenochasmatidae*; 20, *Dsungaripteroidea*; 21, basal dsungaripteroids; 22, *Dsungaripteridae*; 23, *Azhdarchoidea*; 24, *Lonchodectidae*; 25, *Tapejaridae*; 26, tupuxuarids; 27, *Azhdarchidae*.

species, and in so doing, can help to fill in many details on the backdrop to the pterosaur story.

The second way genealogies can be of additional use is more subtle. By combining genealogical data with fossil data (essentially the age and identity of particular fossils), we can begin to trace out the main lines of the pterosaur story, as shown in *Figure 10.2*. Now, if we select one particular lineage—the rhamphorhynchines are a good example—we can see that fossil finds such as *Dorygnathus* from the Early Jurassic, *Rhamphocephalus* from the Middle Jurassic and *Rhamphorhynchus* from the Late Jurassic, provide us with the direct evidence (the tie points) that links this lineage through geological time. Because these pterosaurs all belong to the same lineage, we can join their dots with a single line running from the Early Jurassic to the Late Jurassic. This means that, even though we have not yet found evidence for them (and may never do so, because perhaps it was never fossilized), we can be certain that other rhamphorhynchines must have existed in the time gaps between the dots.

Taking this idea one step further, because genealogical studies have shown that the rhamphorhynchines' closest relatives are the scaphognathines, we can deduce that the lineage of the latter must have existed alongside and for at least as long as the lineage of the former. So, returning to the pterosaur history shown in *Figure 10.2*, this means that, because rhamphorhynchines are known to have existed in the Early Jurassic (where they are represented by *Dorygnathus*), we can infer that the scaphognathine lineage must also have extended back to this point too, even though the earliest record for this clan dates back only to the end of the Middle Jurassic.[7]

What these simple, but effective, concepts do is to equip us with the means to force genealogical data to tell us much more about the history of extinct groups than we can obtain merely by taking the fossil record at face value. If we restrict ourselves purely to fossil evidence, then we must conclude that only three clans of pterosaurs existed in the Early Jurassic. Yet, applying these two ideas reveals that at least another three clans existed at this time and paints a quite different, and probably much truer, picture of pterosaur evolution. If we apply these ideas right across the entire 140-million-year history of pterosaurs, including all the data that we have on pterosaur fossils and the latest results from ongoing analyses of pterosaur interrelationships, we get the tree shown in *Figure 10.2*. It might not be quite as complex or detailed as the genealogy of the kings and queens of England, but it is, without doubt, the most complete and comprehensive picture of the evolutionary history of pterosaurs that has been established so far.

This tree has much to tell us about pterosaurs. Apart from helping us to trace the persistence of clans through time, it can also bring "events" into focus. Notice, for instance, the seemingly simultaneous termination of three pterodactyloid lineages in the early part of the Upper Cretaceous and, in sharp contrast, the apparent explosion in pterosaur diversity at the start of the Upper Triassic. There is also an evolutionary burst of pterodactyloids in the Upper Jurassic. In this chapter, we will encounter these and other major events in the pterosaur story, and examine how they unfolded, who was involved and what their general significance might have been for the Mesozoic world.

We can also take the pterosaur story one stage further. In previous chapters, we saw how some of the key debates about pterosaur anatomy, function and behavior have been resolved, and what we have learned can be combined into a single well-integrated concept for the pterosaur story. What I propose to do in this chapter is to take this concept and show how it can help us to understand and explain some rather curious aspects of pterosaur evolutionary history. Why, for example, did many pterodactyloids achieve large, and even giant, size, but the rhamphorhynchoids did not? And how did birds and pterosaurs manage to live alongside one another for 80 million years? This is the first time such an ambitious account has been attempted. It might not be correct in every detail, and it is certainly not complete, but at least it provides us with a starting point from which we can try to get ever closer to the truth.

Taking to the Air

Presumably at some point in the Triassic, between about 200 million and 250 million years ago, pterosaurs left the security of trees or cliffs and ventured out into an entirely new medium: the air. Unfortunately, we know almost nothing about this major event. As you will doubtless recall from Chapter 4, the exact point at which the pterosaurs branched off from other diapsid reptiles is not at all clear, and intermediate forms between pterosaurs and other reptiles have yet to poke their heads out of the fossil record (or if they have done so, they are keeping their identity well-hidden). So, is there anything we can say about the origin of pterosaurs and their flight ability?

Perhaps. One possibility is to attempt to deduce what pterosaur ancestors were like and how they might have taken wing by extrapolating backward from our knowledge of those species that lie near the base of the pterosaur family tree. So, what do dimorphodontids and anurognathids (the least evolved of all pterosaurs) reveal about pterosaur origins? It seems likely that protopterosaurs were small reptiles that moved around on all fours and spent much, perhaps all, of their time climbing on trees and cliffs, using large, well-developed claws on their fingers and toes. The development of thin flaps of skin fringing the arms and legs, endowing these creatures with some parachuting ability and ensuring them against plunging to their deaths should they miss-judge a leap, is likely to have been the first stage on the road to wings.

The second stage (*Figure 10.3*), the development of these flaps into distinct flight surfaces corresponding in some rudimentary fashion to the propatagium, cheiropatagium and cruropatagium, is also not hard to imagine, especially because plenty of similar examples are found among other backboned animals. It may even be that some of the internal structures, such as wing-fibers, also first appeared at this stage of development.[8]

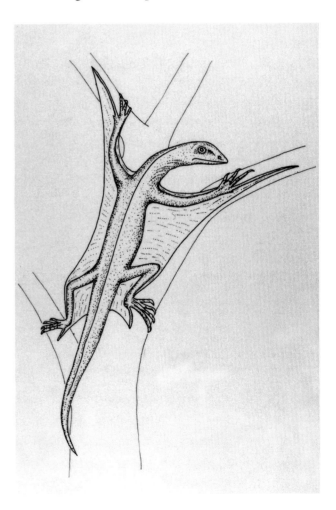

FIGURE 10.3 Restoration of a protopterosaur based on Rupert Wild, 1984. This entirely hypothetical animal represents an early gliding stage, in which the wing-finger has begun to lengthen, and thin sheets of skin, precursors to the wing-membranes of pterosaurs, fringe the body and limbs.

The third stage, the transition to true flight with the development of a flapping ability, might, at first, seem relatively straightforward. Yet, compared with the evolution of a gliding ability, its extreme rarity among animals suggests that it was less easy than we might think. There are two reasons this might be true. First, powered flight required the evolution of a complex system involving the re-engineering of bones and joints to allow the flight stroke to take place, the development of muscles that directed and powered the stroke and the evolution of neural systems and behaviors that controlled and monitored the entire process. Second, and perhaps even more critically, active flight requires much more energy than passive, gliding flight; thus, the evolution of this ability should in some way have given pterosaurs access to more energy, presumably in the form of high-quality food.

Pterosaur teeth and fossil evidence of animals that lived alongside pterosaurs in the Triassic provide some clues as to what might have happened. Let's start with teeth. If we examine early pterosaurs, one feature they all share is the presence of lots of small, sharp-pointed teeth in the rear part of the tooth row. By contrast, teeth toward the front are often modified to form large fangs that were probably used to catch fish, but this is most likely to have been a development that took place *after* pterosaurs had become fully airborne and had developed some considerable sophistication in their flight ability. The small teeth are most likely to have been inherited from pterosaurs' ancestors, which, if correct, means that protopterosaurs probably fed on insects. One early group of pterosaurs, anurognathids, retained this lifestyle. Indeed, some researchers maintain that anurognathids are the earliest of all known pterosaurs, which, if true, would considerably strengthen the argument for insect-eating ancestors.

It seems possible, then, that insects powered pterosaurs to a true flapping flight ability. This is how it might have happened: When pterosaurs took to the air sometime in the Triassic, insects had no aerial predators, other than other insects, such as dragonflies. Protopterosaurs thus gained access to a rich and nutritious harvest, and each improvement in their ability to remain in the air, to maneuver and to catch prey would have been rewarded by even greater access to an energy-rich food source.

Which brings us to a second line of evidence. If pterosaurs really had depended on insects to give them a leg up, shouldn't this be visible in their fossil record? It is. As Ed Jarzembowski, a paleoentomologist (someone who studies fossil insects), pointed out to me many years ago, really big insects are quite common in the Carboniferous and Permian, but disappear in the Late Triassic, at just about the time that pterosaurs are thought to have taken to the air. Coincidence? Perhaps, but at the moment, an insect-powered origin of flight for pterosaurs is the only reasonable theory on the table.[9]

Big Bang or Iceberg? Our first glimpse of the pterosaur story comes in the mid-Late Triassic, about 210 million years ago. Fossils from rocks that originally formed at the bottom of shallow lagoons in what is now central Europe raise the curtain on a stage that, far from being almost empty, as we might expect, was already crowded with pterosaurs and buzzing with their activity. As *Figure 10.2* reveals, at least five different lineages of pterosaurs, each with its own unique membership, ecology and evolutionary history, had already evolved by the time these animals first become visible in the fossil record. Three of them are known directly from fossil material, while two more, ghosts on the stage, can be inferred from our understanding of pterosaur genealogy. Among these players can be found several of the most primitive known pterosaurs—short-winged, deep-headed forms such as *Preondactylus* and the dimorphodontids—but, at the same time, they also include some evolutionarily specialized types, with relatively long wings and a highly modified long, low skull. The most important (not to mention largest) of these was *Eudimorphodon*, which, judging from recent discoveries,[10] had already evolved into several different species.

Such a degree of variety among the earliest known pterosaurs tells us that, evolutionarily speaking, a lot must have happened to the group prior to the mid-Late Triassic. So when exactly did all these currently invisible changes occur and how long did they take? Is it possible, for example, that everything happened rather fast? Perhaps pterosaurs experienced an evolutionary "big bang," such that soon after they first appeared in the late Middle or early Late Triassic, their root stock rapidly branched out into several different lineages and many kinds of pterosaurs (*Figure 10.4*, left). Alternatively, it may be that evolution occurred much more slowly (*Figure 10.4*, right). Pterosaurs might have first taken to the air in the Early Triassic, or perhaps even in the Permian, and then evolved at a rather sedate pace, leaving big gaps between lower branches of the genealogical tree. We might call this the "iceberg" model—even though it's not visible, there is a lot that lies beneath.

The difficulty with the "iceberg" model is that evidence of pterosaurs from pre-Late Triassic rocks seems to be completely and utterly absent: Not

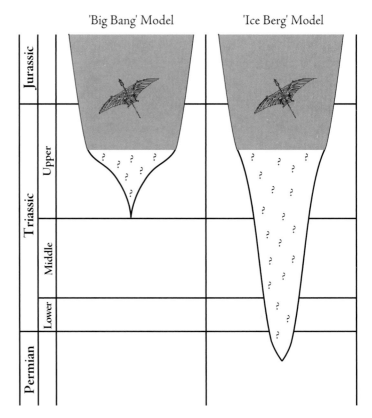

FIGURE 10.4 Two different models for the early evolution of pterosaurs. The "big-bang" model (left) assumes that pterosaurs first appeared in the Middle or even early Late Triassic and evolved rapidly into the five or so lineages known to have existed by the middle Late Triassic. Alternatively, in the "iceberg" model (right) it is assumed that events took place much more slowly and that the first appearance of pterosaurs was in the Early Triassic, or possibly even in the Permian.

a single thin-walled bone, or shard of a wing-finger bone, has yet been found. This is strange, because during periods such as the Jurassic and Cretaceous, when pterosaurs are definitely known to have existed, such fossil fragments regularly turn up. This leaves the "big bang" as the more likely alternative, if only because it fits better with the fossil record. Presumably, in this case, once pterosaurs took to the skies, they found that, apart from the in-flight snacks—insects—they pretty much had the place to themselves. Unfettered by the usual restrictions on evolution, such as competition from incumbents, pterosaurs were free to develop into many different lineages, probably limited only by the speed at which they could evolve and any built-in constraints imposed by their own particular anatomy.[11]

Eudimorphodon, *Prince of the Late Triassic* Incontrovertible evidence of pterosaurs from Triassic rocks was not forthcoming until the early 1970s; since then, a slow but steady stream of discoveries, now amounting to a small treasure chest of fossils,[12] has come to light. Most of these fossils were found in rocks that form the beautiful mountain scenery of northern Italy, but important specimens have also turned up in Austria and even

Greenland, and the picture has been filled out by a row of fragmentary finds, mainly teeth, from Belgium, France, England and the United States. The picture that emerges is a little surprising: Pterosaurs from the lowest part of the genealogical tree, such as dimorphodontids, are there, but, contrary to what we might have expected, they are seemingly rare. By contrast, the most evolutionarily advanced forms around at this time, *Eudimorphodon* and its relatives, dominate the fossil record and appear to have been the most common type of Late Triassic pterosaur.

Eudimorphodontids were the largest Triassic pterosaurs. Some big, old adults could reach over a meter (about 3 feet) between the wing tips, and fragmentary finds suggest that even larger individuals existed.[13] Eudimorphodontids, at least one of which (*Austriadactylus*) bore a well-developed skull crest, had the long snout characteristic of advanced pterosaurs and seem, like many other groups, to have specialized in fishing from the air. Their distinctive multi-pointed teeth would seem to have been well-adapted for gripping slippery prey and, judging by their well-worn tips, were heavily used. The well-developed breastbone, strongly built shoulder girdle and long wings attest to a flight ability that was at least competent enough to permit eudimorphodontids to find and catch their prey while on the wing.

As the restoration in *Figure 10.5* shows, the coastal communities where eudimorphodontids lived were quite unlike those that were to come later in the Mesozoic. Phytosaurs—long-snouted, crocodile-like reptiles—lurked in the shallows, while strange, turtle-shaped placodonts paddled through the depths, pausing to browse on mussels and coral. On land, the first small dinosaurs scampered through the undergrowth, while overhead, a very strange reptile, *Drepanosaurus*, with huge claws on the third finger of each hand and another claw on the end of the tail, swung through the branches.[14]

Above all these zipped and whizzed some of the earliest and most primitive pterosaurs. With big, deep skulls, short wings and relatively powerful hind limbs, *Preondactylus* and the dimorphodontid *Peteinosaurus* give us some idea of what the most primitive pterosaurs probably looked like. *Preondactylus* was about the size of a pigeon and had numerous well-developed teeth, including several large prey-grabbing fangs at the tip of the jaws. This pterosaur also had a fully developed flight apparatus and was almost certainly a competent flier, although its wings seem to have been relatively shorter than in any other pterosaur found so far, and the tail, unlike that of later rhamphorhynchoids, was relatively simple, without the usual stiffening sheath of bony rods.[15] *Peteinosaurus* was in many respects similar, but had the

classic stiffened whip-tail of typical rhamphorhynchoids, and, significantly, the dentistry was rather different. Behind the prey-grabbing fangs was a row of tiny, saw-like teeth that probably served to hang onto food items until they could be swallowed.

The pterosaur community illustrated in *Figure 10.5* is based on fossils from northern Italy, but the first pterosaurs were not confined to this region. In the Late Triassic, there were hardly any natural barriers to prevent them from dispersing far and wide. As *Figure 2.3* shows, the land masses were still linked into the huge super-continent of Pangaea, and there was little to stop pterosaurs from traveling right around its margin or spreading throughout its interior. Unfortunately, rocks of this age are not found everywhere, so we cannot yet map out how pterosaurs colonized Earth, but, by putting together the fossil evidence we do have, it seems that they were already widespread by the end of the Triassic and were distributed worldwide by the Early Jurassic.

Dispersal was not the only challenge, however. Right at the end of the Triassic, Earth experienced some kind of disaster, or disasters, that led to widespread extinctions of animals on land and in the sea.[16] Many animals that had formed the Late Triassic communities to which pterosaurs also belonged, such as the phytosaurs, placodonts and drepanosaurs, died out. Pterosaurs, by contrast, seem to have been far less affected. Most of the main lines survived into the Jurassic, although one group suffered quite badly—the eudimorphodontids. Despite being the most common pterosaur in Late Triassic rocks, no trace of this group, not even a single example of their highly distinctive teeth, has yet been found in the Jurassic.

The Rise and Rise of the Rhamphorhynchids
Perhaps the most striking feature of pterosaur life in the earliest Jurassic was its similarity to that of the Late Triassic. If we could travel back in time to this period and visit the coast of Pangaea, where Europe now lies, the pterosaurs that we are most likely to have encountered were *Campylognathoides* and *Dimorphodon*, direct descendants of the two main Triassic lineages. Like their Triassic relatives, both these pterosaurs seem to have been fish hunters and, when not on the wing, probably spent their ground time clinging to cliffs or tree trunks, dozing or basking in the sun. They were not the only pterosaurs found at the dawn of the Jurassic, but the others, mere shadowy inferences, must await their turn in the story.

FIGURE 10.5 Hangin' out. The northern coast of Tethys in the Late Triassic in what is now northern Italy. Several *Austriadactylus* fly past a *Eudimorphodon* perched on a large conifer. *Drepanosaurus* hangs from a branch, while phytosaurs prowl the waterways. (Painting by Todd Marshall.)

Let's fast-forward a little in time, almost to the end of the Early Jurassic, about 180 million years ago, and turn to the southeast, so that we can observe the large, shallow, warm sea that covered central Europe at about this moment. Rocks that formed from fine mud deposited at the bottom of this sea, now found in several places in Germany and referred to as the *Posidonia* shale,[17] have yielded well over 40 individual pterosaur skeletons, many of them excellently preserved, some of them complete to the tips of their wings. Interestingly, not a single one of these finds is a dimorphodontid, nor do any other pterosaur fossils from this time interval, or from any later point,[18] belong to this group. It looks as if these "big-heads" did not make it to the end of the Early Jurassic.

Campylognathoides, by contrast, is found in some abundance in *Posidonia* shale rocks[19] and has even been reported from as far afield as India.[20] It seems that, unlike the dimorphodontids, its lineage was doing rather well at this point—or was it? In fact, this is the last time *Campylognathoides* and its relatives are to be seen anywhere in the fossil record. There is not a trace of this pterosaur line in younger rocks, so, although campylognathoidids appear to have outlasted the dimorphodontids, it seems that they did not survive beyond the end of the Early Jurassic. Why did these lineages die out? Frankly, we don't know. It might have been pure chance, or perhaps there was some kind of major disaster—volcanic eruptions, a change in the weather patterns—that disrupted the food chain of which they were a part. Yet another possibility is that these fish catchers, specialized as they were, were out-competed by another type of pterosaur even better adapted to this lifestyle—the rhamphorhynchids.

The most common pterosaur from the *Posidonia* shale, featured in *Figure 10.6*, is *Dorygnathus*.[21] With adults that reached nearly a meter (about 3 feet) in wingspan and jaws full of fang-like teeth, *Dorygnathus* is the oldest known member of the biggest and most important branch of all rhamphorhynchoid pterosaurs: rhamphorhynchids—the "prow beaks." *Dorygnathus* is also one of our star witnesses, because it gives away lots of useful information about the pterosaur story in the earliest Jurassic. It might have seemed from the foregoing that this 20-million-year interval was a rather quiet period in pterosaur evolutionary history, filled by lineages doing pretty much the same thing they had done in the Triassic and then quietly expiring in the corner.

In fact, far from it. The location of *Dorygnathus* in *Figure 10.2* demonstrates not only that rhamphorhynchids must have already existed, or first appeared,

FIGURE 10.6 Gone fishing. Out over the Tethys sea in the Early Jurassic, in what is now southern Germany, a flock of *Dorygnathus* are on the lookout for prey. (Painting by Todd Marshall.)

during this interval, but also that the rhamphorhynchines, the particular subgroup to which this group belonged, and its sister subgroup, the scaphognathines, must also have existed or evolved during this time. Moreover, if rhamphorhynchids were present, then their nearest relatives, the line leading to pterodactyloids, must have also existed, and we can be certain that other even earlier branches, such as the anurognathids, must also have been around at this time. Regardless of the exact timing of all these events, we can now be sure that all the main branches of the rhamphorhynchoid part of the pterosaur tree and the branch leading to the pterodactyloids had appeared by the end of the Early Jurassic.

As *Figure 10.2* illustrates, the Early Jurassic, packed with at least six different lineages, was a busy time for pterosaur evolution. But only three of these clans are represented by fossils, so what happened to the others? There are several possible explanations: Maybe the fossil evidence is there, it just has not been found, or recognized, yet. Or, perhaps these groups were rare and hardly ever made it into the fossil record in the first place. A third possibility is that most of these pterosaurs inhabited inland areas and, unlike those that lived in the coastal regions, simply had little chance of becoming fossilized.

Pterosaur communities of the Middle and Upper Jurassic were quite different from those that had gone before. By the end of the Early Jurassic, almost all the early branches of the pterosaur tree had died out, leaving the rhamphorhynchids to take over, which they did, completely. The Middle Jurassic pterosaur fossil record is disappointingly thin, yet, if we aggregate all the finds that have been made so far, the picture that emerges is quite clear—rhamphorhynchids ruled. Early pterodactyloids and anurognathids certainly existed during this 15-million-year interval as well, but, apart from a couple of very fragmentary remains that might represent these groups,[22] there is little direct evidence for them.

Rhamphorhynchids, by contrast, seem to have been omnipresent. Almost every identifiable Middle Jurassic pterosaur fossil belongs to this clan, and, as records from Europe, Asia, Madagascar and South America show, they had spread far and wide. The majority of these finds appear to belong to rhamphorhynchines, a subgroup that seems to have become increasingly adapted to skimming the water surface in the hunt for its prey. Most rhamphorhynchines were probably less than a meter (about 3 feet) in wingspan, although bones of *Rhamphocephalus*, from the Stonesfield Slate of England,[23] indicate that this pterosaur might have reached a whopping 2.5 meters (8 feet) and was one of the largest rhamphorhynchoids.[24]

Scaphognathines must also have existed in the Middle Jurassic but, until recently, direct evidence was frustratingly absent. Encouragingly, new finds from Argentina have borne out genealogical predictions and show that scaphognathines did indeed exist during this interval.[25] The new Argentinean pterosaur, as yet unnamed, has a rather broad snout and powerful, well-spaced teeth, typical scaphognathine features that suggest that members of this clan specialized in bigger prey than their relatives, the rhamphorhynchines.[26]

Another difference between these two groups, which might also help explain why scaphognathines are rarer in the pterosaur fossil record than rhamphorhynchines, concerns their ecology. By and large, fossils of rhamphorhynchines occur in sediments deposited in coastal regions, and it would seem that typically they were found in these areas. Scaphognathines, by contrast, have mainly been discovered in sediments that formed at the bottom of lakes or rivers and seem to have favored inland, rather than coastal, locations. Rocks of the latter type tend to be much rarer than those that represent coastal or shallow sea regions, which perhaps explains the relative rarity of fossil remains of scaphognathines.[27]

Evidently, the lifestyle of these two clans was remarkably successful because both continued without pause through to the end of the Jurassic and, in the case of the scaphognathines, persisted well into the Cretaceous. Rhamphorhynchids seem to have been particularly common in the Late Jurassic. Hundreds of individuals of *Rhamphorhynchus*, ranging from youngsters only a few weeks in age to big, old adults, have been found in the limy mud that accumulated at the bottom of Solnhofen lagoons and in similar places along the northern coast of the Tethys ocean.[28] Moreover, the same pterosaur, or its close relatives, seems to have lived around the edges of many other shallow seas.

At the same time, scaphognathines were snapping up fish from lakes and rivers across what was then Laurasia and almost certainly in Gondwanaland, too. And they were not alone. Finds in the slate beds that formed in the Late Jurassic at the bottom of the Karatau lake in Kazakhstan show, as genealogical studies predict, that anurognathids were there, too (*Figure 10.7*). Flitting around on their relatively short, broad wings, slipping between the trees and sliding around the bushes, anurognathids hunted down the legions of insects—dragonflies, lacewings and beetles—that infested the Karatau lake,[29] snapping up prey in their broad, needle-tooth-lined gape. When not

terrorizing the entomological population, it would seem that anurognathids and their neighbor, the scaphognathine *Sordes*, hung out in the dense lakeside vegetation, suspended from the tips of their long, sharp, strongly curved claws.

Anatomy Is Destiny We have almost reached the end of the story, as far as rhamphorhynchoids are concerned. These long-tailed pterosaurs, the first backboned animals to flap their way through the skies, had survived for more than 75 million years, even though—and this is perhaps one of the most remarkable things about them—evolutionarily speaking, they were highly conservative. Consider, for example, the following: Not a single new rhamphorhynchoid lineage arose after the Early Jurassic. Once they had become established, almost every rhamphorhynchoid clan seems to have stayed in the same kinds of niches for tens of millions of years. Rhamphorhynchoid feeding ecology appears to have been almost entirely restricted to insects and fish or other food that could be caught from the air. Unlike most pterodactyloid lineages, rhamphorhynchoids never achieved large or giant size.

At least part of the explanation for this conservatism is, I think, to be found in the basic anatomy that rhamphorhynchoids inherited from their tree-climbing protopterosaurian ancestors. There can be no doubt that the wings of rhamphorhynchoids were extremely effective and efficient. This is clearly shown by details of their construction and emphasized by the lifestyles pursued by their owners, such as anurognathids and rhamphorhynchids, which required a superb flight ability. So this is not where the problem lay. It is what happened when rhamphorhynchoids were on the ground. As we saw in the previous chapter, with their arms and legs shackled together by the cheiropatagium and a large cruropatagium draped between their legs, rhamphorhynchoids were not at all well-adapted for moving around on land. Moreover, the relative shortness of the forelimb did not help matters. For individuals, this was a problem, but it could be avoided by not visiting the ground at all and sticking, or rather clinging, to trees or cliffs. For rhamphorhynchoids as a whole, it had a profound effect on their evolutionary history.

First of all, it confined these particular pterosaurs to lifestyles that could be pursued on the wing—principally, hunting insects and catching fish or other prey items from the surface of rivers, lakes and lagoons. Other lifestyles, such as wading, swimming, chasing down prey on foot, and many

FIGURE 10.7 Flap, flap, snap. Dusk at the Karatau lake in the early Late Jurassic, in what is now Kazakhstan. In the foreground, *Batrachognathus* hunts for insects. The background shows *Sordes* and more *Batrachognathus*. (Painting by Todd Marshall.)

other possibilities, were excluded either because, as I proposed in the previous chapter, rhamphorhynchoids were physically unable to complete them, or they would have been so inefficient and ineffective that the possibility never evolved in the first place. Another consequence of their life on the wing was that rhamphorhynchoids seldom, if ever, walked around on the ground. This left little, if any, opportunity for leaving tracks, which is surely the main reason why pteraichnites are unknown from the Triassic or much of the Jurassic and why undoubted rhamphorhynchoid tracks have yet to be found.

Rhamphorhynchoids' particular anatomy may also have helped to ensure that they never became giants. There is almost certainly a maximum size beyond which pterosaurs would have found it very difficult to land on, or take off from, a vertical surface. Consequently, in order to evolve into large or giant forms, pterosaurs probably needed some degree of ability to move around on the ground. This would have been a double problem for rhamphorhynchoids. First, they were not well-adapted for life on the ground. Second, being able to run (or rather galumph) fast enough to reach take-off speeds might have been difficult, even for small rhamphorhynchoids, and would have become even more problematic as they got bigger and the necessary take-off speed increased.

What we might conclude from the tale of the rhamphorhynchoids is that the anatomy they started out with was both their destiny and, ultimately, their doom. It got them into the air, but once there, it allowed them a rather narrow set of possibilities. If the story of these wing-fingered fliers had been confined to just the rhamphorhynchoids, it might all have ended sometime in the Early Cretaceous, and pterosaurs would have been little more than a footnote in the annals of animal flight. Fortunately, one group of rhamphorhynchoids evolved into something quite different—a pterodactyloid—and the pterosaur story set off in a whole new direction: downward. Having conquered one medium, the air, now they were going to re-conquer another, the ground.

Act II—Pterodactyloids on Stage Pterodactyloids' appearance in the Late Jurassic is almost as dramatic as the debut of pterosaurs in the Triassic. Frustratingly, we still have no evidence for their ancestors—intermediate forms somewhere between a rhamphorhynchoid and a pterodactyloid—but this is nothing to get excited about. Such "missing links" are notoriously rare. What is more surprising is the almost complete absence of any fossil remains of pterodactyloids in the first 50 million years of the Jurassic,[30] even though the genealogical map illustrated in *Figure 10.2* suggests that they had already appeared by the end of the Early Jurassic and quite possibly well before.

Then, when the curtain did go up in the Late Jurassic, there they were, practically all over the world, from Asia to the Americas and almost everywhere in between. Tracks, too, make a dramatic appearance at about the same time as pterodactyloids. There is hardly a trace of them in Triassic, Early Jurassic or Middle Jurassic rocks, and then suddenly, about halfway through the Late Jurassic, they pop up in France, Spain and large areas of North America.[31]

What are we to make of these curious patterns? As portrayed in *Figure 10.2*, genealogy tells us that the pterodactyloid lineage must have branched off by the time the first of the true prowbeaks (the rhamphorhynchid *Dorygnathus*) appears in the fossil record in the Early Jurassic and that in all probability it branched off earlier, perhaps even in the Triassic. Yet pterodactyloids and their tracks seem to have stayed hiding in the wings until the Late Jurassic. The best explanation I can offer (and in the absence of any good fossil evidence, it is mostly guesswork) is that prior to the Late Jurassic, pterodactyloids were small, they were rare, and they inhabited such regions as plains or upland areas, where the chances of becoming fossilized were even less than the usual millions-to-one against.

Early in the Late Jurassic, pterodactyloids seem to have experienced a dramatic expansion in their ecological and geographic range and moved into a whole range of new habitats associated with lakes, rivers and coasts. This had two results. First, they had a much better chance of ending up in fossil traps, such as the Solnhofen lagoons, so we start to see them in the fossil record. Second, these areas are also suitable for making tracks, and occasionally for preserving them, so the pteraichnite record begins at about this time, too.

It is not known why pterodactyloids moved into these new habitats in the Late Jurassic (perhaps it was related to the development of their walking ability), but we do know who did it: the filter-feeders (ctenochasmatoids) and the clam-crushers (dsungaripteroids). Current understanding of the pterosaur family tree suggests that the four main pterodactyloid clans branched away from each other at about the same time, from which it is easy to deduce that all four must have existed in the Late Jurassic. Curiously, though, two of these four clans, ornithocheiroids and azhdarchoids, do not seem to be present among the hundreds of Late Jurassic fossils found so far, all of which appear to belong either to ctenochasmatoids or dsungaripteroids.

When we first meet them, ctenochasmatoids had already evolved into several different types and, judging by their presence in Asia, Europe, Africa and South America, were already widespread. The least specialized types, such as *Pterodactylus*, floating serenely through *Figure 10.8*, and *Cycnorhamphus*, were about the size of a large seagull and may have been similarly opportunistic, using their relatively simple teeth to dabble for small prey in shallow water, pick up items on the beach or even snap insects from the air. Other ctenochasmatoids were more specialized. *Gnathosaurus*, from Europe, and its Asian relative *Huanhepterus* used their multi-toothed filtering apparatus to sieve for crustaceans or other small water-living animals, while the thinner, more closely spaced teeth of *Ctenochasma* allowed it to harvest even smaller prey.

Early dsungaripteroids seem to have been rather less diverse than their neighbors, the ctenochasmatoids, but, if anything, they were even more widespread. First reported from Europe, they are now known from Asia, Africa and the Americas, suggesting practically a global distribution in the Late Jurassic. Insofar as their anatomy is known (often not far because their fossils remains are so incomplete), most dsungaripteroids seem to have been generally similar to *Germanodactylus*, represented by several almost complete, raven-sized individuals from the Solnhofen Limestones. Although less specialized than their Cretaceous descendants, these early dsungaripteroids already had the winkle-picker jaw tips and robust clam-busting teeth typical of the group. The large number of remains found at fossil localities such as Tendaguru in Tanzania[32] suggest that there, at least, they must have been a familiar sight, stalking in the shallow waters or along the banks of river estuaries and bays, on the lookout for snails, oysters, mussels or perhaps even shrimps.

The presumed lifestyle of ctenochasmatoids and dsungaripteroids also braids rather nicely[33] with two strands of the pterosaur story. The adaptations of these two clans mesh well with the idea that pterodactyloids invaded new habitats—coastal plains and so on—in the Late Jurassic, because we see them living in niches that were rare or absent in plains, forests and mountains. Moreover, the particular lifestyles of ctenochasmatoids and dsungaripteroids, feeding on the banks or in the shallower parts of rivers, lakes and seas, meant that they were moving around in precisely those areas that were most likely to record footprints—in damp mud, silt or sand. Sometimes, against all odds, these beaches, sand banks and mud flats survived and left us with the fossil track sites such as that found at Crayssac today.

The second strand I propose to explore is an even older one—rhamphorhynchoids. Until very recently, it seemed that once pterodactyloids appeared, rhamphorhynchoids went into decline and left the stage soon thereafter, never to return. This led, understandably, to the conclusion that the latter had been driven to extinction by the former, but new details of the pterosaur story show that is quite untrue. It now seems that pterodactyloids and rhamphorhynchoids co-existed for much of the Jurassic and, as recent discoveries in China have shown, continued to live alongside one another well into the Cretaceous. This contradicts the idea that the one replaced the other. This new conclusion is strengthened by the observation that members of the two groups were specialized in quite different ways. Rhamphorhynchoids stuck with the business of skimming for fish or chasing airborne insects, while their long-limbed relatives, the pterodactyloids, radiated into a new set of niches involving sieving and clam-cracking. The final fate of the rhamphorhynchoids has yet to be explained.

Years of the Dragon If ever there was a time when it could be truly said "pterosaurs ruled the skies," it was the Early Cretaceous. Almost half of the slightly more than 100 species described so far come from rocks that formed during this period, and, apart from some early rhamphorhynchoids, almost all the main clans existed during this period, or some part of it. Beyond sheer numerical diversity in this interval, pterosaurs also reached very large sizes for the first time. With wingspans up to 7 meters (23 feet), the largest were more than 15 times the sizes of the smallest species and probably at least 30 times heavier.

Impressive as they are, even these measurements pale somewhat when compared with the sheer anatomical variety of Early Cretaceous pterosaurs. During this period, the relative proportions and shape of the body seem to have varied to a far greater degree than at any time before or after. This is also true for the head, which ranged from short and deep in *Tapejara* to long and low in *J* and bore all sorts of weird and wonderful crests.

Perhaps most tellingly, the shape, size and arrangement of the teeth, and the jaws in which they were set—important pointers to the likely lifestyle—also reached their greatest variety. *Pterodaustro* had the most teeth of any pterosaur, *Dsungaripterus* had the most heavily built set of dentures, and

FIGURE 10.8 "It's not pining, it's passed on. This *Pterodactylus* is no more. It has ceased to be. It's expired and gone to meet its maker. This is a late *Pterodactylus*. It's a stiff. Bereft of life, it rests in peace. It's rung down the curtain and joined the choir invisible. This is an ex-*Pterodactylus*." The carcass of a *Pterodactylus* drifts out into a Solnhofen lagoon in the Late Jurassic in the region of what is now southern Germany. (Painting by Todd Marshall.)

Tapejara had no teeth at all. It would seem that pterosaurs were at their ecological apogee in the Early Cretaceous, living in a broader range of niches,
habitats and environments than they ever had, or ever would again. Not only
were they more widely spread, their ecology had also become more complex.
The finds in the Jehol beds of China, for instance, already include a startling
diversity of pterosaurs that, judging by the construction of their skulls and
teeth, had widely differing ecologies—insect eaters, filter feeders, several
kinds of fishers and possibly even fruit feeders.

The message they send is clear. Pterosaurs were not bit players of the
Mesozoic, restricted to minor roles, decorating patches of empty sky in dinosaur dioramas. They were key players, top predators that, in the Early
Cretaceous at least, seem to have had a significant presence in most habitats
from the middle of the continents to far out over the oceans.

Before we leave this discourse on pterosaur ecology, one more point is
worth making. By and large, pterosaur faunas of the Triassic and Jurassic
seem to have been rather homogenous, the only major difference being between those that lived inland and those that lived in the coastal regions, although even here, the distinction is often rather blurred.

In the Early Cretaceous, they were a lot more complex. There are, for
instance, several cases of Asian and South American pterosaur communities
that apparently consisted of just a single species of dsungaripterid or ctenochasmatid living around the edges of inland lakes. Elsewhere, in what is now
northeast China, extremely diverse pterosaur communities containing anurognathids, scaphognathines, ctenochasmatoids, ornithocheiroids and azhdarchoids were living on densely vegetated coastal plains. Out at sea, it was
a different story again. Here, almost the only pterosaurs to be seen were
ornithocheirids, highly specialized "albatross-like" soaring forms. Melding
together all these facets of local pterosaur communities into a global picture yields an impression of much diversity and complexity—reminiscent in
some respects of bird communities today.

The fates of those pterosaur clans that lived in the Early Cretaceous were
quite varied. Until recently, it was thought that all rhamphorhynchoids had
died out by the end of the Jurassic, but new finds in the Jehol sequence of
northeast China reveal that in this part of the world, they survived until
at least the mid-Early Cretaceous. Curiously, the two types found so far,
anurognathids and scaphognathines, are little different in shape, size and
anatomy from their relatives that lived 20 million years earlier in the Late
Jurassic, and presumably they did much the same thing: hunt for insects

and fish. Some researchers have suggested that these rhamphorhynchoids, and many of the animals and plants that lived alongside them, represent late survivors, clinging on in a refuge that, in contrast to much of the rest of the world, had remained largely unchanged.[34] Perhaps, but the fossil record is not really good enough to be sure, and, in any case, the presence of many typically Cretaceous pterosaurs in the same communities suggests that this refuge was rather porous.

The two pterodactyloid clans that appear to dominate the Late Jurassic—dsungaripteroids and ctenochasmatoids—also seem to have played a major role in the Early Cretaceous. In contrast to the rhamphorhynchoids, both these clans had given rise to larger, more specialized species, while their smaller, less specialized relatives, found worldwide in the Late Jurassic, do not seem to have survived into the Cretaceous.

With their long, chopstick-like jaw tips and battery of crushing teeth, *Dsungaripterus* and its relatives seem to have been highly specialized shellfish feeders that congregated in large flocks around the edges of large shallow lakes, illustrated in *Figure 10.9*. Situated in hot, rather arid regions of central and east Asia, these lakes provided a home for fish and turtles and supported small communities of lizards, dinosaurs and birds. They seem to have been accompanied by just a single species of dsungaripterid, but this pterosaur might have had a larger role than one might expect. Judging from the different fossil remains found, most individuals seem to have been adults of about 2 meters (6 feet) in wingspan, but there were some tiny flaplings there, too, and some big, old animals twice typical adult size. It seems likely that each of these ate different sizes and quite possibly different kinds of shellfish—one species filling several niches at different stages of its life.

Although unspecialized pterodactyloids, such as *Pterodactylus* and its relatives, don't seem to have survived into the Cretaceous, this "seagull" type of role seems to have been successfully taken over by another line of ctenochasmatoids, the lonchodectids. They, too, were only small- to medium-size, the largest reaching perhaps 2 meters in wingspan, and they also had a remarkably conservative body plan that, even when it came to the teeth and jaws, differed little from that of Late Jurassic pterodactyloids like *Pterodactylus*. The lonchodectid skull was long, with numerous short, simple teeth, and jaws whose tips were compressed from top to bottom, rather than from side to side, giving the impression of a pair of sugar tongs with teeth. You could pick up all sorts of things with such a tool, which gives us a clue to their ecology, suggesting that they were generalist feeders, like seagulls today.

The two remaining ctenochasmatoid lineages, both of which at least began the Cretaceous as filter feeders, appear to have evolved in quite different directions. The ctenochasmines stuck with their tooth sieve, which became ever more specialized, and culminated in *Pterodaustro*. Rather like flamingos do today, many thousands of individuals of *Pterodaustro* appear to have gathered to feed on arthropods or other invertebrates that lived in small, shallow, freshwater lakes and pools that dotted the landscape of southern South America in the late Early Cretaceous. Just as for the Asian dsungaripterids, only a single species is known, but it, again, is represented by individuals of all sizes, from flaplings and even an egg with an embryo, up to large adults with wingspans about the size of a condor. So, not only did *Pterodaustro* probably carve the local ecology up into a series of niches, it also seems to have bred near these lakes.[35]

By contrast to ctenochasmatines, some gnathosaurines seem to have evolved in the opposite direction, their teeth becoming larger and larger and steadily fewer in number. These changes point to a shift in diet toward bigger prey items, with a switch at some point from invertebrates back to fish. In this case, the trend culminated in *Cearadactylus*, a large 3- to 4-meter (10- to 13-foot) wingspan gnathosaurine that lived in South America at about the same time as its distant relative, *Pterodaustro*.[36]

Possibly the most important, and certainly the most prominent, of the Early Cretaceous clans were the ornithocheiroids. The first twig on this particular branch of the pterosaur tree—*Istiodactylus*—seems to have frequented coastal plains and river deltas, although exactly how and upon whom it used its cookie-cutter-style dentition (dead dinosaurs?) is not certain. *Istiodactylus* coexisted with its relatives, the ornithocheirids, fossils of which have turned up everywhere and in every period in the Early Cretaceous. These specialist fishers, with murderous looking tooth-grabs splaying from their jaw tips, have been found practically the world over, in sediments that formed on continental plains, in shallow coastal regions and even out in the deep sea. Wherever there were fish worth snagging, ornithocheirids, were there, too. Moreover, working on the principle that if you have a good adaptation, stick with it, the oldest known species found in earliest Cretaceous rocks are little different from those that lived at the end of the Lower Cretaceous, some 40 million years later.

The genealogical tree depicted in *Figure 10.2* suggests that a third ornithocheiroid branch, the pteranodontians, distinguished by the complete loss of teeth, also evolved in the Early Cretaceous. Although the fossil material is

FIGURE 10.9 Individuals of many different sizes (from 19 inches [50 centimeters] up to 10 feet [3 meters] in wingspan) form a vast flock of dsungaripterids feeding on the edges of a large, shallow lake during the Early Cretaceous in the region of what is now Tatal, Western Mongolia. (Painting by Todd Marshall.)

rather poor, several toothless jaw fragments belonging to *Ornithostoma* from the Cambridge Greensand of England (the oldest known pteranodontian), seem to bear this out. Ornithocheiroids were not the only pterosaurs to abandon their teeth, though; azhdarchoids did this, too.

Although their fossil remains have only come to light in the last 20 years, it seems that tapejarids and tupuxuarids (illustrated in *Figure 10.10*), the first two of the three branches of the azhdarchoid lineage, were important members of Early Cretaceous pterosaur faunas. With their relatively short, deep, parrot-like beaks, tapejarids were relatively small and mostly under a couple of meters (6 feet) in wingspan. Their presence in South America and China, and later in Africa, shows that they must have been widespread, although, strangely, they seem to be completely absent from among the thousands of fossils recovered from the Cambridge Greensand.[37] By contrast, tupuxuarids such as *Thalassodromeus*, with their long, narrow, blade-like jaws, are only known from a single location—the Santana nodules of the Araripe plateau in Brazil—and might perhaps have been restricted to such locations as large, relatively calm inland seas where their supposedly water-skimming lifestyle was feasible.[38]

Teeth No More Throughout the later Early Cretaceous, sea levels were on the rise. They reached a high point early in the Late Cretaceous, flooding many continental areas and radically changing the geography of Earth (*Figure 2.3*). On land, these dramatic developments were mirrored by changes in the terrestrial biota: New types of dinosaur—ceratopsians and hadrosaurs—began munching on a flora that was also changing, as flowering plants began to replace typical mid-Mesozoic plants like cycads and ferns. Not to be left out, pterosaurs also experienced some profound and dramatic changes. By about 90 million years ago, early in the Late Cretaceous, many of the typical Cretaceous groups, including all those with teeth, had disappeared and, in all probability, were extinct. Only two clans—pteranodontians and azhdarchids—were left, but their members shared one striking feature: They were toothless.

So, what happened? The broad picture is shown in *Figure 10.2* and reveals that between the mid-Early and mid-Late Cretaceous, there was a complete change in pterosaur faunas. Most of the Early Cretaceous lineages—ctenochasmatoids, dsungaripteroids, ornithocheirids—seem to have died out and were replaced by clans that are hardly known before the Late Cretaceous, pteranodontians and azhdarchids, which, although they

FIGURE 10.10 Feeding frenzy. Although they have been interpreted as skimmers, it is possible that tupuxuarids were not completely restricted to this lifestyle and may have taken advantage of other opportunities, such as the rotting corpse of a dead dinosaur. (Painting by Todd Marshall.)

descended from these earlier lines, seem to have been quite different from them. Unfortunately, the early Late Cretaceous pterosaur fossil record is rather poor, so exact details of this, when and how it happened, are still unknown.

Several clans—anurognathids, scaphognathines, dsungaripterids, gnathosaurines and ctenochasmatines—are completely unknown from Upper Cretaceous rocks and might have already become extinct in the Early Cretaceous. On the other hand, the last records of these groups are scattered across a 25-million-year interval, and we will almost certainly never discover the exact point in time when each of them actually died out.

Three typical Early Cretaceous pterosaur groups did survive, though, and they turn up in early Late Cretaceous rocks. Lonchodectids, seemingly unchanged from earlier forms, have been found at several levels near the base of the Chalk, a rock formed from the shells of countless billions of tiny sea organisms, higher sections of which form, for example, the distinctive white cliffs of Dover in England. Numerous bones and teeth of ornithocheirids have been found in the same sediments and also in rocks that were deposited in the Kem Kem region of Morocco at about the same time. Although still very rare, a few fragments of tapejarids have also been found at the Kem Kem locality. So, to summarize, it seems that lonchodectids, ornithocheirids and tapejarids, little different from those found in the Early Cretaceous, not only survived into the early Late Cretaceous, but were possibly quite common and abundant up until about 90 million to 95 million years ago.

After this, however, there is not one single certain find of either of these clans, or indeed of any other toothed pterosaurs, in the remaining 25 million years of pterosaur history, which goes right up to the end of the Cretaceous. It should be pointed out, though, that the mid-Upper Cretaceous fossil record is very poor, so it is possible that the final disappearance of some, perhaps all, of these groups might have happened a little later. Not much later, though, because late Upper Cretaceous rocks, representing a 20 million year period, have produced many pterosaur fossils, yet these Early Cretaceous clans are definitely not to be found anywhere among them.

Why did toothed pterosaurs, especially the ornithocheirids, previously so abundant, become extinct? One possibility is that some groups, especially the ornithocheirids, were hit hard by a widespread breakdown in oceanic ecosystems that took place about 90 million years ago.[39] Exactly what happened is not clear, but for those species near the top of the food chain, such as pterosaurs, it was probably catastrophic. This event alone does not explain everything, however. While it might have brought about the demise of ornithocheirids and perhaps lonchodectids, it is less likely to have affected groups such as tapejarids, which also lived in typical inland habitats.

Whatever happened in the early Late Cretaceous, at least two pterosaur clans survived. Initially, the most important of these were the pteranodontians, toothless ornithocheiroids that evolved into two quite different forms—*Pteranodon* and *Nyctosaurus*. With its long, gracefully curving jaws and spectacular, bony head crest, *Pteranodon* is one of the most distinctive and arguably most famous of all pterosaurs. Older texts often give the impression that *Pteranodon* was *the* Late Cretaceous pterosaur and could be seen anywhere at any time in this period, but a comprehensive re-study of this pterosaur by Chris Bennett suggests a quite different picture.[40] In fact, *Pteranodon* is only to be found in North America and is known only from a short, 2-million- to 3-million-year time slice that lies almost exactly in the middle of the Late Cretaceous.

Genealogical studies indicate that pteranodontians split away from other ornithocheiroids early in the Cretaceous, yet, apart from *Ornithostoma* and another possible record of the clan from the early Late Cretaceous of North Africa,[41] there is no trace of this lineage before the middle of the Late Cretaceous. Poor as it is, the available evidence suggests that pteranodontians were relatively widespread during the mid-Cretaceous, so its "low profile" probably has more to do with lack of fossilization than any real rarity.

Many of the more than 1,000 finds of *Pteranodon* made so far come from the Niobrara Chalk, which formed in a shallow seaway that stretched down through the middle of North America in the Late Cretaceous. One or two of these individuals reached almost 7 meters (23 feet) in wingspan,[42] although most were little more than half that size. Aerodynamic studies of *Pteranodon* indicate a remarkably efficient flight ability, which presumably enabled this pterosaur to soar over the sea for hours, or perhaps even days, searching for prey.

Nyctosaurus may have been somewhat smaller than *Pteranodon*, and its fossil remains are much, much rarer, but, if anything, it had an even more spectacular cranial crest, as *Figure 5.8* shows. This lineage probably branched away from the line leading to *Pteranodon* before the end of the Early Cretaceous, but the first evidence of its existence is not found until some 15 million years later, in the Niobrara Chalk. After another 15 million year gap, the last record for this line, an incomplete humerus, has been reported from the latest Cretaceous of Brazil.[43] Again, the vagaries of preservation and discovery are probably the main culprits for this exceptionally spotty record.

Azhdarchids Rule The last 15 million years of pterosaur history seems to have differed in at least one way from everything that had gone before. It was almost completely dominated by just a single clan: the long-necked, spear-jawed azhdarchids (*Figure 10.11*). Unlike earlier periods, when at least two or three different clans were always living at the same time, if not always in the same place, almost every find from the last half of the Late Cretaceous appears to belong to an azhdarchid. Pteranodontians practically disappear from the fossil record just beyond the mid-Late Cretaceous, and, after that, it is azhdarchids all the way, the only exception being the single fossil of *Nyctosaurus* from Brazil just mentioned.

The azhdarchid lineage probably first appeared in the Early Cretaceous. Several fossil finds from that interval, supposedly belonging to this clan, might be taken as direct proof of this assumption.[44] They, however, all lack true azhdarchid features and might instead have belonged to *Tapejara* or *Tupuxuara* or their relatives. The oldest reasonably certain records of azhdarchids—bits of toothless jaws and a long neck vertebra—have been found in the early Late Cretaceous Kem Kem beds of North Africa, and further fossils, including the middle-Asian pterosaur *Azhdarcho*, are known from slightly younger mid-Late Cretaceous sediments.[45] *Azhdarcho* was relatively small for an azhdarchid, probably only about the size of a condor, although one or two large bones hint at bigger individuals, but it had typical azhdarchid features, including the characteristically elongated, tube-like neck vertebrae.

Although more common than they were, for example, at the start of the Late Cretaceous, azhdarchids such as *Azhdarcho* and its relatives still shared the mid-Late Cretaceous world with other pterosaurs—pteranodontians. Curiously, at this point in their history, azhdarchids seem to have been restricted to the Old World (at least, fossil azhdarchids of this age have yet to be discovered in the Americas), while pteranodontians are found only in the New World. Whether this pattern is real or merely an artifact of a poor fossil record remains to be seen. In any case, it shows that azhdarchids had yet to achieve world domination, because they are quite unknown, for example, from the Niobrara Chalk.

After the virtual disappearance of pteranodontians about 80 million years ago, azhdarchids had the world to themselves, and they made the most of it. In the late Late Cretaceous, these pterosaurs were found on almost every continent and seem to have lived in a wide variety of habitats, from

FIGURE 10.11 King of the skies. The giant 33 foot (10 meter) wingspan Late Cretaceous azhdarchid Quetzalcoatlus dwarfs a flock of birds. (Painting by Todd Marshall.)

inland plains to coastal regions and even far out over the sea—everywhere you looked, there were azhdarchids.

The surprising thing about azhdarchids is that, although they spread into lots of different habitats, their anatomy, even down to the shape of their beaks, seems to have been pretty much the same for all species. So far as the fossil evidence goes, all of them, from *Azhdarcho* to *Zhejiangopterus*, were toothless, had fishing-rod necks, rather short wings and long, powerful hind limbs. Indeed, the latter are particularly impressive and point to a better ability on the ground than perhaps any other pterosaurs. That azhdarchids took an occasional stroll is shown by the discovery in South Korea of pterosaur tracks that seem to have been made by members of this clan, at least one of which was a giant and left footprints almost a third of a meter (about a foot) long.[46]

Staying with the theme of size, this is one aspect where azhdarchids certainly did vary—a lot. While small forms were only about 1 or 2 meters (3 to 6 feet) in wingspan, most seemed to have reached at least 5 meters (16 feet) or so, and adults of at least three different species grew to 10 meters (33 feet) or more.[47] Moreover, lumping together all the data we have for azhdarchids shows that this remarkable size range was not restricted to just one point in time but, except for the final few million years of the Cretaceous, seems to have been typical for much of the azhdarchid hegemony. If flaplings, juveniles and adults of each species occupied rather different niches, then it would seem that, ecologically speaking, azhdarchids might have covered quite a lot of ground.

In fact, although azhdarchids seem to have been rather homogenous, if we pore over their skeletons with a magnifying glass, differences start to emerge, especially in the construction of the neck bones. Some necks, such as that of a giant azhdarchid from southern Spain, are rather chunky,[48] while in other cases, *Arambourgiania* and *Quetzalcoatlus*, for instance, the shape of individual vertebrae and the arrangement of the joints show subtle, but important, differences. These and other features hint at slightly different ways of doing things. So it might be that the various lifestyles that have been proposed for azhdarchids: scavengers, waders, and airborne fishers, are not either/or options but were indeed pursued by these pterosaurs and formed part of a wide repertoire of adaptations. A nice feature of this idea is that it also helps to explain why azhdarchids have been found in so many and such varied locations, both on land and sea.

Two decades ago, the existence of azhdarchids was hardly suspected. Now, we have evidence of a whole new dynasty of pole-necked pterosaurs that seems to have been as diverse and as successful as any that came before. This dynasty also had one unique feature—it was the last.

Knocking on Heaven's Door There must have been one particular day, about 65 million years ago, when the last pterosaur died, and these extraordinary animals that had graced the skies for so long were no more. Pterosaurs were not the only animals to become extinct at the end of the Mesozoic. Many other groups, including dinosaurs, ammonites, plesiosaurs, and most Mesozoic birds, died out, too. Indeed, the scale of loss was so great that the event is termed a "mass extinction" and numbers among the five largest of such "mega-death" events.[49] Like other mass extinctions, the one that ended the Mesozoic had a profound effect on life on Earth, and most major ecosystems were never quite the same again. Such events are not all bad, though. On the positive side, they also acted as a global reset mechanism, allowing new kinds of animals and plants to evolve and new communities to develop.

Naturally, because extinction, and mass extinction in particular, has played such an important role in the history of life, paleontologists and many other scientists working in related areas are deeply interested in the subject. Unfortunately, it is not all that amenable to investigation. One of the biggest difficulties is that, in most cases, including that of pterosaurs, extinction means that beyond a certain point in the fossil record there are no more data, quite simply because the subject in question has ceased to exist. It's as if one were forced to switch off a detective drama on TV just before the victim was murdered and was then asked to try to work out the circumstances of the murder and who did it. In the case of pterosaurs, it is even more difficult, because the fossil record that we do have is not that good anyway. Imagine only being allowed a quick glance at the TV program every now and then before it was turned off.

Still, the fossil record is all we have. So what does it tell us?

Two things are quite clear. First, pterosaurs appear to have survived right up to the end of the Cretaceous. Both *Arambourgiania* and *Quetzalcoatlus* come from sediments that were deposited very late in the Mesozoic, and fragmentary remains of pterosaurs have been found that date almost to the end of the Cretaceous. Second, pterosaurs did not survive into the Tertiary. Their fossil remains are very easy to spot, and not a single skeleton, or even a tiny fragment of bone, has been found to indicate that somehow one or two managed to outlast the dinosaurs. So, we can be fairly certain that pterosaurs managed to survive long enough to die out with everyone else in the mass extinction.

Who did the dying? The answer seems fairly clear—the azhdarchids and perhaps *Nyctosaurus*, too. Is there anything special about these pterosaurs that might have made them particularly susceptible to extinction? In fact, there is.

An important feature of the very last pterosaurs is that almost all of them seem to have been very large or giant azhdarchids. There are two drawbacks to being large. First, the total population size for "big" species tends to be much lower than for "small" species (think of elephants and rabbits). Second, the rate at which individuals of large species reproduce is generally much slower than the rate for small species (elephant pregnancies can last two years, while in rabbits it is about one month). The situation, then, for the last pterosaurs is that there were fewer different kinds around than at almost any time in their history, and those that still existed were mostly very large and highly specialized, and quite possibly with relatively small populations and slow rates of reproduction.

The conclusion: Pterosaurs were already at a high risk of becoming extinct. They might not have been knocking on heaven's door[50], but they were certainly flying pretty close.

Exactly what happened at the end of the Cretaceous is still not entirely clear.[51] Some scientists have proposed that massive volcanic eruptions disrupted climate and food chains and brought on the mass extinction. Others argue that a huge meteorite impact, which caused global devastation and a short nuclear winter, was the main culprit. And there are other ideas too—lots of them—in fact, more than 100 have been proposed.[52] Happily, we do not need to worry about them, because almost any disruption on a global scale in the latest Cretaceous would probably have been sufficient to finish pterosaurs off.

They were already at risk; there was some kind of major environmental disaster; they became extinct; and that was that.

Well, maybe not quite. In fact, a long way from not quite. The big question is: How had pterosaurs arrived at this point—relatively rare and specialized—in the first place? To try to understand this, we need to analyze their history over a long period, let's say the entire Cretaceous—80 million years—to be on the safe side. The next step is to total how many pterosaurs existed during particular segments of time within this period. Interestingly, it does not seem to matter how we do this—whether we choose to lump pterosaurs together into clans, or count just species, and whether we choose to have a few long segments of time, or lots of short ones—the pattern that emerges, as shown in *Figure 10.12*, is always the same. From about the middle of the Cretaceous onward, pterosaur variety seems to have gone into a terminal dive.

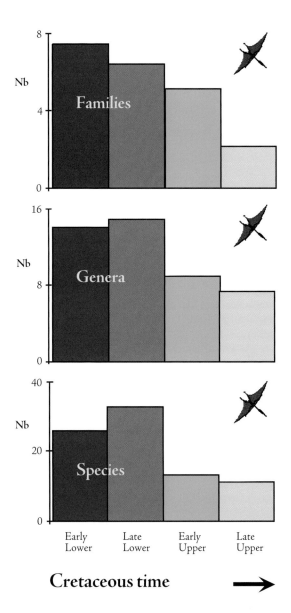

FIGURE 10.12 How pterosaurs fared through the Cretaceous. The number of families, genera and species are shown for four successive divisions of the Cretaceous. Only one conclusion can be drawn from these results: Pterosaurs seem to have become much rarer in the Upper Cretaceous and were rarest of all in the latter half of this interval.

If we examine this decline, it appears to have taken place quite slowly, without any points where pterosaurs suddenly suffered a massive drop in numbers. Quite the contrary, the various clans seem to have disappeared at different times, dsungaripteroids by the mid-Early Cretaceous, ctenochas-matids by the end of the Early Cretaceous, and ornithocheirids in the early Late Cretaceous. That such clans died out is not really surprising. It is a typical feature of the history not only of pterosaurs but of all other animal and plant groups as well: Extinction is the left hand of evolution. What is surprising, however, is that after about the middle of the Cretaceous, no new groups of pterosaurs arose to replace those that had gone before.

So, the decline of pterosaurs may not have so much to do with extinction—stuff happens—but their failure, in the Late Cretaceous, to evolve new clans. Why? A rather striking feature of Late Cretaceous pterosaurs is the almost complete absence of small generalist species such as *Pterodactylus* and *Loncho-dectes*, which seem to have been common in earlier periods. This left ptero-saur communities that were dominated by relatively specialized forms, and, as they became extinct, the numbers of species dwindled until the entire group eventually was threatened with extinction.

So, what happened to the small, unspecialized species? Perhaps this is where birds did have some impact on pterosaur evolution. It might be that, as pterosaur clans died out, the niches and habitats they left behind were not refilled by new pterosaurs, but by birds, which opportunistically moved into these temporarily empty spaces. Prior to the Late Cretaceous, for example, lifestyles such as insect hunting or wading on beaches and river margins were pursued by pterosaurs. Later, however, these roles seem to have been taken over by birds. Ultimately, the effect of this process was to leave ptero-saurs adapted to a relatively narrow range of specialist lifestyles.

Pterosaurs' basic construction also played a key role in shaping their evolutionary history. Even though pterodactyloids seem to have been much more mobile on the ground than rhamphorhynchoids, their fore and hind limbs were still linked together by the flight membranes. Consequently, the possible functional repertoire for the legs, and the range of different shapes and proportions into which they could evolve, was relatively limited.[53] In birds, by contrast, there were no such constraints. Their legs and feet were free to evolve in many different ways, as Late Cretaceous bird fossils confirm. These "early" birds had already diversified into a remarkable variety of forms that included "sea gull-like" generalists and many specialists, such as flight-less runners and divers that inhabited niches into which pterosaurs did not and, in all probability, could not evolve.

The upshot of all these processes was that by the end of the Cretaceous, there were just a few relatively specialized species of pterosaur, whereas birds seem to have been much more diverse and in many cases small, allowing the possibility of large populations and the ability to reproduce rapidly. The lottery that was the end-Cretaceous mass extinction finished off pterosaurs for good. It also extinguished many kinds of birds, but not all. Some got lucky and made it through.

That, perhaps, is why, when you look out the window, the skies of today are filled with birds, and not pterosaurs.

II
POSTSCRIPT

The garden was a wreck. Something had flown straight into the greenhouse, leaving hardly a single pane intact. The visitor was gone, but it had left a calling card—tufts of hair, fixed by sun-dried blood to some of the larger shards of glass. Halfway down the lawn, sprawled out in a little heap, were the half-digested remains of a cat. "Well, at least they didn't manage to keep him down," thought the owner, and it looked like Smilodon's flea problem had finally been solved. Bartlett, the guinea pig, was gone, too, his hutch overturned, and, as for the goldfish pond, pointless even to look—it would be empty, just like last time. Pushing his glasses firmly back onto the bridge of his nose, the man turned to his wife, who had just come through the kitchen doorway, and called to her. "Get the boys up," he said, "and tell them to bring their rifles." He walked over to the shed and started to pull the tarpaulin off a surface-to-air missile launcher they had installed on the roof. "Bloody pterosaurs," he muttered, ruefully. "Why the hell did I think it would be a good idea to bring them back?"

FIGURE 11.1 New to science. A distinctive brown coloration picks out the superbly preserved wing membranes of this new 15 inch (40 centimeter) wingspan example of *Jeholopterus* from the Lower Cretaceous Jehol Group of northeast China. Details of the propatagium of this individual can be seen in Figure 8.4 above left. (Image courtesy of Lü Junchang.)

"Now this is not the end. It is not even the beginning of the end. But it is, perhaps, the end of the beginning."[1] Originally devised and quoted by Sir Winston Churchill, the British prime minister during World War II, these sentences are cited here because it seems to me, at least, that they neatly summarize the current state of development of pterosaurology. Much research remains to be done, of course, and doubtless (to quote another of Sir Winston's famous sayings) it will involve a lot of blood, toil, tears and sweat,[2] but we can at least take heart from the knowledge that, for the first time in 200 years, we have reached a point where the basic anatomy, function and evolutionary history of pterosaurs is widely agreed upon. Moreover, this working model fits together in such a coherent and convincing fashion that it can be used with confidence, both as a basis for improving our knowledge of those aspects of pterosaur biology, such as flight or growth, that we already know something about, and as a starting point for new investigations into areas, such as physiology, of which we still have little or no understanding.

So what problems should be tackled first? This is not so easy to answer. Several issues are particularly vexing, at the forefront of which lies the question of the origin of pterosaurs. Some well-founded evidence as to where pterosaurs fit into the diapsid family tree would be most welcome and, even more importantly, might throw some light on how pterosaurs acquired their particular and peculiar anatomy and how they used it to get into the air.

Beyond the problem of pterosaur origins are plenty of other pressing issues queuing up for attention. How, for example, did pterosaurs breathe, and what was the nature of their physiology? Were they really "hot-blooded" endothermic homoeotherms? If so, how can this be reconciled with prenatal development in an egg buried in the ground: a developmental domain associated with "cold-blooded" exothermic heterotherms like crocodile and lizards?

Aside from the "big questions," there are plenty of seemingly small issues that, once one has gotten the lid off, have a nasty habit of turning into a surprisingly large can of worms. Take pterosaur taxonomy, for instance. Taxonomy is not a particularly complex or difficult science, and the 100 or so species of pterosaur described so far seem pretty small compared with almost 10,000 living species of birds, with at least another couple thousand in the fossil record. Surely, one might think, this corner of pterosaurology should be fairly well sorted out, and so it appears—from a distance.

Up close, though, things look rather different. Here is just one example: Pterosaurs from the Santana and Crato Formations of Brazil have already acquired nigh on 20 different species names, which, as all pterosaurologists agree, is far too many. There are several cases where one pterosaur has two names and there is at least one species, *Coloborhynchus robustus*, which has three, or possibly even four, names. This problem needs addressing, not just because the taxonomist in each pterosaurologist will not rest easy until this tidying up is done,[3] but also because these sorts of studies have an impact on many other aspects of pterosaur biology, such as the meaning of cranial crests, growth patterns and questions of ecology. Extend this idea across all pterosaurs (there are few species that wouldn't benefit from some taxonomic scrutiny), and what started out as a relatively modest piece of research suddenly begins to take on a fundamental significance and threatens to last a lifetime.

It looks as if pterosaurologists are going to need all the blood, toil, tears and sweat that they can lay their hands on. Still, the good news is that even if discoveries of pterosaur fossils ceased tomorrow (which seems highly unlikely, because new finds such as the one shown in *Figure 11.1* are being made almost every week), it should still be possible to answer most of the questions posed here, and many others, on the basis of the fossil collections that we already have. It might seem surprising, but the main constraint on the speed at which our knowledge and understanding of pterosaurs is likely to increase has little to do with the rate of discovery of new finds, or even new technological developments, but is primarily governed by the number of researchers working at any one time. Even including those researchers who only occasionally study pterosaurs, this number is still very small and likely to remain so for the foreseeable future. Inevitably, therefore, progress will be slow, but it will happen.

For now, though, the speed of research is not the most pressing issue. The critical point is that, although it has taken a long time, we have finally reached the first and arguably most important milestone in pterosaur research—a basic understanding of what these animals were and how they worked. Having shuffled the pieces around for a couple of centuries, the general outlines of the picture on the jigsaw puzzle have at last become visible. The task now is to try to fill in some of the big blank spaces that remain and, wherever possible, to improve the resolution of those bits of the puzzle that we already have. The picture will change: It will get bigger and sharper and more detailed and more complex—and as it does, so our wonder and fascination at the sheer extraordinariness of these animals will surely increase, too.

Yet, for all its glory, a picture is only a picture. Nothing will ever be quite like seeing a real, live, breathing pterosaur. And that's completely impossible, isn't it? At present, yes, but, it's worth remembering that ancient DNA has already been recovered from the archaeological record.[4] Admittedly, this DNA is not very old, only a few thousand years, and generally it is very fragmentary. Still, if it can survive a few thousand years, maybe, just maybe, it could survive millions of years. Perhaps a small pterosaur in a very large piece of amber might yield enough fragments of DNA for our increasingly sophisticated genetic techniques to patch it all back together again. Then it should be a relatively simple matter to develop a clone using a chicken's egg, or maybe a crocodile's egg, just to be on the safe side, so that a few months later, out would hatch a hairy little critter with flappy wings, big eyes, and a cute snappy beak.

Impossible? Probably, but I live in hope.

APPENDIX

List of Valid Pterosaur Species

A complete listing of valid species of pterosaur arranged according to the relationships of pterosaurs as they are currently understood (see Chapter 4 for further details).

Pterosauria
    Rhamphorhynchoidea
        Basal forms
           Preondactylus buffarinii
        Dimorphodontidae
           Dimorphodon macronyx
           Peteinosaurus zambellii
        Anurognathidae
           Anurognathus ammoni
           Batrachognathus volans
           Dendrorhynchoides curvidentatus
           Jeholopterus ningchengensis
        Family uncertain
           'Dimorphodon' weintraubi
        Campylognathoididae
           Austriadactylus cristatus
           Campylognathoides indicus
           Campylognathoides zitteli
           Eudimorphodon cromptonellus
           Eudimorphodon ranzii
           Eudimorphodon rosenfeldi

Rhamphorhynchidae
  Rhamphorhynchinae
    Angustinaripterus longicephalus
    Dorygnathus banthensis
    Dorygnathus mistelgauensis
    Dorygnathus purdoni
    Nesodactylus hesperius
    Rhamphocephalus bucklandi
    Rhamphorhynchus 'longiceps'
    Rhamphorhynchus muensteri
  Scaphognathinae
    Cacibupteryx caribensis
    New Cerro Condor pterosaur
    Harpactognathus gentryii
    Pterorhynchus wellnhoferi
    Scaphognathus crassirostris
    Sordes pilosus
Pterodactyloidea
  Ornithocheiroidea
    Istiodactylidae
      Istiodactylus latidens
      Liaoning istiodactylid
    Ornithocheiridae
      Anhanguera blittersdorffi
      Anhanguera cuvieri
      Anhanguera fittoni
      Anhanguera santanae
      Arthurdactylus conandoylei
      Boreopterus cuiae
      Brasileodactylus araripensis
      Coloborhynchus capito
      Coloborhynchus clavirostris
      Coloborhynchus moroccensis
      Coloborhynchus robustus
      Coloborhynchus sedgwickii
      Coloborhynchus wadleighi
      Haopterus gracilis

      New Isle of Wight ornithocheirid
      Liaoningopterus gui
      Liaoxipterus brachyognathus
      Ludodactylus sibbicki
      Ornithocheirus mesembrinus
      Ornithocheirus simus
      Ornithocheirus sp.
    Pteranodontidae
      Ornithostoma sedgwicki
      Pteranodon longiceps
      Pteranodon sternbergi
    Nyctosauridae
      Nyctosaurus gracilis
      Nyctosaurus lamegoi
      New Mexican nyctosaurid
  Ctenochasmatoidea
    Basal ctenochasmatoids
      Cycnorhamphus canjuersensis
      Cycnorhamphus suevicus
      Pterodactylus antiquus
      Pterodactylus kochi
      'Pterodactylus' micronyx
    Ctenochasmatidae
      Ctenochasmatinae
      Beipiaopterus chenianus
      Ctenochasma elegans
      Ctenochasma porocristata
      Ctenochasma roemeri
      Eosipterus yangi
      Pterodaustro guinazui
      Gnathosaurinae
      Cearadactylus atrox
      Gnathosaurus macrurus
      Gnathosaurus subulatus
      Huanhepterus quingyangensis
      Plataleorhynchus streptophorodon
      'Pterodactylus' longicollum

Dsungaripteroidea
  Basal dsungaripteroids
    Germanodactylus rhamphastinus
    Germanodactylus cristatus
    Herbstosaurus pigmaeus
    Kepodactylus insperatus
    Normannognathus wellnhoferi
    Tendaguripterus recki
  Dsungaripteridae
    Domeykodactylus ceciliae
    Dsungaripterus weii
    Lonchognathosaurus acutirostris
    Noripterus complicidens
    'Phobetor' parvus
Azdarchoidea
  Lonchodectidae
    Liaoning lonchodectid
    Lonchodectes compressirostris
    Lonchodectes giganteus
    Lonchodectes machaerorhynchus
    Lonchodectes microdon
    Lonchodectes platysomus
    Lonchodectes sagittirostris
  Tapejaridae
    Chaoyangopterus zhangi
    Jidapterus edentus
    Sinopterus dongi
    Sinopterus guii
    Tapejara imperator
    Tapejara navigans
    Tapejara wellnhoferi
  Basal neoazhdarchids
    Tupuxuara longicristatus
    Tupuxuara leonardii
    Thalassodromeus sethi

Azhdarchidae
  Arambourgiania philadelphiae
  Azhdarcho lancicollis
  Hatzegopteryx thambena
  Montanazhdarcho minor
  Phosphatodraco mauritanicus
  Quetzalcoatlus northropi
  Quetzalcoatlus sp.
  Zhejiangopterus linhaiensis

Possibly valid species of uncertain relationships
  "Araripesaurus castilhoi"
  "Mesadactylus ornithosphyos"
  "Puntanipterus globosus"
  "Rhamphinion jenkinsi"
  "Santanadactylus araripensis"
  "Santanadactylus brasilensis"
  "Santanadactylus pricei"
  "Santanadactylus spixi"

NOTES AND SOURCES

Chapter 1: Dragons of the Air
Pages 2–15

1. The lifestyle of *Quetzalcoatlus*, a giant pterosaur that lived in what is now North America, right at the end of the Mesozoic a little more than 65 million years ago, is uncertain, but with its spear-like jaws and remarkably long neck (see *Figure 10.11*) it would seem to have been well adapted for fishing from the air.

2. Flying reptiles are often referred to as pterodactyls, but the correct collective term for these animals is pterosaur, or Pterosauria, if you wish to be really formal. Pterodactyl is a vernacular name for one particular group of pterosaurs that are more correctly referred to as the pterodactyloids and are mainly distinguished by their short tails. The other major group of pterosaurs, the rhamphorhynchoids, are characterized by the presence of long tails (although, as explained in Chapter 4, one clan of rhamphorhynchoids actually had short tails). Strictly speaking, rhamphorhynchoids are not a completely natural grouping, because their descendants, the pterodactyloids, are artificially excluded from the group. Usually, this state of affairs is shown by always citing the name in quotation marks, but as that is a mere technical nicety, I have left them out in this book. If you are the kind of person who checks to see that the fork drawer contains only forks, you can mentally reinstate the quotation marks as we go along.

3. Among pterosaurs' recent TV appearances, Episode 4 of the award-winning BBC television series *Walking with Dinosaurs*, first screened in 1999, was entirely devoted to *Ornithocheirus*, a large, fish-eating Lower Cretaceous form.

4. Although often assumed to have featured in the first (and best) installment of *Jurassic Park*, pterosaurs only briefly appeared in the second film, and their first substantial part was not until the third, although they made up for their late arrival in a most commendable fashion by trying (unsuccessfully, I am sorry to say) to eat a research student. This was, however, only pterosaurs' most recent role in a long and illustrious career in the movies. One of their earliest appearances (possibly their first on film) was in *King Kong*, first shown in 1933, in which an apparently deaf *Pteranodon* attempted to fly off with leading lady and champion screamer Fay Wray but, tragically, was beaten to death by the hero of the film. A more successful kidnapping was made by another *Pteranodon* in *One Million Years BC*, first shown in 1967, that managed to fly all the way back to its nest carrying a scantily clad Raquel Welch. Sadly, despite pterosaurs' predilection for

young women, even *Quetzalcoatlus* would have found it impossible to lift a full-grown human, especially if there was a giant ape hanging on the other end.

5. Fossil remains of *Pteranodon* show that it most certainly did not have teeth but, undeterred, numerous artists and model makers have equipped this pterosaur with an impressive and very fierce-looking set of dentures. Happily, fact has now caught up with fiction after the description, by the pterosaurologist Dino Frey and colleagues in 2003, of a Lower Cretaceous pterosaur that looks like *Pteranodon*, even down to the large crest sticking up from the back of the skull and, no doubt gladdening the hearts of prehistoric model makers everywhere, jaws that positively bristle with large, sharp-pointed, highly dangerous-looking teeth. A photograph of this pterosaur, the aptly named *Ludodactylus*, meaning "play pterosaur," alluding to the plastic toys that predated the discovery of the real thing, can be seen in *Figure 3.3*.

6. While most pterosaur fossils consist only of hard parts, there are a few rare examples, detailed in Chapters 5 and 6, of fossilized soft tissues and, as related in Chapter 9, lots and lots of pterosaur tracks.

7. See, for example, Stephen J. Gould on *The Mismeasure of Man*, published in 1981, and Richard Dawkins on "Progress" in *Keywords in Evolutionary Biology*, edited by Evelyn Fox Keller and Elisabeth A. Lloyd (1992).

8. Studies by Larry Witmer and colleagues (2003) on the brains of pterosaurs and our latest ideas on the construction of their flight membranes (see Unwin, 2003) suggest that these prehistoric fliers had clever wings that constantly monitored flight conditions and that were capable of responding rapidly to air movement by changing their shape. More details of these extraordinary structures can be found in Chapter 8.

9. "CAT scanning" stands for Computerized Axial Tomography scanning, and it is essentially a way of using a computer to construct a three-dimensional picture of both the inside and outside of a structure using X-rays.

10. Already an eminent scientist by the turn of the 19th century and a leading light of the newly founded Museum d'Histoire Naturelle in Paris, Georges Cuvier was destined to become one of the world's most famous and highly decorated naturalists. He was the first to recognize that pterosaurs were flying reptiles, an idea that was published in 1801, or year 9, as it was then, according to the French Revolutionary Calendar.

11. As will be explained in more detail in Chapter 6, parts of the head and body of pterosaurs were covered with short, fine "hair"—a development unique to these animals and not related in any way to mammal hair.

Chapter 2: Pterosaur Planet
Pages 16–29

1. A good nontechnical description of the Mesozoic world can be found in *Dinosaurs: A Global View* by Sylvia J. Czerkas and Stephen A. Czerkas (1991).

2. Although it actually looks more complicated, cricket, the quintessential English sport, is in fact simpler than quantum physics. Played between two teams of 11 and involving, among other things, the yorker, googlies, silly-mid-on and cake, games can last five days and still end in a tie.

Try BBC Radio 4, Testmatch Special (www.news.bbc.co.uk/sport1/hi/cricket/ test_match_special) for a truly surreal experience.

3. While 20 years per generation is reasonable for today, this period was certainly shorter in the past.

4. The Albian Stage, for example, which occurred at the end of the Early Cretaceous (see *Figure 2.2*) is based on a thick sequence of rocks that crop out in the Albe (or Aube) region of north-eastern France.

5. The proportion of daughter elements to the original undecayed element preserved in, for example, volcanic rocks, can be used to determine the absolute age of the enclosing rock by comparing it with rates of decay measured in the laboratory. It is this approach that sets the end of the Cretaceous at 65 million years ago, give or take a half-million years or so. One problem with isotope dating is that it requires material from volcanic rocks, whereas fossils are normally preserved in sedimentary deposits.

6. Geological stages are split into even finer divisions called biozones, often only a million years or so long, and defined using fossils such as ammonites. As yet, relatively few pterosaur remains can be dated so precisely.

7. Earth's magnetic field, which points in different directions in different latitudes, is "frozen" into certain minerals borne by some rocks, such as lavas, as they cool down and become solid. By determining the age of such rocks using absolute radiometric dating (see note 4) it is possible to establish the original position and orientation of these rocks with regard to other rocks of the same age, and from this to reconstruct the disposition of the continents.

8. The same sediments that produced *Sordes pilosus* (see *Figure 2.1*) also yielded thousands of fossil insects that represent many of the 29 major groups that live alongside us today (see *The History of Insects*, edited by Alexander Rasnitsyn and Donald Quicke, 2001). Indeed, expeditions to Karatau were primarily interested in collecting insects, and the discovery of *Sordes* was essentially a lucky byproduct of this activity.

9. For a recently published general account of these marine monsters, see Richard Ellis' *Sea Dragons: Predators of the Prehistoric Oceans* (2003).

10. A general overview of crocodilians, including the rather strange, possibly vegetarian, notosuchians, can be found in an account by Darren Naish, published in *Geology Today* (2001).

11. A good up-to-date account of mammalian history in the Mesozoic can be found in the recently published third edition of Michael Benton's *Vertebrate Paleontology* (2004).

12. For a brief account of this remarkable discovery, see Hu Yaoming *et al.* (2005).

13. There is a plethora of books on dinosaurs, among which *The Complete Dinosaur*, edited by James Farlow and Michael Brett-Surman, stands out. *The Dinosauria*, edited by David Weishampel, *et al.*, now out in a second edition (2004), is the bible.

14. In 2004, Eric Buffetaut and colleagues reported in the science journal *Nature* on a rather astonishing find: The neck vertebra of a pterosaur pierced right through by the tooth of a large theropod. In another case, mentioned by Philip Currie and Aase Jacobsen (1995), a theropod tooth was found still embedded in the lower leg bone of a pterosaur. Both these fossils could be interpreted as evidence of theropods catching and killing pterosaurs, although it is also possible that the teeth were left behind by theropods scavenging pterosaurs that were already dead.

15. Despite being the first Mesozoic bird to come to light, more than 150 years ago, *Archaeopteryx* is still the oldest known bird. Other contenders for this title include *Protoavis* from the Triassic of Texas (a mish-mash of several reptiles) and *Praeornis*, a feather from the Upper Jurassic of Kazakhstan that, in light of recent discoveries in China (see note 16) may have belonged to a theropod dinosaur.

16. Further details of the spectacular new feathered dinosaurs from the Lower Cretaceous of China, many of which are featured in large color photographs, can be found in *Fossil Treasures from Liaoning* by Wu Qicheng, published in 2002, and in *The Jehol Biota*, published in 2003 and edited by Chang Mee-Mang and colleagues. In *Unearthing the Dragon* (2005), Mark Norell gives an exciting and racy account of the discovery and significance of the Chinese feathered dinosaurs.

17. In a superbly illustrated article in the March 2003 issue of *Scientific American*, Richard Prum and Alan Brush report on the striking similarities between the origin and evolution of feathers as evident, on the one hand, from the fossil record and on the other as deduced from studies of the earliest stages of feather development in chickens.

18. A detailed account of the evolutionary history of birds in the Mesozoic can be found in *Mesozoic Birds: Above the Heads of Dinosaurs*, edited by Luis Chiappe and Larry Witmer (2002).

19. Cretaceous seas were inhabited by a family of large, flightless diving birds, named the hesperornithiforms after the most famous and important member of the clan, *Hesperornis*. Although completely unrelated to modern divers such as loons, *Hesperornis* and its relatives, some of which reached 2 meters (6 feet) or more in length, had a very similar lifestyle to these birds, hunting for fish or squid by propelling themselves along underwater using their powerful hind limbs and large webbed feet. See note 18 for further details of these and other Mesozoic birds.

Chapter 3: Considering Medusa
Pages 30–55

1. This "story" is based on a fossil of *Pterodactylus* collected from the Upper Jurassic Solnhofen Limestones of Bavaria (see note 4) and now found in the collections of the Naturmuseum Senckenberg in Frankfurt-am-Rhein, Germany. The special feature of this fossil is that it has a broken wing-finger, suggesting that it died in a violent event, such as a storm.

2. Actually, it's a bacteria-run planet, but mammals are better at public relations.

3. The wing-finger (equivalent to our fourth or ring finger) was massively enlarged in pterosaurs and supported the outer part of the wing (see also Chapters 5 and 7). See note 1 for further details of the broken-wing *Pterodactylus*.

4. *Solnhofen: A Study in Mesozoic Palaeontology* by K. Werner Barthel, Nicola Swinburne and Simon Conway Morris (1990), provides an excellent introduction to this extraordinary deposit, its world-famous fossil biota and the processes that led to its formation. This work is complemented by a superb collection of photos in *Die Fossilien von Solnhofen*, published by Karl Frickinger in two volumes, the first in 1994 and the second in 1999.

5. Helmut Tischlinger, in conversation with the author, October 2004.

6. See Chapter 1, note 5.

7. A pouch-like throat sac similar to that found in pelicans is preserved in several fish-eating ptero-saurs (see Chapter 5) and was probably also present in *Ludodactylus*.

8. The El Niño phenomenon (or ENSO, El Niño Southern Oscillation, as it is more properly known) has a huge impact on bird populations. The 1983 event, for instance, is blamed for kill-ing more than 11 million seabirds on Christmas Island. Further information can be found in Barber and Chavez (1983) and at the following Website: www.pices.int/publications/scientific_reports/Report10/default.aspx.

9. Details of the Zhejiang locality can be found in Cai Zhenquan and Wei Feng (1994) and David Unwin and Lü Junchang (1997).

10. Pterosaur carcasses carried a cargo of bacteria with them that, unless halted by burial and the processes of fossilization, slowly but surely destroyed their hosts' soft tissues.

11. This extraordinary fossil, which consists of a small piece of wing membrane attached to bones of the forelimb, was first described by David Martill and David Unwin (1989) and later discussed by Alex Kellner (1996).

12. David Martill, a paleontologist based at Portsmouth University in England, once suggested to me that this extraordinarily high concentration of phosphate, which is rather hard to explain, may have resulted from a very large pile of guano (for which local pterosaurs may well have been responsible) falling or sliding into the lagoon. Despite much field work in the Araripe area, he has yet to discover any serious evidence in favor of this incredibly attractive hypothesis.

13. The "Medusa Effect" was first written about by David Martill in *Geology Today* in 1989.

14. This fossil, the only known example of *Austriadactylus*, first described by Fabio Dalla Vecchia and colleagues in 2000, now resides in the collections of the Staatliches Museum für Naturkunde in Stuttgart, Germany.

15. Crinoids, marine organisms related to sea lilies, are still alive today, but are much rarer than they were in the past.

16. The *Posidonia* shales, named after a shell that is commonly found in these beds (by a taxonomic quirk the shell is now called *Bositra*) and quarried in the region of Holzmaden in southern Ger-many, are famous for yielding thousands of superbly preserved fossils (some with fossilized soft parts) of animals such as ichthyosaurs, plesiosaurs, crocodiles, fish, ammonites, clams, ancient squid, sea lilies and pterosaurs. The *Posidonia* shales originally formed in the Early Jurassic about 190 million years ago at the bottom of a warm, shallow sea that covered much of what is now Western Europe. A concise account of this famous fossil-yielding rock sequence can be found in the section by Rupert Wild on Holzmaden in the first edition of *Palaeobiology* (1993).

17. In the 1850s, strip mining for phosphate for use in the agricultural and chemical industries, began in the area around Cambridge, England (see, for example, *The Cambridgeshire Coprolite Mining Rush*, published by Richard Grove in 1976). The diggers were after a layer called the Cambridge Greensand (the green coming from crystals of the mineral glauconite), because this was the source of the phosphate, in the form of lumpy phosphatic nodules, widely, but quite wrongly thought to be coprolites—fossilized dung. Operating on the principle that "where there's muck, there's brass," much of the area around Cambridge and to the south was complete-ly excavated in the search for "coprolites." Along with the phosphate nodules, the diggers also

found many thousands of fossils, mostly ammonites, bivalves and belemnites (relatives of squid), but among them were rarities, including the bones of marine reptiles, dinosaurs, turtles, an early flightless diving bird and pterosaurs. Such items fetched a few pennies from the dons, and before long a huge collection had been built up in the Woodwardian (now Sedgwick) Museum of Cambridge University.

18. Good overviews of this deposit and the closely related Crato Limestone Formation can be found in *Santana Fossils: An Illustrated Atlas* by John Maisey (1991) and in *Field Guides to Fossils: Fossils of the Santana and Crato Formations, Brazil* by David Martill (1993).

19. The Jehol Group consists of a thick sequence of sediments that were deposited between about 139 and 124 million years ago and is divided into the Yixian Formation, beneath, and the Jiufotang Formation, on top. These rocks have yielded many tens of thousands of fossils, including pterosaurs, birds and feathered dinosaurs, that form the Jehol Biota. See also Chapter 2, note 15. According to the latest assessments of Chinese paleontologists, more than 100 pterosaur fossils have been recovered from the Jehol Group.

20. Steve Etches, a plumber by trade, is one of the select few to have found and collected pterosaur fossils. Thirty years of scouring local exposures of the Kimmeridge Clay, a sequence of Upper Jurassic rocks that mainly crop out along the south coast of England, have yielded thousands of fossilized remains of the animals and plants that inhabited the seas in which the Kimmeridge Clay settled out. Among Steve's treasured finds are a handful of pterosaur bones belonging to at least two different species, and I know that one day he is going to find a complete skeleton.

21. This pterosaur was first described under the name *Dsungaripterus parvus* by Natasha Bakhurina in 1982, but as Bakhurina and David Unwin (1995) later noted, details of the skull show that it belongs to a new and quite distinct genus of dsungaripteroid pterosaur.

22. This example of *Pterodactylus kochi* was first described by the German paleobiologist Othenio Abel in 1925 and, although some of the fossilized soft tissues were lost during preparation, it still has superbly preserved remains of the flight membranes.

23. The so-called Zittel wing was first described by the German paleontologist Karl Alfred von Zittel in 1882, and more recently by Peter Wellnhofer (1985) and Kevin Padian and Jeremy Rayner (1993).

24. A detailed account of how he produces fantastic photographs of fluorescing fossils is given by Helmut Tischlinger in the 2002 issue of the journal *Archaeopteryx*.

25. The CAT-scanning facility run by Tim Rowe at the University of Austin, in Texas, has been tremendously successful at discovering the otherwise invisible internal anatomy of fossils. Several spectacular sequences, including a trip through the skull of *Rhamphorhynchus*, can be found on the facility's "digimorph" Website at http://digimorph.org/. The work specifically referred to here concerns a study by Larry Witmer and colleagues, published in *Nature* in 2003, on well-preserved skulls of *Rhamphorhynchus* and *Anhanguera*. As detailed in Chapter 5, CAT-scanning and digital-imaging techniques enabled these scientists to reconstruct the shape and orientation of the brains of these two pterosaurs in remarkable detail.

26. While still a student, a former colleague of mine, Roland Goldring, who taught paleontology for many years at Reading University in England, had the good luck to be present at the meeting

in Wilhelmshaven and actually witnessed the flight of Holst's *Rhamphorhynchus*. Recounting his memories of the event, he recalled, "Both the flight and the wing beats were remarkably smooth, and the whole thing was most impressive."

27. In their classic 1974 paper, Cherrie Bramwell and George Whitfield reported on their experiments on the head crest of *Pteranodon*, while more recently Matt Wilkinson of Cambridge University, England, has carried out some exciting new work on pterosaur flight by testing different wing models in a wind tunnel.

28. I do most of my writing on a 10-year-old Apple Macintosh LC II. It's slow and it's noisy, but it still goes, and I love it.

29. A good introduction to phylogenetic systematics and how it is applied to fossils can be found in *Systematics and the Fossil Record: Documenting Evolutionary Patterns* by Andrew Smith (1994), but there are many other works in this area, including the great-granddaddy of them all, Willi Hennig's *Phylogenetic Systematics*, the English translation of which was published in 1966.

30. Sometimes, the exact condition of a character in a particular species may be unknown, for example, because that region of the body is not preserved in fossil material. Such "missing data" are indicated by a question mark, which, to paleontologists' chagrin, is often common in data tables based on fossils.

31. A basic assumption of phylogenetic systematics is that the pattern of relationships between species, based on the most harmonious arrangement of the data, requires the least amount of change (i.e., evolution) possible. This presumes that evolution is parsimonious, an assumption that may or may not be justified.

32. The two principal data tables for pterosaurs, devised by David Unwin and Alex Kellner in the 1990s, can both be found in *Evolution and Palaeobiology of Pterosaurs* by Eric Buffetaut and Jean-Michel Mazin (2003).

33. If you wish to gain some idea of what the complete fossil record looks like, and the contribution that pterosaurs make to the whole, *The Fossil Record II*, edited by Michael Benton and published in 1993, is a good place to start.

34. See David Unwin (1997) for the most recent summary of the pterosaur track record.

Chapter 4: A Tree for Pterosaurs
Pages 56–87

1. The evolutionary pathway to flight in pterosaurs almost certainly involved a phase when they took to the air from a high point, such as a tree or a cliff. Probably the hardest step involved the development of flapping, partly because this is a complex activity, but mainly because, from an energy point of view, it is very expensive. One possibility is that pterosaurs got the extra energy they needed by feeding on insects.

2. Species names, derived from Latin or Greek, are made up of two components: one for the genus—the group to which the species belongs, and one for the particular species. Thus, *Homo* is the genus name, while *sapiens* is the species or trivial name. Other species in the genus, for example, *Homo erectus* and *Homo habilis* (both now extinct), have the same genus name, but different species names.

3. The Biological Species Concept was first clearly defined by Ernst Mayr in his seminal 1942 work *Systematics and the Origin of Species,* and in its most basic form states that species are "groups of actually or potentially interbreeding natural populations, which are reproductively isolated from other such groups." There are many other species concepts, but Mayr's remains the most widely known and widely taught.

4. In fact, the vast majority of animal species, living and extinct, are recognized this way.

5. There are at least 200 published species names for pterosaurs but, at the time of writing, there are only about 100 valid species. Lest 100 should sound like a lot, bear in mind that there are about 10,000 species of living birds and at least another 2,000 extinct species in the bird fossil record, which, coincidentally, spans about the same length of time (150 million years) as that of pterosaurs.

6. Naming a new pterosaur, or indeed any animal, living or extinct, is a complicated process and bound by a strict set of rules, now in their fourth edition (see Ride, *et al.,* 1999) and titled the *International Code of Zoological Nomenclature* (www.iczn.org). *Describing Species,* by Judith Winston, published in 2000, is a good introduction to how these rules work, how one actually goes about naming a new species, and how to cope with many other arcane aspects of taxonomy.

7. As the attentive reader will already have noticed, in this book I tend to use just the genus name, for example, *Pteranodon,* rather than the full species name, such as *Pteranodon longiceps.* This is partly for reasons of readability, but mainly because in many cases the genus contains only a single species, which means that the specific and generic names apply to the same thing. Even where the genus contains two or more species, they are usually so similar that, except in a few special cases, it is not necessary to distinguish between them, and the generic name can be used with impunity.

8. The rules and recommendations of the International Code of Zoological Nomenclature (ICZN), mentioned in note 6, specify how to deal with synonymy, homonymy and lots of other potentially embarrassing taxonomic problems that end in -ymy. Like vultures that have spotted a wearying camel, those with a penchant for legalistic minutiae are strongly attracted by the interminable discussions that tend to arise when there are far more names than species. The rest of us, of course, have a life.

9. Chris Bennett's first paper on the taxonomy of *Rhamphorhynchus* appeared in 1995. In it, he proposed that all known individuals of this pterosaur from the Solnhofen Limestone (now totaling more than 200 specimens) belonged to a single species represented by an extensive growth series, rather than to five distinct species, as was previously thought. Further details of this controversial but, in my opinion, convincing idea were published in a second paper in 1996.

10. See Chris Bennett's 1993 paper on *Pteranodon,* the pterosaur upon which his doctoral research, completed in 1991, was primarily based.

11. In 1975, Peter Wellnhofer published a genealogy of rhamphorhynchoid pterosaurs, which depicted several direct relationships, including the descent of *Rhamphorhynchus* from *Dorygnathus.* This was eventually incorporated into a general pterosaur evolutionary tree, which appeared in the pterosaurologists' bible, the *Handbuch der Paläoherpetologie, Part 19, Pterosauria,* published in 1978.

12. The two main studies, one by David Unwin and the other by Alex Kellner, are to be found in the pterosaur volume edited by Eric Buffetaut and Jean-Michel Mazin (2003). The very latest analysis of pterosaur relationships was published by Unwin in 2004.

13. Amniotes, which include reptiles, birds and mammals, are characterized above all else by the presence of specialized membranes in the egg (absent in fish and amphibians) that enclose the embryo and provide it with protection, water and a well-developed blood system, through which gases can be exchanged.

14. The first pterosaur egg, complete with embryo, was reported from the Lower Cretaceous Jehol Group of Liaoning Province, China, in a June 2004 issue of *Nature* by the Chinese paleontologists Wang Xiao-Lin and Zhou Zhonghe. Two further eggs with embryos, one from the Lower Cretaceous of Argentina, described by Luis Chiappe and Laura Codorníu, and a second from the Jehol Group, reported by Ji Qiang and colleagues, appeared in *Nature* just five months later. What an eye-opener: no eggs for 200 years and then three in six months!

15. Thomas von Soemmering published one of the earliest papers on pterosaurs, which appeared in 1812, at a time when only a single specimen was known. Fortunately for Soemmering, this pterosaur, a superb example of *Pterodactylus antiquus*, had fetched up in the natural history collection of the Bavarian Academy of Sciences, of which he was the director. A short but excellent account of Soemmering can be found in Peter Wellnhofer's *Encyclopaedia of Pterosaurs*.

16. Edward Newman's restoration appeared in the first issue of the *Zoologist*, which came out in 1843.

17. Already an eminent scientist by the turn of the 19th century and a leading light of the newly founded Museum d'Histoire Naturelle in Paris, Georges Cuvier was to become one of the world's most famous and highly decorated naturalists. His most important publication, *Recherches sur les Ossemens Fossiles*, first appeared in 1812, with revised editions in 1824, 1825 and 1836.

18. In mammals, the quadrate is no longer involved in the jaw joint, but is reduced to a tiny bone that forms one of the ear ossicles (the incus or anvil). The extraordinary history of the transformation of the reptilian jaw joint into the mammalian condition (which took place at about the same time as pterosaurs were taking to the air), and the fate of the various bones involved, including their incorporation into our ears, can be found in Jim Hopson's chapter on synapsid evolution, which appeared in 1994 in *Major Features of Vertebrate Evolution*, edited by Don Prothero and Robert Schoch.

19. The most comprehensive biography of Harry Govier Seeley, although still lamentably short, appeared in the June 1907 issue of *The Geological Magazine*.

20. Richard Lydekker's review of *Dragons of the Air* appeared in *Nature* in 1901.

21. In a paper published in 1993, Stafford Howse and Andrew Milner showed that the fossil material upon which *Ornithodesmus* was originally based, a series of sacral vertebrae, belonged to a small theropod dinosaur. Following taxonomic practices laid out in the ICZN rules, Howse and colleagues published a new name, *Istiodactylus*, in 2001 for the pterosaur remains that were formerly included under the name *Ornithodesmus*.

22. The traditional approach to classifying amniotes is based on the construction of the skull region behind the eye and, in particular, the presence or absence of perforations, termed temporal fenestrae. Those with no openings at all, for example, are termed anapsids and include groups

such as turtles. Diapsids, as their name reveals, have two temporal openings, while another group, the synapsids, which includes all mammals, living and extinct, and some of their close relatives, have a single temporal opening located in a relatively low position.

23. Some scientists think turtles may also be diapsids, but this idea is still highly controversial.

24. *Scleromochlus*, known from several imprints of its skeleton found in Upper Triassic rocks near Elgin in Scotland, seems to have been a small, lightly built animal about 30 centimeters (10 inches) in length, with a large head, short arms and long hind limbs. Unfortunately, the fossil remains of *Scleromochlus* are not well-preserved, and when I examined this material, I found it almost impossible to distinguish anything other than the coarsest of details.

25. Almost immediately after *Scleromochlus* was first discovered and described at the start of the 20th century, the paleontologist Friedrich von Huene suggested, in a paper published in 1914, that it might be a pterosaur ancestor. Over the years, this idea drifted into obscurity, only to be revived by Kevin Padian in 1984, with a second contribution by him in 1997, in which he proposed that *Scleromochlus* be united with pterosaurs in a group called the Pterosauromorpha. Other paleontologists, such as Michael Benton of Bristol University, England, have argued that even though they are both ornithodirans, pterosaurs and *Scleromochlus* share no special relationship with one another (see Benton, 1999).

26. Chris Bennett proposed in a paper published in 1996 that pterosaurs were not ornithodirans at all, but more closely related to archosauriforms such as *Euparkeria* (see note 25). In the same paper, he showed that almost all the ornithodiran characteristics to be found in pterosaurs are restricted to their hind limbs and, in a later paper that appeared in 1997, argued that these features could have evolved as adaptations for springing into the air from trees or cliffs, rather than being inherited from an ancestral ornithodiran.

27. *Euparkeria* from the Lower Triassic of South Africa was described in detail by Rosalie Ewer in 1965, and further information can be found in papers by Johann Welman (1995), David Gower and Erich Weber (1998) and Phil Senter (2003).

28. General information and references to the main literature on prolacertiforms can be found in papers by Susan Evans (1988), Michael Benton and Jackie Allen (1997), David Dilkes (1998) and Johannes Müller (2003).

29. The main description of *Tanystropheus* appeared in a monographic work published by Rupert Wild in 1973.

30. A brief description, together with a full introduction to the literature on *Sharovipteryx* (previously known as *Podopteryx*), can be found in the chapter by David Unwin and colleagues in *The Age of Dinosaurs in Russia and Mongolia* (2000).

31. I am grateful to Professor Robert Reisz of the University of Toronto, Canada, for keeping me up to date on the latest findings regarding *Sharovipteryx*.

32. Research on the genealogy of diapsid reptiles by David Hone, based in the Department of Earth Sciences of Bristol University in England, indicates that pterosaurs do not share a close relationship with prolacertiforms.

33. In the early 19th century, long, slender, hollow bones that were thought to belong to birds were found in the Lower Jurassic of Dorset, the Middle Jurassic of Oxfordshire, the Lower

Cretaceous of Sussex and the Upper Cretaceous of Kent, all in England. All of them eventually proved to belong to pterosaurs.

34. *The Dinosaur Hunters* by Deborah Cadbury, published in 2000, is a highly readable account of Gideon Mantell and the paleontological milieu of early 19th century Britain.

35. Owen's paper was read before the Geological Society of London on Dec. 17, 1845, and published in volume 2 of the society's *Quarterly Journal* in 1846.

36. Entry for Dec. 17, 1845, in *The Journal of Gideon Mantell*, edited by E. Cecil Curwen and published in 1940.

37. Birds are now known from rocks of Wealden age, most notably in the Jehol Biota of China, so Gideon Mantell could have been right, but, dogged as usual by his bad luck, he was 5,000 miles too far west and 170 years too early.

38. This specimen was eventually described as a bird by the Chinese paleontologist Dong Zhiming in 1993.

39. The term "dimorphodontids" is derived from Dimorphodontidae, the family to which *Dimorphodon* and its relatives, such as *Peteinosaurus*, belong.

40. Details of several different functional aspects of the skull and skeleton of *Dimorphodon* can be found in a 1983 paper by Kevin Padian, professor of paleontology at the University of California, Berkeley.

41. *Peteinosaurus* was first described in a superb and highly influential monograph on Triassic pterosaurs published in 1978 by Rupert Wild. Further details of this pterosaur have been uncovered by Fabio Dalla Vecchia, an Italian pterosaurologist based in Monfalcone, Italy, and appeared in the 2003 pterosaur volume edited by Eric Buffetaut and Jean-Michel Mazin.

42. *Preondactylus* was first described by Rupert Wild in 1993 and further details were published by Fabio Dalla Vecchia in 1998 and 2003. A second incomplete skeleton of *Preondactylus*, thought to be a pellet of bones spat out by a fish after it had caught and partially consumed this pterosaur, was described by Dalla Vecchia and colleagues in 1989.

43. Two specimens of *Batrachognathus*, an adult and a juvenile, were recovered by Russian paleontologists from the Upper Jurassic Karabastau rocks of the Karatau ridge in Kazakhstan (see Ryabinin 1948 and Natasha Bakhurina and David Unwin 1995). The Lower Cretaceous Jehol Group of Liaoning Province, China, has yielded a fairly complete skeleton of the pterosaur *Dendrorhynchoides*, first described by Ji Shu-an and Ji Qiang in 1998 and identified by Unwin and colleagues (2000) as an anurognathid.

44. The extraordinarily well-preserved remains of *Jeholopterus* were first described by Wang Xiao-Lin and colleagues in 2002.

45. All recent studies of pterosaur relationships that have included rhamphorhynchoids, for example, by David Unwin (1995, 2003, 2004) and Alex Kellner (2003), have concluded that anurognathids are not directly related to dimorphodontids, and the apparent similarities between these two clans were probably present in all early pterosaurs and do not indicate any special relationship.

46. As for dimorphodontids, the term "campylognathoidids" is derived from the family name Campylognathoididae.

47. The main descriptions of *Campylognathoides* were published by Felix Plieninger in 1895 and Peter Wellnhofer in 1974. *Campylognathoides* was also reported from India in 1974 by Sohan L. Jain, but this record is unconfirmed.

48. The first finds of *Eudimorphodon* were reported on in detail by Rupert Wild in his 1978 monograph and later, in 1993, he described a well-preserved, although headless, individual with impressions of the flight membranes. Fabio Dalla Vecchia contributed further information on Italian finds of *Eudimorphodon* in papers published in 1995, 2001 and 2003. Multi-cusped teeth, seemingly similar to those of *Eudimorphodon*, have been reported from several European locations (summarized by Dalla Vecchia in his 2003 paper). The fossil from Greenland, discovered by a paleontological expedition led by Farish Jenkins of Harvard University and described in 2001 by Jenkins and colleagues, was also found in Triassic sediments. It is relatively small and seems to represent a very young individual.

49. See Chapter 3, note 13.

50. Gustav von Arthaber (1919), Carl Stieler (1922) and A. Salée (1928) published general descriptions of *Dorygnathus*, and Ferdinand Broili reported in 1939 on a specimen in the collections of the Bayerische Staatssammlung für Paläontologie und Geologie in Munich, Germany, that has some patches of fossilized soft tissues.

51. If pterosaurs had used superlatives, then surely the degree of beakiness would have been one of them.

52. Helmut Tischlinger from Stammham in Bavaria, Germany, kindly provided me with this and much other unpublished information regarding the Solnhofen pterosaurs.

53. The single most important description of the skeletal anatomy of *Rhamphorhynchus* was published by Peter Wellnhofer in his landmark monograph of 1975. This work contains a comprehensive listing of earlier studies.

54. Again, Peter Wellnhofer's 1975 monograph (see previous note) is the principal reference, but there have been several important studies published more recently. These include a detailed account of the Zittel wing by Kevin Padian and Rayner in 1993, several beautifully illustrated papers on the Dark-Wing *Rhamphorhynchus* by Helmust Tischlinger and Dino Frey in 2001 and 2002 and by Dino Frey and colleagues in 2003, and a most impressive account of the external morphology of the brain, based on a CAT scanning analysis by Larry Witmer and colleagues, published in *Nature* in 2003.

55. The Cerro Cóndor site hit the headlines in 2002, when it produced the first South American Jurassic mammal, described by Oliver Rauhut and colleagues in a report in *Nature* that is most notable for containing the first mention of pterosaurs from this site. A paper documenting the exciting new pterosaur finds, which appear to belong to at least two quite different rhamphorhynchoids, is now in preparation.

56. The scaphognathine *Harpactognathus* was described by Ken Carpenter and colleagues in 2003 in the pterosaur volume edited by Eric Buffetaut and Jean-Michel Mazin.

57. The description of *Pterorhynchus* by Stephen Czerkas and Ji Qiang in 2002 formed a chapter in the volume *Feathered Dinosaurs and the Origin of Flight*, edited by Sylvia Czerkas.

58. First described by Andrei Sharov in 1971, further details of *Sordes* appeared in a *Nature* article by David Unwin and Natasha Bakhurina in 1994 and in a review by Bakhurina and Unwin in 1995.

59. *Scaphognathus* was one of the first pterosaurs to be found and studied, its description forming part of a monograph that was published in 1831 by Georg Goldfuss, a professor at the University of Bonn. Goldfuss thought he could see evidence of soft tissues, including fine hair-like structures, an idea that was dismissed by later researchers, but, as Helmut Tischlinger recently showed using ultraviolet light studies, Goldfuss was quite correct. A second juvenile individual of *Scaphognathus* was briefly described by Peter Wellnhofer in his 1975 monograph, and a third near-adult individual is now under study by Chris Bennett.

60. Dino Frey and colleagues proposed, in a contribution to the pterosaur volume published in 2003, that the construction of the shoulder girdle in ornithocheiroids corresponds to a "top-decker" type of aircraft. In this type of construction, the wings are located well above the center of gravity, conferring greater stability but less maneuverability than in middle-decker or bottom-decker designs.

61. *Istiodactylus* was first described by the Rev. Walter Hooley in 1913 (under the name of *Ornithodesmus*—see note 20) and further details were published by Stafford Howse and Andrew Milner in 1993 and Howse and colleagues in 2001.

62. David Unwin's 2001 paper on the Cambridge Greensand pterosaurs contains the most comprehensive modern review of the remains of *Ornithocheirus* from this deposit. A superbly preserved complete skull of *Ornithocheirus* from the Santana Formation of Brazil was described by Peter Wellnhofer in 1987, under the name of *Tropeognathus mesembrinus*.

63. Among the Cambridge Greensand ornithocheirids, there are certainly two valid forms: *Ornithocheirus* and *Coloborhynchus*. A third form, *Anhanguera*, has also been identified, but is quite probably based on fossil material that actually belongs to *Coloborhynchus*. *Ornithocheirus* and *Coloborhynchus* are also to be found in the Santana Formation and are probably senior synonyms (that is, older names that take priority over more recently proposed names) for many of the ornithocheirids that have been described over the last 40 years.

64. As Michael Fastnacht of the University of Mainz in Germany showed in his paper of 2001, *Coloborhynchus* can easily be identified by the highly distinctive blunt-ended shape of the jaw tips and the size and arrangement of the teeth. This pterosaur has also been reported from North America (Yuong-Nam Lee, 1994), Europe (Richard Owen, 1874), Africa (Bryn Mader and Alex Kellner, 1999) and Mongolia (Natasha Bakhurina and David Unwin, 1995).

65. *Haopterus*, first described by Wang Xiao-Lin and Lü Jungchang in 2001, was initially thought to be a member of the Pterodactylidae, but details of its dentition give it away as an ornithocheirid. Another large ornithocheirid from the same region, *Liaoningopterus*, was recently published by Wang Xiao-Lin and Zhou Zhonghe (2003) while a third, small ornithocheirid, *Boreopterus*, also from the Jehol Biota, has just been described in 2005 by Lü Jungchang and Ji Qiang.

66. See note 13.

67. *Pteranodon* is one of the best known and best understood of all pterosaurs, after a series of milestone papers by Chris Bennett on its anatomy (1987 and 2001), sexual dimorphism (1992), growth (1993) and taxonomy (1994).

68. A paper published in 2003 by Chris Bennett details the extraordinary antler-like crests of *Nyctosaurus* and also lists all the pertinent references to this pterosaur.

69. The principal description of *Pterodactylus* is to be found in Peter Wellnhofer's 1970 monograph, and several important exceptionally well-preserved specimens, described by Dino Frey and David Martill (1998) and Frey and Helmut Tischlinger (2000), have come to light in the last few years. Frustratingly, and despite several recent attempts by Chris Bennett (1996, 2002) and Stéphane Jouve (2004) to resolve this problem, the exact number and identity of species of *Pterodactylus* is still unclear.

70. That the flamingo pterosaur *Pterodaustro* has true teeth, rather than bristle-like structures made out of keratin, was shown in a brief note in *Nature* by Luis Chiappe and Anusuya Chinsamy in 1996.

71. Accounts of the fossil site of Loma del *Pterodaustro* can be found in papers by Luis Chiappe and colleagues published in 1998. Chiappe and colleagues also presented a revised account of the skull anatomy of *Pterodaustro* in 2000, and Laura Codorniú and Chiappe described two hatchlings of *Pterodaustro* in 2004. The egg with embryo was described by Chiappe and colleages in a November 2004 issue of *Nature*.

72. The main accounts of Solnhofen germanodactylids can be found in papers by Felix Plieninger (1901) and Peter Wellnhofer (1970). Other records are described by Eric Buffetaut and colleagues (1998), David Unwin (1988) and David Unwin and Wolf-Dieter Heinrich (1999).

73. *Dsungaripterus* was detailed by Young Chung Chien (Yang Zhong-jian) in two papers (1964 and 1973). The Tatal pterosaur was first described by Natasha Bakhurina in 1982 and further details can be found in the 1995 review by Bakhurina and David Unwin.

74. Good accounts of the skull anatomy of *Tapejara* can be found in papers by Peter Wellnhofer and Alex Kellner (1991) and Dino Frey and colleagues (2003).

75. The Lower Cretaceous Jehol Group of Liaoning Province, China, has yielded several tapejarids, including *Chaoyangopterus*, described by Wang Xiao-Lin and Zhou Zhonghe in 2003; *Jidapterus*, described by Dong Zhiming and colleagues in 2003; and two species of *Sinopterus*, the first described by Wang Xiao-Lin and Zhou Zhonghe in 2003 and the second by Li Jianjun and colleagues in 2003.

76. Two species of *Tupuxuara* have been named, both by Alex Kellner and Diogenes Campos (in 1988 and 1994) and both based on skull fragments. A complete skull, featured in a paper in *Science* and published in 2002 by the same authors, was given the name *Thalassodromeus sethi*.

77. The two principal papers on *Zhejiangopterus* were published by Cai Zhenquan and Wei Feng in 1994 and by David Unwin and Lü Jungchang in 1997.

78. Still not fully described, even though it was found in the mid 1970s, some details of *Quetzalcoatlus* can be found in Douglas Lawson's original 1975 paper, in a general account by Wann Langston that appeared in the February 1981 issue of *Scientific American* and in a paper on the skull of this pterosaur published in 1996 by Alex Kellner and Langston.

79. Descriptions of two new azhdarchids and details of other members of this family can be found in two papers, one by Eric Buffetaut and colleagues and the second by Pereda-Suberbiola and colleagues in the 2003 pterosaur volume edited by Buffetaut and Jean-Michel Mazin.

Chapter 5: The Head Inside Out
Pages 88–109

1. The display behavior described here is highly speculative but, as the last section of this chapter relates, there is evidence, direct and indirect, that supports its main themes.

2. Those of us brought up in Britain in the 1960s learned from the classic children's program, "Blue Peter" (www.bbc.co.uk), that the primary requirement for making almost anything is sticky-back plastic.

3. This approach, in which knowledge of the soft-tissue anatomy of living animals is used to infer the likely anatomy of extinct forms, dates back more than 200 years and was first used by the great-grandfather of comparative anatomy and vertebrate paleontology, Georges Cuvier. Originally referred to as uniformitarianism, the modern version of this technique, called the Extant Phylogenetic Bracket, has been enthusiastically promoted by Larry Witmer, who made use of it, for example, to pinpoint the position of the nostrils of *Tyrannosaurus* and other dinosaurs (see Witmer, 2001).

4. A horny covering of the snout, referred to in birds as the rhamphotheca, has been reported in *Rhamphorhynchus* by Peter Wellnhofer in 1975, *Tapejara* by David Martill and Dino Frey in 1998, and *Pterodactylus* by Frey and Helmut Tischlinger in 2000.

5. Known as the antorbital fenestra, the possible functions of this opening in pterosaurs and other diapsids—most probably as housing for an air sac connected to the lungs that served to lighten the skull—have been discussed in detail by Larry Witmer (1997).

6. This is the anapsid condition found in early reptiles and some living groups, such as turtles.

7. The 12 cranial nerves are the bane of medical students' lives. Numerous mnemonics exist to prompt the memory; few are printable.

8. According to a recent analysis of the jaw musculature in *Pteranodon*, published by Chris Bennett in 2001, the main jaw-closing muscle, the mandible adductor, originated from the cheek region and, together with additional muscles, was part of the pterygoideus muscle complex that originated from the back of the palate, inserted onto each mandible just in front of the jaw hinge. These muscles closed the jaw. It was opened by another muscle, the mandible depressor, which originated from the outer edge of the back of the skull and inserted on the mandible just behind the jaw articulation. More or less the same muscles were probably to be found in other pterosaurs, although their relative sizes and strengths are likely to have differed considerably.

9. Fossil evidence of throat sacs has been found in *Rhamphorhynchus* and in *Pterodactylus*.

10. *Istiodactylus*, from the Lower Cretaceous of the Isle of Wight, has narrow, blade-like teeth with sharp edges front and back. Tightly packed in a narrow arcade, they could deliver a cookie-cutter bite capable of slicing off chunks from their prey (see Stafford Howse, *et al.*, 2001).

11. The best example of stomach contents in a pterosaur is to be found in a specimen of *Rhamphorhynchus* described by Peter Wellnhofer in 1975. This "greedy guts" appears to have swallowed a fish that was so large that it may well have led to its downfall.

12. A few species of birds feed by skimming, notably the skimmer *Rhynchops*, which also has an elongated knife blade-like prow on the lower jaw. Accidents do happen, and there are reports of skimmers cartwheeling over and eventually into the water after hitting large, slightly submerged objects.

13. A good introduction to nightjars, their anatomy, and behavior can be found in *Nightjars: A Guide to Nightjars and Related Nightbirds* by Nigel Cleere, published by Pica Press (1998).

14. The way azhdarchids fed has been much debated, with suggestions ranging from vulture-like scavenging to stork-like wading, although fishing from the air seems most likely (see David Martill, 1998).

15. Further discussion of the jaws of *Tapejara* and the possible diet of this pterosaur can be found in Peter Wellnhofer and Alex Kellner's 1991 paper.

16. Casts of the brain are preserved in several different pterosaurs, including *Dorygnathus* from the Lower Jurassic, *Rhamphorhynchus* and *Pterodactylus* from the Upper Jurassic, *Tapejara*, *Anhanguera* and *Ornithocheirus* from the Lower Cretaceous and *Pteranodon* and an azhdarchid from the Upper Cretaceous.

17. See papers by Larry Witmer, *et al.*, and David Unwin in the October 30, 2003 issue of *Nature*.

18. See, for example, papers by Tilly Edinger (1927, 1941) and Jim Hopson (1979).

19. In some well-preserved pterosaur skulls, a small opening low in the eye socket indicates the point where the optic nerve entered the orbit.

20. See, for example, Kenneth Kardong (1995).

21. The relationship of the semicircular canals to head posture in pigeons and further references to this phenomenon can be found in the paper by Erichsen and colleagues published in 1989.

22. Larry Witmer and colleagues have argued on the basis of their analysis of pterosaur skulls and brains that such an orientation would have permitted some stereoscopic vision in *Anhanguera*, highly important for judging distance, especially when attempting to snatch something up from the surface of the water while flying at speed.

23. Interestingly, *Archaeopteryx*, the oldest known bird (from the Upper Jurassic Solnhofen Limestones of Germany) had a brain mass-to-body mass ratio that was about the same as that for pterosaurs.

24. See Chapter 3, note 13.

25. See Chapter 4, note 54. One possibility with a crest made of skin is that its color could be altered by flooding capillaries in the skin with blood.

26. A head crest was tentatively identified by Karl Wanderer in 1908 in a well-preserved example of *Rhamphorhynchus*, but has yet to be verified. Excitingly, ultraviolet light studies have recently led to the discovery of soft tissue crests, again almost certainly constructed from skin, in pterosaurs such as *Pterodactylus*, which otherwise lack any evidence of bony crests.

27. The extraordinary crests of *Tapejara* have been described by Diogenes Campos and Alex Kellner (1997) and David Martill and Dino Frey (1998).

28. The idea that pterosaur crests may have been used as sails was suggested recently by Dino Frey, Helmut Tischlinger and colleagues in a chapter of the 2003 pterosaur volume, edited by Eric Buffetaut and Jean-Michel Mazin.

29. This idea was proposed by Peter Wellnhofer in 1988 for *Ornithocheirus mesembrinus*.

30. This proposal was put forward by Cherrie Bramwell and George Whitfield in 1974 in their paper on the biomechanics of *Pteranodon*.

31. The radiator model for pterosaur head crests was suggested by Alex Kellner and Diogenes Campos (1988).

32. See Chris Bennett (1993).

33. See, for example, the chapter on mating systems and sexual selection in *Herpetology*, edited by E. Harvey Pough and colleagues and published in 1998.

34. Charles Darwin, 1871, *The Descent of Man and Selection in Relation to Sex.*

35. Strictly speaking a "lek" is defined as an aggregation of displaying males that females attend primarily for the purpose of mating. For more on this type of behavior, see *Leks* by Jacob Hoglund and Rauno Alatalo, published by Princeton University Press (1995).

Chapter 6: The Body Inside Out
Pages 110–139

1. There is good fossil evidence for "hair" in pterosaurs, as recounted later in this chapter, and a mane is preserved in both *Sordes* and *Pterodactylus*. Whether pterosaur hair varied in color is not known, so I played it safe here by sticking to a monochromatic scheme.

2. Pterosaur evolution is, in some respects, a history of reduction, unification and loss. Among the main events were loss of the teeth (in pteranodontians and azhdarchoids), fusion or loss of the neck ribs (in anurognathids and pterodactyloids), loss of tail vertebrae (in anurognathids and pterodactyloids), reduction of metacarpals one to three (in large pterodactyloids), loss of an intermediate bone in the third finger of the hand (in anurognathids), loss of fingers one to three in the hand (*Nyctosaurus*), loss of the final bone of the wing-finger (in at least one anurognathid and *Nyctosaurus*), reduction of the fibula (many different groups) and loss of the fifth toe in the foot (most pterodactyloids).

3. One reason for assuming a correlation between lack of variation in the skeleton and lack of variation in soft-tissue structures concerns the nervous system. The main nerves that fed into the nerve networks and monitored and controlled the major organs were connected to the spinal cord via gaps between individual vertebrae. Because the number and arrangement of vertebrae between the neck and the base of the tail is highly conservative in pterosaurs, it can be assumed that the pattern of enervation, together with the organs and tissues that it served, showed a similar degree of conservatism.

4. Each vertebra was built on the same basic plan. A spool-shaped body called the centrum formed the base and articulated with its neighbors via a dish-shaped facet at the front and a matching ball-shaped facet at the back, the "proceolous" condition, literally meaning "with the dish at the front." On top of the centrum sat the neural arch, pierced along its length by the neural canal, which transmitted the spinal cord, and drawn out, front and back, into paired processes (the zygapophyses) that articulated with those of the vertebrae in front and behind. The top of the neural arch usually rose up into a large, blade-like process (the neural spine) and, in many vertebrae, flat, wing-like processes stuck out on either side of the neural arch and acted as attachment points for the ribs.

5. The presence of nine cervical vertebrae in all known pterosaurs has been established by Chris Bennett, who reported on it at the Society of Vertebrate Paleontology meeting in Denver in 2004.

6. A detailed description of this superb fossil was published in 2000 by Alex Kellner and Yuki Tomida.

7. Uncrushed, undistorted neck vertebrae are rare and only known in a few pterosaurs, including *Anhanguera*, *Arambourgiania*, *Coloborhynchus*, *Tupuxuara* and *Azhdarcho*.

8. Best match is defined as the position in which the articular surfaces of two bones that form a joint achieve the greatest degree of contact.

9. This specimen, held in the collections of the seminary (Philosophisch-Theologische Hochschule) in Eichstätt, Germany, was first described by Peter Wellnhofer in 1970.

10. Gastralia, or belly ribs, are found in many lizards and crocodiles, where their function is to support the abdomen. Further details can be found in *Vertebrates: Comparative Anatomy, Function, Evolution* by Kenneth Kardong, published in 1995.

11. In some pterosaurs, the last one or two dorsal vertebrae seem to have lacked ribs and have occasionally been referred to as lumbar vertebrae.

12. A notarium is also found in some birds, but, unlike in pterosaurs, there is no direct bone-to-bone contact with the shoulder girdle.

13. This remarkable example of *Rhamphorhynchus*, "Mr. Greedy Guts," is part of the Karl Strobl collection, but can be seen in the Jura Museum in Eichstätt, Germany. It was first described and illustrated by Peter Wellnhofer in 1975 in his monograph on the rhamphorhynchoids from the Solnhofen Limestone.

14. A good account of skeletal pneumatization in birds, and the literature on this subject, can be found in a paper by Pat O'Connor, published in 2004 in the *Journal of Morphology*.

15. The most recent descriptions and discussions of pneumatization in pterosaurs can be found in papers by Chris Bennett on *Pteranodon* (2001) and Nils Bonde and Per Christiansen (2003) on *Rhamphorhynchus*.

16. Nils Bonde and Per Christiansen reached a similar conclusion in their 2003 paper, based on a study of pneumatization in the vertebrae of an extraordinarily well-preserved and well-prepared example of *Rhamphorhynchus* in the collections of the Geological Museum in Copenhagen, Denmark.

17. A good introduction to the breathing apparatus of birds can be found in *Birds: Their Structure and Function* by Anthony King and John McLelland, published in 1984.

18. Until recently, it was thought that all long tails were constructed in the same way, but recent finds suggest that the bundles of bony spars were absent in *Eudimorphodon* and perhaps also in *Preondactylus*, although the possibility that they had failed to ossify, and so were simply not preserved, cannot be ruled out.

19. Remains of a tail flap are preserved in the scaphognathines *Sordes* and *Pterorhynchus*. In both cases (see Alexander Sharov, 1971; Stephen Czerkas and Ji Qiang, 2002) the flap has a long, low rhombus shape, rather like the feathers on an arrow.

20. Anurognathids were long believed to have been short-tailed, because the first example to be found, a specimen of *Anurognathus* from the Solnhofen Limestone, appeared to have a stumpy little tail, similar to that of pterodactyloids. Surprisingly, the well-preserved skeleton of a Chinese anurognathid, *Dendrorhynchoides*, first described in 1998 by Ji Shu-an and Ji Qiang appears to have a long tail, which led some pterosaurologists to believe that anurognathids were like other rhamphorhynchoids, after all. According, however, to more recent information, published by David Unwin and colleagues in 2000, it now seems that much of the tail of *Dendrorhynchoides* was faked. Other new discoveries of anurognathids, including a second specimen of *Anurognathus* and another Chinese form, *Jeholopterus*, quite clearly have short tails.

21. Further details of the peculiar tail of *Pteranodon* can be found in Chris Bennett's monograph on this pterosaur, published in 2001.

22. An excellent account of the architecture of the shoulder girdle and its associated musculature in a rhamphorhynchoid (*Campylognathoides*) and a pterodactyloid (*Anhanguera*) was published by Chris Bennett in 2003 in the pterosaur volume edited by Eric Buffetaut and Jean-Michel Mazin.

23. The shoulder joint and its operation in pterosaurs has been the subject of several studies; one of the earliest was reported by Ernest Hankin and David Watson in 1914. Cherrie Bramwell and George Whitfield presented the first detailed analysis of this joint and its function in 1974, and further accounts were published by Kevin Padian in 1983, Chris Bennett in 2001, and Dino Frey, Marie-Celeste Buchy and David Martill in 2003.

24. This new information on the pterosaur sternum appeared in a paper on *Eudimorphodon*, published by Rupert Wild in 1993.

25. Immature pterosaurs show that the first, or proximal, syncarpal was made up of two carpal bones called the radiale and the ulnare, while the second, or distal, syncarpal was made up of three distal carpals.

26. The medial carpal is so named because it lies on what is formally referred to as the medial side of the wrist. Until quite recently, this carpal was known as the lateral carpal, a most inappropriate name that stems from Harry Seeley's complete misunderstanding of pterosaur forelimbs. At an early point in his studies, he got the pterosaur hand backward, confusing fingers one to four as the fourth to the first. He only realized his mistake after visiting Germany, where he was confronted by numerous well-preserved Solnhofen pterosaurs, in which the orientation of the hand was quite clear.

27. During the 19th century, it was widely believed that the pteroid represented the first finger (in humans, the thumb) and, consequently, that the wing-finger was equivalent to the fifth finger. After much debate in the early 20th century, it was concluded that the pteroid was a completely new bone, so the wing-finger was, in fact, the fourth digit, while the fifth had been lost. In a paper, however, published in 1998, Dino Frey and David Martill returned to the 19th century idea that the pteroid was a remnant of the first finger. So far, this proposal has garnered no support, because, starting from the typical reptilian condition, it requires more evolutionary restructuring of the fingers than the currently accepted interpretation, in which the wing-finger corresponds to the fourth digit.

28. Matt Wilkinson of Cambridge University, England, has shown, using scale models tested in wind tunnels, that a pteroid-supported forewing, forming a leading edge flap, would have been highly effective during flight. This idea is discussed further in Chapter 8.

29. Tetrapods, including humans, have a five-finger hand, although, as paleontologists Jennifer Clack and Michael Coates have discovered, early forms may have had as many as eight fingers, echoing our fishy ancestry (see Clack, *Gaining Ground: The Origin and Evolution of Tetrapods*, 2002). This aside, at some point, the ancestors of pterosaurs must have lost the fifth (little) finger.

30. In fingers one to three and toes one to four of pterosaurs, the bone preceding the claw is relatively elongated, a mechanical design that increases the leverage that can be exerted by muscles that attach to the claw. A similar construction is found in animals that live in trees or on rock faces and seems to be an adaptation for climbing.

31. Figures and descriptions of the claw sheathes of pterosaurs have recently been published by Dino Frey, Helmut Tischlinger and colleagues in the 2003 pterosaur volume edited by Eric Buffetaut and Jean-Michel Mazin.

32. The apparent absence of fingers one, two and three in *Nyctosaurus* was first discovered by Chris Bennett, who reported on it at the Society of Vertebrate Paleontology meeting in Mexico in 2000.

33. Curiously, a second specimen of *Anurognathus* also has only three bones in the wing-finger, although other anurognathids seem to have the usual complement of four.

34. The three bones of the pelvis consisted of the blade-like ilium above, the buttress-like pubis below and to the front, and the plate-like ischium below and to the back.

35. The role of the pterosaur prepubis in breathing was discussed in an abstract by Leon Claessens that appeared in 2004.

36. In many reptiles, the head of the thigh bone sits directly on top of the shaft in what is known as a terminal position. The result is that, as one can easily see in lizards and crocodiles, their legs stick out sideways from the body. In birds and dinosaurs, by contrast, the head is turned almost at a right angle and directed inward, which enabled the legs to be brought into a vertical position, directly beneath the body. Pterosaurs, ever unconventional, had a thigh bone with a sloping head that was directed inward and upward, usually at about 45 degrees, although in ornithocheiroids, for example, it was much steeper, reaching about 70 to 80 degrees.

37. The two bones in the first row of tarsals consisted of the astragalus on the outside and the calcaneum on the inside.

38. Toes one to four contained two, three, four and five bones, respectively.

39. Details of the foot webs were published by Dino Frey, Helmut Tischlinger and colleagues in 2003 in a paper on pterosaur soft tissues that appeared in the volume edited by Eric Buffetaut and Jean-Michel Mazin.

40. This specimen, in the collections of the Staatliches Museum für Naturkunde in Karlsruhe, Germany, was described in the paper mentioned in the previous note.

41. See Chapter 3, note 11.

42. The claim by Georg Goldfuss that he could see hair in the original fossil of *Scaphognathus* was viewed with skepticism by his contemporaries, including Herman von Meyer, who presented a detailed summary of what was known about these animals in his epic work, *Der Fauna der Vorwelt*. Even after it was established that pterosaurs were hairy, the presence of hair in the

Goldfuss Scaphognathus continued to be doubted right up until about a year ago, when an ultraviolet light study of the fossil by Helmut Tischlinger (reported in 2003 in an issue of the journal *Globulus*) showed that Goldfuss had been right after all. In a paper that came out in 1908, Karl Wanderer published the first convincing evidence of hair in a specimen of *Rhamphorhynchus* that, although thought to be lost, is still found in the collections of the Museum für Naturkunde in Dresden, Germany. Ferdinand Broili championed the idea of pterosaur hair in the 1920s and 1930s, publishing a series of papers (in 1927, 1938 and 1939) describing pterosaurs from the Solnhofen Limestone and later from the *Posidonia* Shales of Holzmaden in which he was certain that traces of hair could be seen.

43. Hair follicles were first described by Karl Wanderer in his 1908 paper and further reported by Ferdinand Broili in his publications (see previous note). A full description of these intriguing structures was presented by Peter Wellnhofer in his 1975 monograph on Solnhofen rhamphorhynchoids.

44. A summary of the fuzzy and feathered Jehol theropods can be found in *The Jehol Biota*, edited by Chang Mee-Mann and published in 2003.

45. In their paper on *Jeholopterus*, published in 2002, Wang Xiao-Lin and colleagues suggested that pterosaur hair and dinosaur fuzz may be the same thing, an idea that has also been supported by other Chinese paleontologists, such as Ji Qiang and Yuan Chongxi, in their brief comments on a second specimen of *Jeholopterus*, also published in 2002.

46. A detailed treatment of Harry Seeley's arguments in favor of warm-blooded pterosaurs can be found in his final work on pterosaurs, *Dragons of the Air*, published in 1901. Baron von Nopcsa addressed this issue in a 1916 paper titled "Zur Körpertemperatur der Pterosaurier," while Bob Bakker promoted his view of hot-blooded dinosaurs and pterosaurs in several books, including *The Dinosaur Heresies*, which first came out in 1986.

47. Detailed accounts of the microscopic structure of pterosaur bones can be found in papers published in 2000 by Armand de Ricqlès and colleagues, in 2004 by Kevin Padian and colleagues, and, most recently, in a Ph.D. thesis completed by Lorna Steele at the University of Portsmouth, in Portsmouth, U.K., in 2005.

48. See *The Hot-Blooded Insects: Mechanisms and Evolution of Thermoregulation* by Bernd Heinrich, published in 1993, for an introduction to endothermy in insects.

49. The conclusion that pterosaur physiology was essentially like that of birds and bats is intuitively attractive, but could it be wrong? There are alternatives. One possibility is that pterosaurs had metabolic rates that were intermediate between those of reptiles and birds and acquired part of their heat energy from the sun, using their wings rather like solar panels, an option open to pterosaurs, because the wing surfaces were made of living tissue, unlike, for example, the "dead" wings of birds. With a physiology somewhere between endothermy and ectothermy, heterothermy and homeothermy, pterosaurs would not have needed to eat as often as birds and bats, and with their relatively large, efficient wings, they might not have needed to carry out sustained flapping for long periods—quite a different physiological model from that found in birds and bats, but possibly as, or even more, efficient.

Chapter 7: Babes on the Wing
Pages 140–163

1. The scenario presented here is speculative. That said, as detailed in this chapter, there is a rapidly growing body of evidence to show that pterosaurs incubated their eggs in the ground, that they were highly precocial and capable of walking and even flying soon after they hatched, and that they might not have needed much, if any, parental care.

2. A summary of the fossil record of vertebrate embryos and neonates (individuals that have just been born or hatched) can be found in the first chapter of *Reptilian Incubation: Environment, Evolution and Behaviour*, edited by D. Charles Deeming, published in 2004.

3. See the paper in the Dec. 2, 2004, issue of *Nature*, by Luis Chiappe and colleagues.

4. See the paper in the June 10, 2004, issue of *Nature*, by Wang Xiao-Lin and Zhou Zhonghe.

5. The Happy New Year card was sent to me by Wang Xiao-Lin and Zhou Zhonghe. I shall treasure it always.

6. The so-called pterosaur eggs from the Stonesfield Slate, first described by Professor James Buckman (1860) and later by William Carruthers (1871), are now thought to have been laid by turtles.

7. The example of *Pterodactylus kochi*, with a supposed egg preserved in the body cavity, was figured in *Die Fossilien von Solnhofen*, published by Karl Frickinger in 1999. This faint, egg-shaped structure, located toward the front of the body (not where one would expect it to be), is probably an impression of part of the skeleton, possibly the breast bone.

8. See, for example, the chapter by Karl Hirsch, titled "The Fossil Record of Vertebrate Eggs" in *The Palaeobiology of Trace Fossils*, edited by Stephen Donovan and published in 1994.

9. As reported, for example, by Jens Franzen in his chapter on horses in the volume on Messel, edited by Stephan Schaal and Willi Ziegler (1992), there are several individuals of *Propalaeotherium*, a fox terrier-size early horse from the Tertiary fossil locality of Messel, Germany (about 50 million years old), in which young are preserved inside the body cavity of the mother.

10. The bird egg with embryo from the Jehol beds of Liaoning Province, China, was reported in the Oct. 22, 2004, issue of *Science* by Zhou Zhonghe and Zhang Fucheng.

11. See the paper in the Dec. 2, 2004, issue of *Nature* by Ji Qiang and colleagues.

12. See note 3.

13. See the chapter by Michael Thompson and Brian Speake in *Reptilian Incubation: Environment, Evolution and Behaviour*, edited by D. Charles Deeming, published in 2004.

14. See, for example, *The Megapodes* by Darryl Jones and colleagues published by Oxford University Press in 1995.

15. See *Dinosaur Eggs and Babies*, edited by Kenneth Carpenter and colleagues, published in 1994.

16. Full details of these pterosaurs can be found in Peter Wellnhofer's monographs on Solnhofen Limestone pterosaurs, published in 1970 and 1975.

17. See Chapter 4, note 68.

18. Ji Qiang and colleagues in their *Nature* paper of December 2004 argue that the proportions of the second Chinese embryo closely match those of the ctenochasmatid *Beipiaopterus*, also from the Jehol Group of Liaoning Province, China. *Beipiaopterus*, however, is almost certainly the same thing as *Eosipterus*. Whether the embryo also belongs to this pterosaur is not clear.

19. These *Pterodaustro* flaplings were described in a paper published by Laura Codorniú and Luis Chiappe in 2004.

20. Wing-loading estimates for a flapling and an adult of *Pterodactylus kochi*, and for several other pterosaurs, were published by Grant Hazlehurst and Jeremy Rayner in *Paleobiology* in 1992.

21. Mentioned in a paper published by Lev Nesov in 1991.

22. See note 3.

23. See, for example, the paper on the growth of *Tyrannosaurus*, by Greg Erickson and colleagues, in the Aug. 12, 2004, issue of *Nature*.

24. See the paper by Chris Bennett on *Rhamphorhynchus*, published in 1995.

25. See note 13.

26. Parent pterosaurs feeding their young have been the subject of numerous illustrations; one of the best appeared on the cover of the American edition of Peter Wellnhofer's *The Illustrated Encyclopedia of Pterosaurs*, published in 1991.

27. See, for example, *Crocodiles and Alligators of the World*, by David Alderton, published in 1991.

28. The mere fact that flaplings and juveniles could fly considerably improved their chances of becoming fossilized by taking them near, or over, potential fossil traps, such as lakes or lagoons. That the young of birds and bats are usually unable to fly surely played an important role in their extreme rarity in the fossil record.

29. Chris Bennett and Stefan Jouve have argued, convincingly, in papers published in 1996 and 2004, respectively, that fossils assigned to *Pterodactylus elegans* and *Ctenochasma gracile* all belong to the same pterosaur: *Ctenochasma elegans*, which, as a result, is now represented by a row of juveniles (some of near flapling size) and adults.

30. See *Herpetology*, by F. Harvey Pough and colleagues, published in 1998.

31. See, for example, *The Origin and Evolution of Birds*, Second Edition, by Alan Feduccia, published in 1999.

32. This type of developmental pattern is technically referred to as "determinate growth."

33. Technically referred to as "indeterminate growth."

34. Good detailed accounts of features of the skeleton that can be used to distinguish adult from nonadult pterosaurs can be found in several papers by Chris Bennett, published in 1993, 1995 and 1996.

35. It might seem obvious that such a large pterosaur was an adult, but because pterosaurs could, in theory, have grown even larger (perhaps up to 15 or even 20 meters in wingspan), size is no guarantee of adulthood, even in this case.

36. Such examples have been termed "sub-adults" and can be defined, loosely, as pterosaurs that are as large as or larger than the smallest known adult, but whose skeleton still shows evidence of immaturity.

37. Details of *Quetzalcoatlus* can be found in papers by Douglas Lawson (1975) and Wann Langston (1981) and in *The Illustrated Encyclopedia of Pterosaurs* by Peter Wellnhofer (1991). *Hatzegopteryx* was described by Eric Buffetaut and colleagues in two papers (2002 and 2003), and the giant Spanish pterosaur was briefly reported on by Julio Company and colleagues in 2001.

38. The fossil remains of *Quetzalcoatlus*, *Hatzegopteryx* and the giant Spanish pterosaur were all thought, at first, to be the bones of dinosaurs and were only later identified as pterosaurs.

Chapter 8: High Fliers
Pages 164–195

1. *Coloborhynchus*, a large ornithocheirid pterosaur up to 6 meters (20 feet) in wingspan, is known from several well-preserved fossil remains from the Santana Formation of Brazil, a sequence of sediments deposited in a large, almost land-locked, sea, sometime in the late Early Cretaceous. A spectacular fossil, consisting of several articulated neck vertebrae of an ornithocheiroid (possibly *Coloborhynchus*) one of which is almost completely pierced by a theropod tooth, was reported on in the July 1, 2004 issue of *Nature* by Eric Buffetaut and colleagues. The theropod tooth is thought to have belonged to a spinosaur, medium to large predatory dinosaurs, that have also been reported from the Santana Formation.

2. Cosimo Alessandro Collini, superintendent in the late 1700s of the Naturalienkabinett (literally a cabinet of natural objects, a direct forerunner of modern natural history museums) in the Mannheim palace of Karl Theodor, Elector of Palatine, was the first person to describe a pterosaur scientifically, in 1784. He concluded, erroneously, that it was some kind of sea creature—an idea that was also supported by Johann Wagler, a zoologist from Munich, who went even further and published a restoration in 1830, showing how pterosaurs had used their forelimbs as flippers.

3. See Chapter 4, note 17.

4. A good introduction to gliding mammals can be found in *The Biology of Gliding Mammals*, edited by Ross Goldingay and John Scheibe, published in 1999.

5. When air meets a cambered (curved) airfoil or an inclined surface it has to move faster across the upper surface than it does across the lower surface. The result is that a relatively greater pressure is exerted on the lower surface than on the upper surface and the difference between the two is experienced by the wing as lift. Further details on aerodynamics, particularly as it pertains to animal flight, can be found in *A Practical Guide to Vertebrate Mechanics* by Chris McGowan and published in 1999.

6. Paragliders are made of the same materials as parachutes, but, instead of being umbrella-shaped, they take the form of a wing that is inflated by air passing though it. This wing is much more steerable than a parachute and has another advantage—it produces lift. Under the right conditions, for example, locations where air is moving quite quickly and also rising, such as near the crest of a hill or ridge, paragliders can stay airborne for long periods.

7. While this is certainly true of animal gliders, it is not necessarily the case for manmade gliders. Gliding aircraft, hang-gliders and paragliders all descend relative to the air, but they can maintain height or even climb if they are in air (for example, in a thermal) that is rising as fast or faster than they are descending.

8. Ornithopters are machines that fly by flapping their wings. Visit www.ornithopter.org for more information, including accounts of successful flights by man-powered ornithopters (a demonstration is planned for the 2006 Winter Olympics in Turin, Italy) and details of where you can purchase your own ornithopter.

9. Interestingly, there are many examples of small islands that lack any significant predators and that are also inhabited by several kinds of flightless birds. Unfortunately, things often go badly for them if predators do manage to reach their island, the Dodo being perhaps the most infamous example. *The Song of the Dodo. Island Biogeography in an Age of Extinction* by David Quammen (1997) is an excellent introduction to this subject.

10. See Chapter 4.

11. Abel's views on pterosaur flight were laid out in one of his earliest works: *Grundzüge der Palaeobiologie der Wirbeltiere* [Basics of the Palaeobiology of Vertebrates], published in 1912, where they were elegantly summarised in the section heading "Der Drachenflug der Rhamphorhynchiden und der Flatterflug der Pterodactyliden" [The gliding-flight of the rhamphorhynchids and the flapping-flight of the pterodactylids].

12. Harry Seeley made it quite clear in his *Dragons of the Air*, published in 1901, that he believed pterosaurs to have been energetic fliers that flew by flapping their wings. Baron Franz von Nopcsa put forward a similar view in a 1924 paper that was mainly devoted to savaging his arch rival, G. von Arthaber, and his ideas about pterosaurs.

13. The Magnificent Frigate bird, for example, has a wingspan of over seven feet (2.29 metres) while the body weighs only just over 3 pounds (about 1.5 kilos) and less than 10 percent of this, a mere 4 or 5 ounces (100-125 grammes), is accounted for by the skeleton.

14. If you bend a solid column or a tube with a narrow central cavity, such as a ballpoint pen, it will eventually break (fail) by snapping. But, if you bend a thin-walled tube such as a drinking straw, it will fail by collapsing inward, also known as buckling. The long bones of most small to medium sized pterosaurs probably snapped before they buckled (see Chapter 3, note 1), but in large and giant species the walls were so thin that they were at much greater risk of failing by buckling.

15. Cuvier died in 1832 and the first pterosaur with clear evidence of wing membranes, a specimen of *Pterodactylus kochi* from the Solnhofen Limestones, was not found until the 1840s. In fact, traces of wing membrane were also preserved in the first example of *Scaphognathus* described by Goldfuss the year before Cuvier died, but they were not recognized as such until quite recently (see Chapter 6, note 42).

16. The first two major finds of pterosaurs with fossil evidence of wing membranes, examples of *Rhamphorhynchus* from the Solnhofen Limestone, were described in the same year, 1882, by Karl Alfred von Zittel, Professor of Geology and Palaeontology in Munich, and Othniel C. Marsh, Professor of Paleontology at Yale University.

17. A detailed list of pterosaurs with evidence of the wings can be found in a detailed account of pterosaur wings and their preservation by Kevin Padian and Jeremy Rayner that appeared in 1993.

18. There are two main reasons why the wing membranes are poorly preserved in the region of the body. First, the sheer thickness of the body may have prevented the flight membranes from coming into contact with the surrounding sediment and leaving some kind of impression. Second,

the region in and around the body is often heavily prepared to expose features such as the ribs, thus, even if evidence of the membranes had been fossilised it was often removed accidentally during preparation of the specimen.

19. Fossilized evidence of wing membranes in pterosaurs from the Lower Cretaceous of China have been described by Wang Xiao-Lin *et al.*, (2002) and Lü Junchang (2002), from the Crato Limestone of Brazil by Dino Frey, *et al.*, (2003) and in the Dark-Wing *Rhamphorhynchus* by Frey and Helmut Tischlinger (2003) and Frey, *et al.*, (2003).

20. Bats and some gliding mammals have a uropatagium, a term derived by combining the Greek word "oura" meaning tail with the Latin word "patagium," which means the edging on a tunic. The patagium between the hind limbs of pterosaur has also been referred to as the uropatagium, although this is not the most appropriate term because, in pterosaurs, this patagium has nothing to do with the tail. (Chris Bennett argued, in a paper that appeared in 1987, that the cruropatagium did in fact attach to the tail, but despite all the recent discoveries of well preserved fossils, such as the Dark-Wing *Rhamphorhynchus*, there is still no evidence for this idea.) It does, however, connect to the legs, the technical word for which is crus. Consequently, I prefer to use the term "cruropatagium."

21. Back in the mid-1990s, I presented this idea at a workshop on pterosaurs and was somewhat taken aback at the strong reaction it provoked. Everyone in the room, except for me, seemed to be completely certain it was wrong. This, according to the biologist J. B. S. Haldane, writing in the *Journal of Genetics* (volume 58, page 464) in 1963, is the first stage in the acceptance of a scientific idea: It is "worthless nonsense." The second stage is "this is an interesting, but perverse, point of view." Then comes the third stage, "this is true, but quite unimportant," while the final stage is "I always said so." At the time of writing, we seem to have reached stage two.

22. This relationship was discovered by Grant Hazlehurst while carrying out doctoral studies on pterosaur flight at Bristol University in the early 1990s and appears in his Ph.D. dissertation, titled *The Morphometric and Flight Characteristics of the Pterosauria*. Hazlehurst found that in bats, where the arms and legs are also both involved in the wings, there was a similar correlation of the two, but not in birds where the legs are completely separate from the wings.

23. Photographs and drawings of what was quite clearly a cruropatagium were published by Peter Wellnhofer in 1970, in his description of *Pterodactylus*, and by Alexander Sharov in 1971, in his description of *Sordes*. Despite this, authors such as Kevin Padian and Jeremy Rayner continued to write, as late as 1993, that "No incontrovertible evidence for an interfemoral membrane or a uropatagium has been advanced for any pterosaur."

24. Pterodactyloids with cruropatagia like those of *Pterodactylus* include the ctenochasmatoid *Eosipterus* (Lü Junchang, 2003), a tapejarid from the Crato Limestone of Brazil (Dino Frey, *et al.*, 2003) and the azhdarchid *Zhejiangopterus* (David Unwin and Lü Junchang, 1997).

25. See, for example, page 28 of Peter Wellnhofer's *Handbuch der Paläoherpetologie*, published in 1978.

26. This extraordinarily well-preserved piece of wing membrane, which belonged to a medium-size pterosaur from the Lower Cretaceous Santana Formation of Brazil, was first described by David Martill and David Unwin in the July 13, 1989 issue of *Nature*.

27. Detailed accounts of the fibres found in pterosaurs' wing, which have been given the formal name of aktinofibrillen by Peter Wellnhofer in his 1987 paper on the beautiful Vienna specimen

of *Pterodactylus*, can be found in papers by Kevin Padian and Jeremy Rayner (1993), David Unwin and Natasha Bakhurina (1994) and Chris Bennett (2000).

28. This idea was proposed by Colin Pennycuick in his review of pterosaur wings, which appeared in 1988.

29. A comprehensive listing of the fossil evidence for pterosaur wings can be found in the 1993 paper by Kevin Padian and Jeremy Rayner.

30. These details were first published in a paper by David Unwin and Natasha Bakhurina in a paper on *Sordes pilosus* published in the September 1, 1994, issue of *Nature*.

31. See, for example, the 1993 paper by Kevin Padian and Jeremy Rayner.

32. Chris Bennett's proposals regarding the function of pterosaur wing fibres appeared in his paper on pterosaur flight, published in 2000.

33. See Cherrie Bramwell and George Whitfield's classic "Biomechanics of *Pteranodon*" paper from 1974.

34. See Dino Frey and Helmut Tischlinger, 2003, and Frey, *et al.*, 2003.

35. The mattress-lie structure was first described by David Martill and David Unwin in their 1989 paper in *Nature* (see note 27).

36. Striated muscle, found, for example, in your biceps and triceps, usually operates the skeletal system and is under voluntary control. Smooth muscle, by contrast, is almost entirely concerned with visceral functions, such as the digestive tract, blood vessels and lungs. The next time you throw up, despite every effort to do otherwise, blame it on your smooth muscle.

37. This idea was first proposed by Larry Witmer, *et al.*, in a paper on pterosaur brains published in *Nature* in 2003, and by David Unwin in the same issue of *Nature*, in a commentary on the Witmer, *et al.*, paper.

38. A good general introduction to animal flight mechanics can be found in a review paper by Ulla Norberg (2002).

39. A detailed account of the musculature of the shoulder girdle region of two completely different pterosaurs, the long-tailed *Campylognathoides* and the short tailed *Anhanguera* can be found in Chris Bennett's chapter in the 2003 pterosaur volume, edited by Eric Buffetaut and Jean-Michel Mazin.

40. See, for example, *Figure 10.36* in *Vertebrates:Comparative Anatomy, Function, Evolution*, by Kenneth Kardong, published in 1995.

41. As Chris Bennett points out in his account of the shoulder musculature detailed in note 40, there is no evidence that the supracoracoideus muscle of pterosaurs had an attachment site that extended onto the breastbone, nor is the anatomy of the shoulder girdle suited to redirecting the action of this muscle so that it could raise the humerus, as in birds.

42. The relatively restricted range of movement at the shoulder joint in ornithocheiroids was first noted by Cherrie Bramwell and George Whitfeld in their 1974 study of *Pteranodon* and subsequently elaborated on by Grant Hazlehurst and Jeremy Rayner in a paper published in 1992 on an ornithocheirid from the Lower Cretaceous of Brazil.

43. Steven Winkworth's model of *Pteranodon*, complete with big feet, is illustrated on page 39 of Peter Wellnhofer's *The Illustrated Encyclopedia of Pterosaurs* (1991).

44. See Paul MacCready *The Great Pterodactyl Project* (1985).

45. The lift generated by an aerofoil depends on its degree of camber, how fast it is moving through the air and its surface area. Because a pterosaur could not increase its wing area and had to slow down to land, the only way to maintain lift was to increase the degree of camber.

46. Chris Bennett has shown in his analysis of pterosaur hind limbs, published in 1995, that they were well-adapted for leaping.

47. The earliest attempts to understand pterosaur aerodynamics date back to the early 20th century with the pioneering work of Ernest Hankin and David Watson (1914) and Howard Short (1914), but apart from a few isolated studies, by D. von Kripp (1943) and Erich von Holst (1957), little happened until the 1970s. Cherrie Bramwell and George Whitfields' seminal work on the aerodynamics of *Pteranodon*, published in 1974, sparked a series of studies of this and other pterosaurs by W. B. Heptonstall (1971), Ross Stein (1975, 1976) and James Brower (1983), culminating in the work by Grant Hazlehurst (see note 22), some of which was published in the early 1990s (e.g., by Grant Hazlehurst and Jeremy Rayner in 1992). Our understanding of pterosaur wings has changed dramatically in the last decade but, so far, only a single researcher, Matt Wilkinson of Cambridge University who completed a Ph.D. on the aerodynamics of *Anhanguera* in 2003, has attempted to analyze pterosaur wings in the light of these findings.

48. Reptiles have a density of about 0.9 to 1.0 grams per cubic centimeter, which is fairly typical of vertebrates as a whole. Birds, however, may have a density as low as 0.73 grams per cubic centimeter, and pterosaurs may have had a similar density, or perhaps even less, in the case of such heavily pneumatized species as *Pteranodon*.

49. Some of the most reliable mass estimates for pterosaurs are to be found in Ph.D. theses by Grant Hazlehurst, completed in 1991 (see note 22) and Matt Wilkinson, whose thesis *Flight of the Ornithocheirid Pterosaurs* was completed in 2003.

50. Bird-like restorations of wing shape in pterosaurs were proposed by Kevin Padian in papers published in 1983, 1985, and 1987 and in 1993 with Jeremy Rayner.

51. This was established by Grant Hazlehurst in his doctoral research on pterosaur flight and appeared in a paper by him and Jeremy Rayner in 1992.

52. Some details of the extraordinary flight ability of frigate birds is described in a short paper by Henri Weimerskirch and colleagues which appeared in the January 23, 2003, issue of *Nature*.

53. For a detailed account of *Pteranodon*, see Chris Bennett's 2001 monograph on this pterosaur.

54. This conclusion is also supported by the results of aerodynamic analyses of pterosaurs reported on by Grant Hazlehurst and Jeremy Rayner in 1992, and Sankar Chatterjee and Jack Templin in 2004.

55. See, for example, *The Origin and Evolution of Birds*, Second Edition by Alan Feduccia, published in 1999.

56. The problem of how to generate enough power to sustain flapping flight as size increases is well explained in Knut Schmidt-Nielsen's *Scaling. Why Is Animal Size So Important?* (1984). In their

1974 study of *Pteranodon*, Cherrie Bramwell and George Whitfield calculated that a 6.95 metre wingspan individual with an estimated weight of 16.6 kilos would have just about been able to generate enough power for flapping flight, but would not have been able to climb at a steep angle or hover. Since *Quetzalcoatlus* was considerably larger and heavier, but must also have had at least some flapping ability, this suggests that Bramwell and Whitfield's calculations may have been overly pessimistic.

Chapter 9: Grounded
Pages 196–223

1. This story is based on details of the pterosaur track site at Crayssac in southwest France. Detailed accounts of the pterosaur tracks and circumstances under which they were formed can be found in chapters by Jean-Michel Mazin and colleagues and by Jean-Paul Billon-Bruyat and Mazin in the pterosaur volume (2003), edited by Eric Buffetaut and Mazin.

2. The article was eventually published in *Nature* in 1987, under the heading *Pterosaur locomotion: joggers or waddlers?*

3. In Figure 51 of Harry Seeley's *Dragons of the Air*, *Dimorphodon* stands on all fours, while in Figure 52, it is up on its hind limbs.

4. See, for example, three major works by Peter Wellnhofer: the monographs on Solnhofen pterosaurs (1970, 1975) and the pterosaur *Handbuch* (1978).

5. Figure 52 in Cherrie Bramwell and George Whitfeld's *Biomechanics of* Pteranodon (1974).

6. Kevin Padian's ideas were laid out in his Ph.D. Thesis, *Studies of the structure, evolution and flight of Pterosaurs (Reptilia: Pterosauria)*, Yale University; and two papers both published in 1983, one in *Palaeobiology*, the other in *Postilla*.

7. Walking pterosaurs reconstructed with a steeply tilted body can be found in several works, including Harry Seeley's *Dragons of the Air* (1901) and in papers by Chris Bennett on the terrestrial ability of pterosaurs (1990) and the anatomy of *Pteranodon* (2001).

8. According to Kevin Padian's writing in *Palaeobiology* in 1983 (page 229), "one function commonly attributed to pterosaurs was impossible: They could not have walked quadrupedally by moving the forelimbs parasagitally over the ground."

9. The new fossil remains included several superbly preserved skeletons of *Anhanguera*, a large 4- to 5-meter-wingspan ornithocheiroid from the Lower Cretaceous of Brazil, the best example of which is to be found in the paleontological collections of the Museum of Natural History in Tokyo (see Alex Kellner and Yuki Tomida, 2000). Several other museums, including the Coal and Fossil Museum at Iwaki, also in Japan, the American Museum of Natural History in New York, the Bavarian State Museum for Paleontology in Munich, Germany, and the Staatssammlung für Naturkunde in Karlsruhe, Germany, also have well-preserved fossil remains of this and other Santana pterosaurs, such as *Tapejara* and *Tupuxuara*, which have been instrumental in understanding how pterosaurs stood and walked.

10. The paper on the pterosaur from Mexico, by Jim Clark and colleagues, appeared in the Feb. 26, 1998, issue of *Nature* and was deemed so important that a full-color photograph of the fossil appeared on the cover of the journal.

11. Squirrels, birds and some lizards that are specialized for climbing also have very narrow, deeply curved, sharply pointed claws, as Derek Yalden explained in his 1985 paper on the claws of *Archaeopteryx*. Animals that use their claws to grab and hold onto prey tend to have thicker, wider claws so that they don't snap if the prey tries to break free.

12. Bicycle clips are thin metal or plastic bands that prevent a bicycle rider's trousers from getting caught up in the chain.

13. You can view the Department of Earth Sciences at Bristol at http://www.gly.bris.ac.uk/. A version of Robodactylus can be found on the same site at: http://palaeo.gly.bris.ac.uk/dinosaur/animation.html under the link Anhanguera - stick model.

14. Speaking of his bestseller *A Brief History of Time*, Stephen Hawking noted that "each equation in a book halves its sales." Don Henderson very kindly calculated for me that inclusion of the full set of 36 equations will reduce sales of this book to the full stop at the end of this sentence.

15. Both Colin Pennycuick, in a paper published in 1986, and David Unwin, in the *Nature* paper of 1987, have pointed out that pterosaurs are front-heavy and would have found it difficult to balance on their legs alone.

16. The center of mass is the point in any object (living or not) through which gravity can be assumed to operate. This is not a theoretical point, it really does exist, and any time you balance something, you are making use of this point. With regard to pterosaurs, it is the point within an individual where the mass toward the front exactly balances the mass toward the rear and, when viewed from the side, usually lies about level with the shoulders.

17. See note 7.

18. The fossil remains on which Roborhamphus was based were first described in the 1975 monograph on *Rhamphorhynchus* by Peter Wellnhofer.

19. William Stokes' paper, the first to describe a track that had definitely been made by a pterosaur, came out in the *Journal of Paleontology* in 1957, although a brief abstract mentioning the new find had appeared in the *Bulletin of the Geological Society of America* some three years earlier.

20. The paper by Kevin Padian and Paul Olsen, reporting on their experimental investigation of *Pteraichnus*, appeared in 1984 in the same organ (*Journal of Paleontology*) as the first description of this pterosaur track.

21. Detailed accounts of the Clayton Lake trackway and other aspects of pterosaur tracks can be found in *Dinosaur Tracks and Traces*, edited by David Gillette and Martin Lockley (1989) and in further papers by Chris Bennett (1992, 1997) and David Unwin (1997).

22. The paper by Martin Lockley and colleagues, with the emphatic title, "The fossil trackway *Pteraichnus* is pterosaurian, not crocodilian" appeared in *Ichnos*, a journal devoted to fossil tracks and traces of all kinds. The paper by Jean-Michel Mazin and colleagues appeared in *Comptes rendus de l'Academie des Sciences Serie II, Earth and Planetary Sciences*.

23. See *Dinosaurios: Rutas por el Jurásico de Asturias*, by José Carlos Martínez García-Ramos and colleagues, published in 2002.

24. The 1997 paper on *Purbeckopus*, written by Jo Wright and colleagues, appeared in the *Proceedings of the Geologists' Association*.

25. The first detailed description of the pterosaur tracks from South Korea, a multinational effort including researchers from Korea, Germany and the United States and led by Hwang Koo-Geun from the Chonnam National University in Kwangju, Korea, appeared in the *Geological Magazine* in 2002.

26. In a 1922 paper, Carl Stieler speculated that rhamphorhynchoids sprang into the air propelled by a powerful kick from the legs, perhaps assisted by an additional push from the tail. Evidence that supports this idea, for example from trackways, has yet to be found.

27. See Chapter 2, note 13.

28. Typically, when climbing steep snow or ice, mountaineers use two ice axes (one commercial brand being called "pterodactyls"), one in each hand, and a set of crampons on each boot, at the front end of which project two sharp spikes that one kicks into the snow or ice.

29. Bat-like restorations of pterosaurs, hanging upside-down from trees, date back at least to the works of Othenio Abel (1925). Cherrie Bramwell and George Whitfield (1974) depicted *Pteranodon* hanging from a cliff (it does not look very comfortable) and even Peter Wellnhofer's *The Illustrated Encyclopedia of Pterosaurs* (1991) shows some pterosaurs hanging by their feet.

30. "Lucien" is described in a paper, *Ichnological evidence for quadrupedal locomotion in pterodactyloid pterosaurs*, by Jean-Michel Mazin and colleagues in the pterosaur volume edited by Eric Buffetaut and Mazin (2003).

31. Track evidence for swimming pterosaurs was first described by Jo Wright and Martin Lockley in an abstract published in 1999 in the *Journal of Vertebrate Paleontology*.

32. A duck-like swimming pose for pterosaurs was first proposed by Carl Stieler (1922), based on his study of the Lower Jurassic pterosaur *Dorygnathus*, and was taken up recently by Martin Lockley and Jo Wright in their paper, *Pterosaur swim tracks and other ichnological evidence of behavior and ecology*, in the volume edited by Eric Buffetaut and Jean-Michel Mazin (2003).

Chapter 10: The Pterosaur Story
Pages 224–265

1. This is only one of many possible scenarios in which pterosaurs might have become extinct. Several others are discussed in this chapter. Parasites that live on or in the skin are common in birds and bats, and pterosaurs almost certainly had them, too. A Russian paleoentomologist, Alexander Ponomarenko, suggested in papers that appeared in 1976 and 1986 that a fossil insect called *Saurophirus*, found in Lower Cretaceous lake sediments of Mongolia and Siberia, may have been a pterosaur parasite, but this proposal has not been enthusiastically embraced by pterosaurologists (see, for example, Natasha Bakhurina and David Unwin, 1995).

2. There are one or two exceptions. In a 2003 paper on Triassic pterosaurs, Fabio Dalla Vecchia and co-authors discussed the early evolution of pterosaurs; in the 2003 symposium volume edited by Eric Buffetaut and Jean-Michel Mazin, David Unwin outlined a new history of the group; in a somewhat earlier work published in 1996, Buffetaut and colleagues mused on their extinction.

3. The pterosaur from Greenland, a new species of *Eudimorphodon* recovered from rocks of Upper Triassic age, was described by Farish Jenkins and colleagues in 2001. The New Zealand pterosaur,

a fragmentary bone from the Upper Cretaceous collected by Joan Wiffen, was published by her and Ralph Molnar in 1988. The record from Antarctica, a single pterosaur bone from the Upper Jurassic, was recovered by Mike Hammer in the early 1990s and reported on by him in 1996.

4. See Chapter 4, note 52.

5. See Chapter 3, note 18.

6. The Middle Jurassic, a 20-million-year period from which just a mere handful of pterosaur fossils have been recovered, is undoubtedly the darkest age of pterosaur history. Other notoriously gloomy intervals include the middle Early Cretaceous and the early Late Cretaceous, although both these have been lightened by new finds in recent years.

7. Part of the clan history that is inferred to have existed before the oldest fossil remains are encountered is known technically as the "ghost lineage." Animals or plants that have a poor or patchy fossil record may have several long ghost lineages, whereas if the fossil record is better, the ghost lineages tend to be fewer and shorter.

8. Chris Bennett argued in a paper published in 1997 that the long, powerful legs of early pterosaurs equipped their ancestors with the ability to leap in a powerful fashion from tree to tree or out into the air at the beginning of a glide. While this is possible, it seems more likely that lengthening of the legs was related to the advantage to be gained, by an incipient flier, from increasing the area of the cheiropatagia and the cruropatagium.

9. It has occasionally been suggested (see, for example, the paper by Kevin Padian, 1984, and its references) that pterosaurs might have taken to the air directly from the ground—running, jumping, gliding, flapping and eventually getting airborne. Such a scenario seems inherently unlikely, especially for early pterosaurs, which had a relatively poor running ability and whose limbs were encumbered by flight membranes.

10. Three different species of *Eudimorphodon* have been described so far: *Eudimorphodon ranzii* by Rocco Zambelli in 1973, *Eudimorphodon rosenfeldi* by Fabio Dalla Vecchia in 1995 and *Eudimorphodon cromptonellus* by Farish Jenkins and colleagues in 2001.

11. Rapid evolution of the type proposed here is not uncommon. During the Early Cenozoic, for example, after the mass extinctions at the end of the Cretaceous, survivors such as mammals and modern birds seem to have experienced a rapid phase of evolution, during which it took just a few million years for them to split into most of the major groups that are seen today.

12. See the chapter by Fabio Dalla Vecchia in the pterosaur volume edited by Eric Buffetaut and Jean-Michel Mazin (2003) for a comprehensive review of Triassic pterosaurs.

13. An individual pterosaur with a wingspan in excess of 2 meters (6 feet) has been proposed by Fabio Dalla Vecchia (2000) on the basis of a large wing-finger bone from the Upper Triassic of Italy.

14. The leading authority on drepanosaurids is Silvio Renesto (http://dipbsf.uninsubria.it/paleo/), who has published several papers on these remarkable animals, the latest in 2004 in the *Eudimorphodon* volume, produced by the Natural History Museum in Bergamo, Italy. An analysis of the relationships of drepanosaurids to other reptiles (apparently, they were early diapsids and related to creatures such as *Coelurosauravus*, a small gliding form from the Permian) can be found in a recent paper by Phil Senter, published in 2004.

15. The absence of the sheath of bony rods in the long tails of some rhamphorhynchoids was first noticed by Fabio Dalla Vecchia, who reported on it in his 1998 and 2003 papers.

16. The end-Triassic mass extinction was one of the five largest events of this type. While it saw the decimation or outright disappearance of several major lineages of animals and plants on land and a severe reduction of such important marine groups as bivalves and ammonoids, the pattern and timing of these extinctions and their possible causes remain unclear (see, for example, the paper by Paul Olsen and colleagues in the May 17, 2002, issue of *Science*).

17. See Chapter 3, note 16.

18. An incomplete and, unfortunately, headless skeleton of a pterosaur from the Early, or possibly Middle, Jurassic of Mexico, first described by Jim Clark and colleagues in 1994, is thought to belong to *Dimorphodon*. If this identification is correct, it would indicate that the dimorphodontid clan survived for rather longer than I have suggested here. The proportions, however, of the limb bones of the Mexican pterosaur are different from those of dimorphodontids, and it would seem to belong to one of the more advanced rhamphorhynchoid clans (see, for example, its position in the list of pterosaur names in this book's appendix), although exactly which one is difficult to determine.

19. The best specimens of *Campylognathoides* come from the famous *Posidonia* Shale locality of Holzmaden, in Baden-Würtemberg, Germany (see Chapter 3, note 16), and were described in detail by Felix Plieninger (1895) and Peter Wellnhofer (1974).

20. *Campylognathoides indicus*, represented by a few teeth and limb bones, is the only record of a pterosaur from the Indian subcontinent (Sohan L. Jain, 1974).

21. First found as early as 1830 by Carl Theodori in the Early Jurassic of northern Bavaria, the most important specimens of *Dorygnathus* were collected from several locations in Würtemberg, most notably around Holzmaden and Ohmden. Felix Plieninger (1907), Gustav von Arthaber (1919) and A. Saleé (1928) all published detailed descriptions of the skeleton; Carl Stieler (1922) reconstructed *Dorygnathus* using a partial, but very well-preserved skeleton from Braunschweig; and Ferdinand Broili (1939) identified fossilized soft parts, including hair.

22. The earliest record of anurognathids, the only one from the Middle Jurassic, was first reported by Natasha Bakhurina and David Unwin in 1995 and comes from Bakhar in Central Mongolia, which, interestingly, bearing in mind the supposed diet of anurognathids, has also yielded a considerable number of fossil insects. The pterosaur remains consist solely of impressions of several wing bones, and, while they cannot be certainly identified as anurognathid, they do show several typical features of the group. *Herbstosaurus*, based on fragmentary remains of a pelvis and hind limbs from the Lotena Formation of Neuquén Province, Argentina, was thought by its first describer (Rodolfo Casamiquela, 1975) to be a dinosaur, but was later recognized as a pterosaur (see John Ostrom, 1978, and José Bonaparte, 1978). Recent studies indicate that it is a pterodactyloid and, at present, the only reasonably certain Middle Jurassic record for the group (Unwin, 1996).

23. The Stonesfield Slates, found in Oxfordshire, England, were originally formed along the coastline of a shallow island dotted-sea and seem to represent beach sands that were deposited, together with the remains of animals and plants, by violent storms in the Middle Jurassic. Quarrying of the slates in the early 19th century led to a flood of fossils, including some of the earliest finds of dinosaurs, the first evidence of mammals from Mesozoic rocks and, most importantly, pterosaurs.

24. See the review of Middle Jurassic pterosaurs by David Unwin (1996), for the most recent account of *Rhamphocephalus*.

25. See Chapter 4, note 52.

26. This idea was first proposed by Chris Bennett in 2004 and is based on his study of new material of *Scaphognathus* from the Upper Jurassic Solnhofen Limestones of Bavaria.

27. Further comments on the possible ecology of scaphognathines and rhamphorhynchines can be found in a paper by Ken Carpenter and colleagues that appeared in the 2003 pterosaur volume edited by Eric Buffetaut and Jean-Michel Mazin.

28. Further details of the fossil material of *Rhamphorhynchus* can be found in the 1975 monograph by Peter Wellnhofer and also in the *The Illustrated Encyclopedia of Pterosaurs* by the same author.

29. The Karabastau beds of the Karatau ridge in Kazahkstan are one of the most important localities for Jurassic fossil insects. More than 18,000 specimens belonging to 19 orders have already been recognized, and further details, together with a good introduction to the literature on Karatau insects, can be found in *History of Insects*, edited by Aleksander Rasnitsyn and Donald Quicke in 2000.

30. See note 24.

31. Details of Late Jurassic pterosaur tracks can be found in several papers published in the *Palaeobiology of Pterosaurs*, edited by Eric Buffetaut and Jean-Michel Mazin (2003).

32. An excellent account of the expeditions to Tendaguru can be found in *African Dinosaurs Unearthed* by Gerhard Maier, published in 2004.

33. Whenever facts fit nicely together, scientists should suspect the worst.

34. See the paper by Makoto Manabe and colleagues in *Nature* (2000).

35. *Pterodaustro* was first described by José Bonaparte (1970, 1971), then again by Teresa Sanchez (1973), and most recently by Luis Chiappe and colleagues in 2000 and 2004.

36. For detailed accounts of *Cearadactylus*, see Guiseppe Leonardi and G. Borgomanero (1985) and David Unwin (2003).

37. The Cambridge Greensand and its pterosaurs was recently reviewed by David Unwin (2001).

38. Alex Kellner and Diogenes Campos proposed that *Thalassodromeus* and other tupuxuarids were skimmers, in a paper that appeared in *Science* in 2002.

39. There seems to have been a major crisis in the early Late Cretaceous that led to extinctions among backboned animals, including the ichthyosaurs and some kinds of crocodiles and snakes, and also among the invertebrates, including groups of clams, shellfish and echinoids (see, for example, the 1994 paper by Natalie Bardet in *Historical Biology*).

40. Chris Bennett published several important papers on *Pteranodon* in 1991, 1994 and 2001.

41. The main account of these north African pterosaurs can be found in a paper by Peter Wellnhofer and Eric Buffetaut (1999).

42. See note 14.

43. This fossil, which bears the name *Nyctosaurus lamegoi* and was the first pterosaur to be found in South America, was described by Llellwyn Price in 1953.

44. Several fragmentary pterosaur fossils including a humerus from the Lower Cretaceous of Texas (described by Murry and colleagues in 1991), a partial skeleton from the Crato Formation of Brazil (mentioned by David Martill and Dino Frey in a 1998 paper) and a neck vertebra from the Upper Jurassic of Tanzania, Africa (noted by Julia Sayão and Alex Kellner in an abstract that appeared in 2001), are claimed to represent early records of azhdarchids, but could also represent azhdarchoids, such as tapejarids or lonchodectids, and might, in the case of the vertebra, even belong to ctenochasmatoids.

45. Several papers (e.g. by Lev Nesov 1995 and David Archibald and colleagues in 1998) have discussed the age of the sediments at the fossil locality of Dzharakhuduk, where *Azhdarcho* was first found, but the consensus (e.g. Hans-Dieter Sues and Alexander Averianov, 2004) now seems to be that it is mid-Late Cretaceous.

46. See Chapter 9, note 25.

47. The latest estimates of the wingspan of *Quetzalcoatlus northropi*, using data from more complete azhdarchids such as *Zhejiangopterus*, yield a most likely wingspan of 10 meters (33 feet) and a weight of perhaps 50 to 70 kilos (110 to 154 pounds).

48. Remains of a "small" 5-meter (16-foot) wingspan azhdarchid from southeast Spain were first described by Julio Company and colleagues in 1999. Fragmentary fossil remains of much larger individuals, at least 10 meters (33 feet) in wingspan were only recognized later (see Julio Company and colleagues, 2001).

49. For a good introduction to the topic of mass extinctions, try *When Life Nearly Died* by Michael Benton (2003).

50. The song "Knockin' on Heaven's Door" was originally recorded by Bob Dylan, and appeared on the soundtrack for the 1973 film *Pat Garrett and Billy the Kid*.

51. Another excellent account of mass extinctions and, just as importantly, what happened afterward, can be found in *Mass Extinctions and Their Aftermath* by Tony Hallam and Paul Wignall (1997).

52. In his paper *Scientific Methodologies in Collision: The History of the Study of the Extinction of the Dinosaurs*, Michael Benton (1990) gives a good account of the more than 100 ideas proposed so far to explain the extinction of the dinosaurs at the end of the Mesozoic.

53. The subject of ongoing work by Danny Elvidge and David Unwin, some hints of our results can be gleaned from abstracts for scientific conferences such as, for example, the one published for the annual meeting of the Society of Vertebrate Paleontology in Minnesota in 2003.

Chapter 11: Postscript
Pages 266–270

1. This quotation formed part of a speech given by Sir Winston Churchill at the Lord Mayor's luncheon in the Mansion House, London, on Nov. 10, 1942, after the battle of El Alamein in Egypt, widely regarded as one of the major turning points in World War II.

2. This quotation comes from a speech made to the House of Commons by Winston Churchill on May 13, 1940, three days after he became British prime minister.

3. The urge to put forks in the fork drawer can be very strong, especially in taxonomists.

4. Ancient DNA (aDNA) is known from the archaeological record and has been reported from fossils up to 9,000 or 10,000 years old. In the 1990s, several labs published reports of DNA in much older fossils, even including dinosaurs, although, so far, not one of these claims has proved to be reliable. Theoretically, though, if DNA can survive for a few thousand years, there is no reason why it shouldn't last for longer.

BIBLIOGRAPHY

Abel, O. 1912. *Grundzüge der Palaeobiologie der Wirbeltiere*. E. Schweizerbart'sche Verlagsbuchhandlung, Stuttgart.

Abel, O. 1925. On a skeleton of *Pterodactylus antiquus* from the lithographic shales of Bavaria, with remains of skin and musculature. *American Museum Novitiates*, **192**, 1-12 .

Alderton, D. 1991. *Crocodiles and Alligators of the World*. Sterling Publishing.

Anon. 1907. Eminent Living Geoloists: Professor H. G. Seeley, F.R.S., F.L.S., F.G.S., F.Z.S., F.R.G.S. *Geological Magazine*, (5) **4**, 241-253.

Archibald, J.D., Sues, H.-D., Averianov, A.O., King, C., Ward, D.J., Tsaruk, O.A., Danilov, I.G., Rezvyi, A.S., Veretennikov, B.G., and Khodjaev, A. 1998. Precis of the Cretaceous paleontology, biostratigraphy and sedimentology at Dzharakuduk (Turonian-Santonian), Kyzylkum Desert, Uzbekistan. In: Lucas, S.G., Kirkland, J.I., and Estep, J.W. (eds.), *Lower and Middle Cretaceous Terrestrial Ecosystems*. New Mexico Museum of Natural History and Science, Bulletin, **14**, 21-27.

Arthaber, G. von 1919. Studien über Flugsaurier auf Grund der Bearbeitung des Wiener Exemplares von *Dorygnathus banthensis* Theod. sp. *Denkschrift der Akademie der Wissenschaften Wien, mathematisch-naturwissenschaftliche Klasse*, **97**, 391-464.

Bakhurina, N.N. 1982. Pterodaktil' iz nizhnego mela Mongolii. *Paleontologicheskii Zhurnal Akademii Nauk SSSR, Moskva*, 104-108. [A pterodactyl from the Lower Cretaceous of Mongolia. *Paleontolog. Journal*, 105–109.]

Bakhurina, N.N. and Unwin, D.M. 1995. A survey of pterosaurs from the Jurassic and Cretaceous of the former Soviet Union and Mongolia. *Historical Biology*, **10**, 197-245.

Bakker, R. 1986. *The Dinosaur Heresies*. Morrow, New York.

Barber, R.T. and Chavez, F.P. 1983. Biological Consequences of El Nino. *Science* **222**, 1203–1210.

Bardet, N. 1994. Extinction events among Mesozoic marine reptiles. *Historical Biology*, **7**, 313-324.

Barthel, K.W., Swinburne, N.H.M., and Morris, S.C. 1990. *Solnhofen: a study in Mesozoic palaeontology*. Cambridge University Press, Cambridge.

Bennett, S.C. 1987. New evidence on the tail of pterosaur *Pteranodon* (Archosauria: Pterosauria). *Fourth Symposium on Mesozoic Terrestrial Ecosystems*, Short Papers, 18-23.

Bennett, S.C. 1989. A pteranodontid pterosaur from the Early Cretaceous of Peru, with comments on the relationships of Cretaceous pterosaurs. *Journal of Paleontology*, **63**, 669-677.

Bennett, S.C. 1990. A pterodactyloid pterosaur pelvis from the Santana Formation of Brazil: implications for terrestrial locomotion. *Journal of Vertebrate Paleontology*, **10**, 80-85.

Bennett, S.C. 1991. *Morphology of the Late Cretaceous pterosaur* Pteranodon *and systematics of the Pterodactyloidea*. Unpublished PhD. Thesis, University of Kansas.

Bennett, S.C. 1992. Sexual dimorphism of *Pteranodon* and other pterosaurs with comments on cranial crests. *Journal of Vertebrate Paleontology*, **12**, 422-434.

Bennett, S.C. 1992. Reinterpretation of problematic tracks at Clayton Lake State Park: not one pterosaur but several crocodiles. *Ichnos*, **2**, 37-42.

Bennett, S.C. 1993. The ontogeny of *Pteranodon* and other pterosaurs. *Paleobiology*, **19**, 92–106.

Bennett, S.C. 1994. Taxonomy and systematics of the Late Cretaceous pterosaur *Pteranodon* (Pterosauria, Pterodactyloidea). *Occasional Papers of the Natural History Museum, The University of Kansas, Lawrence, Kansas*, **169**, 1-70.

Bennett, S.C. 1995. An arboreal leaping origin of flight and the relationships of pterosaurs. *Journal of Vertebrate Paleontology*, **15**, 19A.

Bennett, S.C. 1995. A statistical study of *Rhamphorhynchus* from the Solnhofen Limestone of Germany - year-classes of a single large species. *Journal of Paleontology*, **69**, 569-580.

Bennett, S.C. 1996. The phylogenetic position of the Pterosauria within the Archosauromorpha. *Zoological Journal of the Linnean Society*, **118**, 261-308.

Bennett, S.C. 1996. Year-classes of pterosaurs from the Solnhofen Limestone of Germany: taxonomic and systematic implications. *Journal of Vertebrate Paleontology*, **16**, 432-444.

Bennett, S.C. 1997. Terrestrial locomotion of pterosaurs: a reconstruction based on *Pteraichnus* trackways. *Journal of Vertebrate Paleontology*, **17**, 104–113.

Bennett, S.C. 1997. The arboreal leaping theory and the origin of pterosaur flight. *Historical Biology*, **12**, 265–290.

Bennett, S.C. 2000. New information on the skeleton of *Nyctosaurus*. *Journal of Vertebrate Paleontology*, **20** (suppl. to nb. 3), 27A.

Bennett, S.C. 2001. The osteology and functional morphology of the Late Cretaceous pterosaur *Pteranodon*. *Palaeontographica, Abteilung A*, **260**, 1-153.

Bennett, S.C. 2002. Soft tissue preservation of the cranial crest of the pterosaur *Germanodactylus* from Solnhofen. *Journal of Vertebrate Paleontology*, **22**, 43-48.

Bennett, S.C. 2003. Morphological evolution of the pectoral girdle of pterosaurs: myology and function. In: Buffetaut, E. and Mazin, J.-M. (eds.), *Evolution and Palaeobiology of Pterosaurs*. Geological Society of London, Special Publication, **217**, 191-215.

Bennett, S.C. 2003. New crested specimens of the late Cretaceous pterosaur *Nyctosaurus*. *Paläontologische Zeitschrift*, **77**, 61-75.

Bennett, S.C. 2004. New information on the pterosaur *Scaphognathus crassirostris* and the pterosaurian cervical series. *Journal of Vertebrate Paleontology*, **24** (suppl. to nb. 3), 38A.

Benton, M.J. 1990. Scientific methodologies in collision: the history of the study of the extinction of the dinosaurs. *Evolutionary Biology*, **24**, 371-424.

Benton, M.J. 1993. *The Fossil Record 2*. Chapman and Hall, London.

Benton, M.J. 1999. *Scleromochlus* and the origin of dinosaurs and pterosaurs. *Philosophical Transactions of the Royal Society of London, Series* B, **354**, 1423-1446.

Benton, M.J. 2003. *When Life Nearly Died: The Greatest Mass Extinction of All Time*. Thames and Hudson.

Benton, M.J. 2004. *Vertebrate Palaeontology*, 3rd Edition, Blackwell Publishing Oxford.

Benton, M.J. and Allen, J.L. 1997. *Boreopricea* from the Lower Triassic of Russia, and the relationships of the prolacertiform reptiles. *Palaeontology*, **40,** 931-953.

Billon-Bruyat, J-P. and Mazin, J.M. 2003. The systematic problem of tetrapod ichnotaxa: the case study of *Pteraichnus* Stokes, 1957 (Pterosauria, Pterodactyloidea). In: Buffetaut, E. & Mazin, J.-M. (eds), *Evolution and Palaeobiology of Pterosaurs*. Geological Society of London, Special Publication, **217**, 315-324.

Bonaparte, J.F. 1970. *Pterodaustro guinazui* gen. et sp. nov. pterosaurio de la Formacíon Lagarcito, Provincia de San Luis, Argentina y su significado en la geología regional (Pterodactylidae). *Acta Geologica Lilloana*, **10**, 207-226.

Bonaparte, J.F. 1971. Descripcion del craneo y mandibulas de *Pterodaustro guinazui* (Pterodactyloidea - Pterodaustriidae nov.) de la Formacion Lagarcito, San Luis, Argentina. *Publicaciones del Museo Municipal de Ciencias Naturales de Mar del Plata*, **1**, 263-272.

Bonaparte, J.F. 1978. *El mesozoico de America del Sur y sus Tetrapodos*. Opera Lilloana **26**. Tucuman, Argentina.

Bonde, N. and Chistiansen, P. 2003. The detailed anatomy of *Rhamphorhynchus*: axial pneumaticity and its implications. In: Buffetaut, E. and Mazin, J.-M. (eds.)., *Evolution and Palaeobiology of Pterosaurs*. Geological Society of London, Special Publication, **217**, 217-232.

Bramwell, C.D. and Whitfield, G.R. 1974. Biomechanics of *Pteranodon*. *Philosophical Transactions of the Royal Society of London, Series* B, **267**, 503–581.

Broili, F. 1927. Ein *Rhamphorhynchus* mit Spuren von Haarbedeckung. *Sitzungsberichte der Bayerischen Akademie der Wissenschaften, mathematisch-naturwissentschaftliche Abteilung*, 49-67.

Broili, F. 1938. Beobachtungen an Pterodactylus. Sitzungsberichte der Bayerischen Akademie der Wissenschaften, mathematisch-naturwissentschaftliche Abteilung, 139-154.

Broili, F. 1939. Ein Dorygnathus mit Hautresten. Sitzungsberichte der Bayerischen Akademie der Wissnschaften, mathematisch-naturwissentschaftliche Abteilung, 129–132.

Brower, J.C. 1983. The aerodynamics of *Pteranodon* and *Nyctosaurus*, two large Pterosaurs from the Upper Cretaceous of Kansas. *Journal of Vertebrate Paleontology*, **3**, 84–124.

Buckman, J. 1860. On some fossil reptilian eggs from the Great Oolite of Cirencester. *Quarterly Journal of the Geological Society of London London*, **16**, 107–110.

Buffetaut, E., Clarke, J.B. and LeLoeuff, J. 1996. A terminal Cretaceous pterosaur from the Corbières (southern France) and the problem of pterosaur extinction. *Bulletin de la Société géologique de France*, **167**, 753–759.

Buffetaut, E., Grigorescu, D., and Csiki, Z. 2002. A new giant pterosaur with a robust skull from the latest Cretaceou sof Romania. *Naturwissenschaften*, **89**, 180-184.

Buffetaut, E., Grigorescu, D., and Csiki, Z. 2003. Giant azhdarchid pterosaurs from the terminal Cretaceous of Transylvania (western Romania). In: Buffetaut, E. and Mazin, J.-M. (eds.), *Evolution and Palaeobiology of Pterosaurs*. Geological Society of London, Special Publication, **217**, 91-104.

Buffetaut, E., Lepage, J.-J., and Lepage, G. 1998. A new pterodactyloid pterosaur from the Kimmeridgian of the Cap de la Hève (Normandy, France). *Geological Magazine*, **135**, 719-722.

Buffetaut, E., Martill, D. and Escuillié, F. 2004. Pterosaurs as part of a spinosaur diet. *Nature*, **430**, 33.

Buffetaut, E. and Mazin, J.-M. (eds.). 2003. *Evolution and Palaeobiology of Pterosaurs*. Geological Society of London, Special Publication, **217**.

Cadbury, D. 2000. The dinosaur hunters: a story of scientific rivalry and the discovery of the prehistoric world. Fourth Estate, London.

Cai, Zhenquan and Feng, Wei. 1994. On a new pterosaur (*Zhejiangopterus linhaiensis* gen. et sp. nov.) from Upper Cretaceous in Linhai, Zhejiang, China. *Vertebrata Palasiatica*, **32**, 181-194.

Campos, D.A. and Kellner, A.W.A. 1997. Short note on the first occurrence of Tapejaridae in the Crato Member (Aptian), Santana Formation, Araripe Basin, Northeast Brazil. *Anais da Academia Brasileira de Ciências*, **69**, 83-87.

Carpenter, K. Hirsch, K.F., and Horner, J.R. 1994. Dinosaur eggs and babies. Cambridge University Press, New York.

Carpenter, K., Unwin, D.M., Cloward, K., Miles, C., and Miles, C. 2003. A new scaphognathine pterosaur from the Upper Jurassic Morrison Formation of an unusual death of an unusual pterosaur. In: Buffetaut, E. and Mazin, J.-M. (eds.), *Evolution and Palaeobiology of Pterosaurs*. Geological Society of London, Special Publication, **217**, 45-54.

Carruthers, W. 1871. On some supposed vegetable fossils. *Quaterly Journal of the Geological Society of London*, **27**, 443–449.

Casamiquela, R.M. 1975. *Herbstosaurus pigmaeus* (Coeluria, Compsognathidae) n. gen. n. sp. del Jurasico medio del Nequéri (Patagonia septentrionali). Uno de los mas pequenos dinosaurios conocidos. *Acta primero Congreso Argentino Paleontologia et Biostratigrafia*, **2**, 87-102.

Cecil Curwen, E. (ed.). 1940. *The Journal of Gideon Mantell*. Oxford University Press, London.

Chang, Mee-Mann. 2003. *The Jehol Biota*. Shanghai Science and Technology Press, Shanghai.

Chatterjee, S. and Templin, R.J. 2004. The terrestrial and aerial locomotion of pterosaurs. *Geological Society of America, Special Paper*, **376**, 1-64.

Chiappe, L.M. and Chinsamy, A. 1996. *Pterodaustro*'s true teeth. *Nature*, **379**, 211–212.

Chiappe, L.M., Codorniú, L., Grellet-Tinner, G., and Rivarola, D. 2004. Argentinian unhatched pterosaur fossil. *Nature*, **432**, 571.

Chiappe, L.M., Kellner, A.W.A., Rivarola, D., Davila, S., and Fox, M. 2000. Cranial morphology of *Pterodaustro guinazui* (Pterosauria: Pterodactyloidea) from the Lower Cretaceous of Argentina. *Contributions in Science, Natural History Museum of Los Angeles County*, **483**, 1-19.

Chiappe, L.M., Rivarola, D., Cione, A., Frenegal-Martinez, M., Sozzi, H., Buatois, L., Gallego, O., Laza, J., Romero, E., Lopez-Arbarello, A., Buscalioni, A., Marsicano, C., Adamonis, S., Ortega, F., McGehee, S., and Di Iorio, O. 1998. Biotic association and palaeoenvironmental reconstruction of the "Loma del *Pterodaustro*" fossil site (Early Cretaceous, Argentina). *Geobios*, **31**, 349–369.

Chiappe, L.M. and Witmer, L.M. 2002. *Mesozoic Birds: Above the Heads of Dinosaurs*. University of California Press, Berkeley.

Clack, J.A. 2002. *Gaining Ground: The Origin and Evolution of Tetrapods*. Indiana University Press, Bloomington.

Claessens, L.P.A.M. 2004. The identity and function of the pterosaur prepubis. *Journal of Vertebrate Paleontology*, **24** (Suppl. to **3**), 46A.

Clark, J.M., Fastovsky, D.F., Montellano, M., Hopson, J.A., and Hernandez, R. 1994. Additions to the Middle Jurassic vertebrate fauna of Huizachal Canyon, Tamaulipas, Mexico. *Journal of Vertebrate Paleontology*, **14** *(Suppl. to **3**)*, 21A.

Clark, J.M., Hopson, J.A., Hernandez, R.R., Fastovsky, D.E., and Montellano, M. 1998. Foot posture in a primitive pterosaur. *Nature*, **391**, 886-889.

Cleere, N. 1998. *Nightjars: A Guide to Nightjars and Related Nightbirds*. Pica Press.

Codorniú, L. and Chiappe, L.M. 2004. Early juvenile pterosaurs (Pterodactyloidea: *Pterodaustro guinazui*) from the Lower Cretaceous of central Argentina. *Canadian Journal of Earth Sciences*, **41**, 9-18.

Collini, C. 1784. Sur quelques Zoolithes du Cabinet d'Histoire naturelle de S. A. S. E. Palatine and de Bavière, à Mannheim. *Acta Acad. Theodoro-Palatinae*, pars physica, **5,** 58–103.

Company, J., Ruiz-Omeñaca, J. I., and Pereda-Suberbiola, X. 1999. A long-necked pterosaur (Pterodactyloidea, Azhdarchidae) from the Upper Cretaceous of Valencia, Spain. *Geologie en Mijnbouw*, **78**, 319-333.

Company, J., Unwin, D.M., Ruiz-Omenaca, J.I., and Pereda-Suberbiola, X. 2001. A giant azhdarchid pterosaur from the latest Cretaceous of Valencia, Spain, the largest flying creature ever? *Journal of Vertebrate Paleontology*, **21** (Suppl. to **3**), 41A-42A.

Currie, P.J. and Jacobsen, A. R. 1995. An azhdarchid pterosaur eaten by a velociraptorine theropod. *Canadian Journal of Earth Sciences*, **32**, 922–925.

Cuvier, G. 1801. [Reptile volant.] In: Extrait d'un ouvrage sur les espèces de quadrupèdes dont on a trouvé les ossemens dans l'intérieur de la terre. *Proc. Class. Sci., Math. et Phys, Inst. Nationale*, 26 Brumaire, l'an 9.

Cuvier, G. 1809. Sur le squelette fossile d'un reptile volant des environs d'Aichstedt, que quelques naturalistes ont pris pour un oiseau, et dont nous formons un genre de sauriens, sous le nom de Petro-Dactyle. *Annales du Musee d'Histoire naturelle de Marseille*, **13**, 424-437.

Cuvier, G. 1812. Rechereches sur les ossemens fossiles. 1st edn, Paris.

Cuvier, G. 1824 Recherches sur les ossemens fossiles. 2nd edn, Paris.

Cuvier, G. 1825 Recherches sur les ossemens fossiles. 3rd edn, Paris.

Cuvier, G. 1834–1836. Recherches sur les ossemens fossiles, 4th edn, chap. III, Article VI, Sur un genre du sauriens... In: D'Ocagne, E. Paris, **10**.

Czerkas, S.A. and Ji Qiang. 2002. A new Rhamphorhynchoid with a headcrest and complex integumentary structures. In: S. J. Czerkas (ed.), *Feathered Dinosaurs and the Origin of Flight*. The Dinosaur Museum of Blanding, Utah. The Dinosaur Museum Journal, **1**.

Czerkas, S.J. and Czerkas, S.A. 1991. *Dinosaurs: A Global View*. Mallard Press, New York.

Dalla Vecchia, F.M. 1995. A new pterosaur (Reptilia, Pterosauria) from the Norian (Late Triassic) of Friuli (Northeastern Italy). Preliminary note. *Gortania - Atti Mus. Friul. St.Nat.*, **16**, 59-66.

Dalla Vecchia, F.M. 1998. New observations on the osteology and taxonomic status of Preondactylus buffarinii Wild, 1984 (Reptilia, Pterosauria). Boll. Soc. Pal. Italy., **36** (3, 1997),355-366.

Dalla Vecchia, F.M. 2001. Triassic pterosaurs: unravelling the puzzle. *Two hundred years of pterosaurs – A symposium on the anatomy, evolution, palaeobiology and environments of Mesozoic flying reptiles, Tolouse, France, September 5-8, 2001. Strata*, **11** (Série 1), 33-35.

Dalla Vecchia, F.M. 2003. New morphological observations on Triassic pterosaurs. In: Buffetaut, E., and Mazin, J.-M. (eds.), *Evolution and Palaeobiology of Pterosaurs.* Geological Society Special Publication, **217**, 23-44.

Dalla Vecchia, F.M., Muscio, G., and Wild, R. 1989. Pterosaur remains in a gastric pellet from the Upper Triassic (Norian) of Rio Seazza Valley (Udine, Italy), Gortania. Atti del Mus. Friuliano di Storia Nat., Udine, **10** (1988),121-132.

Dalla Vecchia, F. M., Wild, R., Hope, H., and Reitner, J. 2000. A crested rhamphorhynchoid pterosaur from the Late Triassic of Austria. *Journal of Vertebrate Paleontology*, **22**, 196-199.

Darwin, C. 1871. *The Descent of Man, and Selection in Relation to Sex.* Appleton, New York. 2 vols.

Dawkins, R. 1992. "Progress." In: Fox Keller, E. and Loyd, E. (eds.), *Keywords in Evolutionary Biology.* Harvard University Press, Cambridge.

Deeming, D.C. (ed.) 2004. *Reptilian Incubation: Environment, Evolution and Behaviour.* Nottingham University Press, Nottingham.

Dilkes, D.W. 1998. The Early Triassic rhynchosaur Mesosuchus browni and the interrelationships of basal archosauromorph reptiles. *Philosophical Transactions of the Royal Society of London*, Biological Sciences, **353**, 501-541.

Dong Zhiming. 1993. A Lower Cretaceous enantiornithine bird from the Ordos Basin of Inner Mongolia, People's Republic of China. *Canadian Journal of Earth Sciences*, **30**, 2177-2179.

Dong Zhiming and Lü Junchang. 2005. A new ctenochasmatid pterosaur from the Early Cretaceous of Liaoning Province. *Acta Geologica Sinica*, **79**, 164-167.

Dong Zhiming, Sun Yuewu, and Wu Shaoyuan. 2003. On a new pterosaur from the Lower Cretaceous of Chaoyang Basin, western Liaoning, China. *Global Geology*, **22**(1), 1-7.

Edinger, T. 1927. Das Gehirn der Pterosaurier. *Z. Anat. EntwGesch.*, **83**, 105–112.

Edinger, T. 1941. The brain of *Pterodactylus. Am. J. Sci.*, **239**, 665–682.

Ellis, R. 2003. *Sea Dragons: Predators of the Prehistoric Oceans.* University Press of Kansas, Lawrence, Kansas.

Elvidge, D.J. and Unwin, D.M. 2003. Locomotor modules, linkage and morphological disparity in pterosaurs and other flying vertebrates. *Journal of Vertebrate Paleontology*, **23** (Suppl. to **3**), 48A.

Erichsen, J.T., Hodos, W., Evinger, C., Bessette, B.B., and Phillips, S.J. 1989. Head orientation in pigeons: postural, locomotor and visual determinants. *Behav. Evol.*, **33**, 268-278.

Erickson, G.M., Makovicky, P.J., Currie, P.J., Norell, M.A., Yerby, S.A., and Brochu, C.A. 2004. Gigantism and comparative life-history parameters of tyrannosaurid dinosaurs. *Nature*, **430**, 772–775.

Evans, S.E. 1988. The early history and relationships of the Diapsida. In: Benton M.J. (ed.), *The Phylogeny and Classification of the Tetrapods.* Clarendon, Oxford, **1**, 221–260.

Ewer, R.F. 1965. The anatomy of the thecodont reptile Euparkeria capensis Broom. *Philosophical Transactions of the Royal Society of London*, Series B: Biological Sciences, **248** (751), 379-435.

Farlow, J.O. and Brett-Surman, M.K. 1997. *The Complete Dinosaur*. Indiana Univ. Press, Bloomington.

Fastnacht, M. 2001. First record of *Coloborhynchus* (Pterosauria) from the Santana Formation (Lower Cretaceous) of the Chapada do Araripe, Brazil. *Paläontologische Zeitschrift*, **75**, 23-36.

Feduccia, A. 1999. *The Origin and Evolution of Birds*, Second Edition. Yale University Press, New Haven.

Franzen, J.L. 1992. The Messe horse show, and other odd-toed ungulates. In: Schaal, S. and Ziegler, W. (eds.), *Messel: An Insight into the History of Life and of the Earth*. Clarendon Press, Oxford, 241-247.

Frey, E., Buchy, M-C. and Martill, D.M. 2003. Middle- and bottom-decker Cretaceous pterosaurs: unique deigns in active flying vertebrates. In: Buffetaut, E. and Mazin, J.-M. (eds.), *Evolution and Palaeobiology of Pterosaurs*. Geological Society of London, Special Publication, **217**, 267-275.

Frey, E. and Martill, D. M. 1998. Late ontogentic fusion of the processus tendinis extensoris in Cretaceous pterosaurs from Brazil. *Neues Jahrbuch für Geologie und Paläontologie Monatsheft*, **10**, 587–594.

Frey, E. and Martill, D. M. 1998. Soft tissue preservation in a specimen of *Pterodactylus kochi* (Wagner) from the Upper Jurassic of Germany. *Neues Jahrbuch für Geologie und Paläontologie Abhandlung*, **210**, 421-441.

Frey, E., Martill, D.M. and Buchy, M.-C. 2003. A new crested ornithocheirid from the Lower Cretaceous of northeast Brazil and the unusual death of an unusual pterosaur. In: Buffetaut, E. and Mazin, J.-M. (eds.), *Evolution and Palaeobiology of Pterosaurs*. Geological Society of London, Special Publication, **217**, 55-64.

Frey, E., Martill, D.M. and Buchy, M.-C. 2003. A new species of tapejarid pterosaur with soft tissue head crest. In: Buffetaut, E. and Mazin, J.-M. (eds.), *Evolution and Palaeobiology of Pterosaurs*. Geological Society of London, Special Publication, **217**, 55-72.

Frey, E. and Riess, J. 1981. A new reconstruction of the pterosaur wing. *Neues Jahrbuch für Geologie und Paläontologie Abhandlung*, **161**, 1–27.

Frey, E. and Tischlinger, H. 2000. Weichteilanatomie der Flugsaurierfuße und Bau der Scheitelkämme: Neue Pterosaurierfunde aus den Solnhofener Schichten (Bayern) und der Crato-Formation (Brasilien). *Archaeopteryx*, **18**, 1-16.

Frey, E. and Tischlinger, H. 2003. Am Puls der Flugsaurier. *Fossilien*, **4**, 234-240.

Frey, E., Tischlinger, H., Buchy, M-C., and Martill, D.M. 2003. New specimens of Pterosauria (Reptilia) with soft parts with implications for pterosaurian anatomy and locomotion. In: Buffetaut, E. and Mazin, J.-M. (eds.), *Evolution and Palaeobiology of Pterosaurs*. Geological Society of London, Special Publication, **217**, 233-266.

Frickhinger, K.A. 1994. *Die Fossilien von Solnhofen*. Goldschneck Verlag.

Frickhinger, K.A. 1999. Die Fossilien von Solnhofen. 2. Neue Funde. Goldschneck Verlag.

García-Ramos, J.C.M., Lires, J. and Piñuela, L. 2002. *Dinosaurios. Rutas por el Jurásico de Asturias*. Grupo Zeta, Intermark Comunicacion.

Gillette, D.D. and Lockley, M.G. 1989. *Dinosaur Tracks and Traces*. Cambridge University Press, Cambridge, New York.

Goldfuss, G. A. 1830. *Pterodactylus crassirostris. Isis von Oken, Jena*, 552–553.

Goldfuss, G. A. 1831. [Pterodactylus macronyx at Banz.] N. Jb. Miner. Geogn. Geol. Petrefakt., **2**, 298.

Goldingay, R. and Scheibe, J. (eds.) 1999. *The Biology of Gliding Mammals*. Filander Verlag Fürth.

Gould, S.J. 1981. *The Mismeasure of Man*. W. W. Norton and Co.

Gower, D.J. and Weber, E. 1998. The braincase of Euparkeria, and the evolutionary relationships of birds and crocodilians. *Biological Reviews*, **73**, 367-411.

Grove, R., 1976. *The Cambridgeshire Coprolite Mining Rush*. Oleander Press, Cambridge.

Haldane, J. B. S. 1963. *Journal of Genetics*, **58**, 464.

Hallam, A. and Wignall, P.B. 1997. *Mass extinctions and their aftermath*. Oxford University Press.

Hammer, W.R. 1996. Implications of an Early Jurassic vertebrate fauna from Antarctica. In: Michael Morales (ed.), *The Continental Jurassic*. Museum of Northern Arizona, **60**, 215-218.

Hankin, E.H. and Watson, D.M.S. 1914. On the flight of pterodactyls. *Aero. J.*, **18**, 324–335.

Hawking, S. 1988. *A Brief History of Time: From the Big Bang to Black Holes*. Bantam, Rei Edition.

Hazlehurst, G. 1991. *The morphometric and flight characteristics of the Pterosauria*. Ph.D. Thesis, University of Bristol.

Hazlehurst, G. and Rayner, J.M.V. 1992. An unusual flight mechanism in the Pterosauria. *Palaeontology*, **35**, 927–941.

Hazlehurst, G. and Rayner, J.M.V. 1992. Flight characteristics of Jurassic and Triassic Pterosauria: an appraisal based on wing shape. *Paleobiology*, **18**, 447-463.

Heinrich, B. 1993. *The Hot-Blooded Insects: Strategies and Mechanisms of Thermoregulation*. Harvard University Press, Cambridge.

Hennig, W. 1966. *Phylogenetic Systematics*. University of Illinois Press, Urbana.

Heptonstall, W.B. 1971. An analysis of the flight of the Cretaceous pterodactyl *Pteranodon ingens*. *Scott. J. Geol.*, **7**, 61–78.

Hirsch, K.F. 1994. The fossil record of vertebrate eggs. In: S. K. Donovan (ed.), *The Palaeoecology of Trace Fossils*. John Wiley and Sons, Chichester, 269-294.

Hoglund, J. and Alatalo, R. 1995. *Leks*. Princeton University Press, Princeton.

Holst, E. von. 1957. Der Saurierflug. *Paläont. Z.*, **31**, 15–22.

Hooley, R.W. 1913. On the skeleton of *Ornithodesmus latidens* from the Wealden of Atherfield (Isle of Wight). *Quarterly Journal of the Geological Society of London*, **69**, 372-422.

Hopson, J.A. 1979. "Paleoneurology." In: Gans, C. (ed.), *Biology of the Reptilia*. Neurology. Academic Press, New York, **9**, 39-146.

Hopson, J.A. 1994. Synapsid evolution and the radiation of non-eutherian mammals. In: Prothero, D.R. and Schoch, R.M. (eds.), *Major Features of Vertebrate Evolution*. Short Courses in Paleontology. Paleontological Society, Knoxville, **7**, 190-219.

Howse, S.C.B. and Milner, A.R. 1993. *Ornithodesmus* – a maniraptoran theropod dinosaur from the Lower Cretaceous of the Isle of Wight, England. *Palaeontology*, **36**, 425-437.

Howse, S.C.B., Milner, A.R., and Martill, D.M. 2001. Pterosaurs. In: Martill, D. M. and Naish, D. (eds.), *Dinosaurs of the Isle of Wight*. The Palaeontological Association, London, 324-355.

Huene, F. von. 1914. Beiträge zur Geschichte der Archosaurier. A. Beiträge zur Kenntnis und Beurteilung der Pseudosuchier. 1. Neue Beiträge zur Kenntnis von *Scleromochlus taylori* A. S. Woodward. *Geol. paläont. Abh. N. F.,* **13**, 4–22.

Hwang, Koo-Geun, Huh, M., Lockley, M.G., Unwin, D.M., and Wright, J.L. 2002. New pterosaur tracks (Pteraichnidae) from the Late Cretaceous Uhangri Formation, SW Korea. *Geological Magazine,* **139** (4), 421-435.

Jain, S.L. 1974. Jurassic pterosaur from India. *J. geol. Soc. India,* **15**, 330–335.

Jenkins, F.A., Jr., Shubin, N.H., Gatesy, S.M., and Padian, K. 2001. A diminutive pterosaur (Pterosauria: Eudimorphodontidae) from the Greenlandic Triassic. In: Jenkins, F.A. Jr., Shapiro, M.D., and Owerkowicv, T., *Studies in Organismic and Evolutionary Biology in Honor of A. W. Crompton. Bulletin of the Museum of Comparative Zoology,* Harvard University Press, Cambridge. **156**, 151-170.

Ji Qiang, Ji Shu-an, Cheng Yen-nien, You Hailu, Lü Junchang, Liu Yingqing, and Yuan Chongxi, 2004. Pterosaur egg with a leathery shell. *Nature,* **432**, 572.

Ji Qiang and Yuan Chongxi. 2002. Discovery of two kinds of protofeathered pterosaurs in the Mesozoic Daohugou Biota in the Ningcheng Region and its stratigraphic and biologic significances. *Geological Review,* **48**, 221-224.

Ji Shu-an and Ji Chiang. 1997. Discovery of a new pterosaur in Western Liaoning, China. *Acta Geologica. Sinica.,* **71**(1), 1-6.

Ji Shu-an and Ji Chiang. 1998. A new fossil pterosaur (Rhamphorhynchoidea) from Liaoning. *Jiangsu Geology* **22**, 199-206.

Jones, D.N., Dekker, R.W.R.J. and Roselaar, C.S. 1995. *The Megapodes.* Oxford University Press, Oxford.

Jouve, S. 2004. Description of the skull of a *Ctenochasma* (Pterosauria) from the Latest Jurassic of Eastern France, with a taxonomic revision of European Tithonian Pterodactyloidea. *Journal of Vertebrate Paleontology,* **24** (3), 542-554.

Kardong, K.V. 1995. *Vertebrates – Comparative: Anatomy, Function, Evolution.* Wm. C. Brown Publishers, Dubuque.

Kellner, A.W.A. 1989. A new edentate pterosaur of the Lower Cretaceous from the Araripe Basin, Northeast Brazil. *Anais da Academia Brasileira de Ciências,* **61**, 439-446.

Kellner, A.W.A. 1996. Reinterpretation of a remarkably well-preserved pterosaur soft tissue from the early Cretaceous of Brazil. *Journal of Vertebrate Paleontology,* **16**, 718–722.

Kellner, A.W.A. 2003. Pterosaur phylogeny and comments on the evolutionary history of the group. In: Buffetaut, E. and Mazin, J.-M. (ed.), *Evolution and Palaeobiology of Pterosaurs.* Geological Society Special Publication, **217**, 105-137.

Kellner, A.W.A. and Campos, D.A. 1988. Sobre un novo pterossauro com crista sagital da Bacia do Araripe, Cretáceo Inferior do Nordeste do Brasil. (Pterosauria, *Tupuxuara,* Cretaceo, Brasil). *Anais da Academia Brasileira de Ciências,* **60**, 459-469.

Kellner, A.W.A. and Campos, D.A. 1994. A new species of *Tupuxuara* (Pterosauria, Tapejaridae) from the Early Cretaceous of Brazil. *Anais da Academia Brasileira de Ciências,* **66**, 467-473.

Kellner, A.W.A. and Campos, D.A. 2002. The function of the cranial crest and jaws of a unique pterosaur from the early Cretaceous of Brazil. *Science,* **297**, 389-392.

Kellner, A.W.A. and Langston, W. 1996. Cranial remains of *Quetzalcoatlus* (Pterosauria, Azhdarchidae) from Late Cretaceous sediments of Big Bend National Park, Texas. *Journal of Vertebrate Paleontology*, **16**, 222-231.

Kellner, A.W.A. and Tomida, Y. 2000. Description of a new species of Anhangueridae (Pterodactyloidea) with comments on the pterosaur fauna from the Santana Formation (Aptian-Albian), northeastern Brazil. *National Science Museum Monographs, Tokyo*, **17**, 1-135.

King, A.S. and McLelland, J. 1984. *Birds: Their Structure and Function*. Bailliere Tindall, London.

Kripp, D. von. 1943. Ein Lebensbild von *Pteranodon ingens* auf flugtechnischer Grundlage. *Nova Acta Acad. Caesar. Leop. Carol.*, **12** (83), 217–240.

Langston, W., Jr. 1981. Pterosaurs. *Scientific American*, February, 122-136.

Lawson, D.A. 1975. Pterosaur from the latest Cretaceous of West Texas: discovery of the largest flying creature. *Science*, **187**, 947-948.

Lee, Y.N. 1994. The Early Cretaceous pterodactyloid pterosaur *Coloborhynchus* from North America. *Palaeontology*, **37**, 755-763.

Leonardi, G. and Borgomanero, G. 1985. *Cearadactylus atrox* nov. gen., nov. sp.: novo Pterosauria (Pterodactyloidea) da Chapada do Araripe, Ceara, Brasil. *Coletânea de Trabalhos Paleontológicos, Série Geologia, Brasilia*, **27**, 75-80.

Li, Jianjun, Lü Junchang, and Zhang, Baokun. 2003. A new Lower Cretaceous Sinopterid pterosaur from the Western Liaoning, China. *Acta Palaeontologica Sinica*, **42**, 442-447.

Lockley, M.G., Logue, T.J., Moratalla, J.J., Hunt, A.P., Schultz, R.J., and Robinson, J.W. 1995. The fossil trackway *Pteraichnus* is pterosaurian, not crocodilian: implications for the global distribution of pterosaur tracks. *Ichnos*, **4**, 7–20.

Lockley, M.G. and Wright, J.L. 2003. Pterosaur swim tracks and other ichnological evidence of behavior and ecology: In: Buffetaut, E., and Mazin, J.-M. (ed.) *Evolution and Palaeobiology of Pterosaurs*. Geological Society Special Publication, **217**, 297-313.

Lü Junchang. 2002. Soft tissue in an early Cretaceous pterosaur from Liaoning Province, China. *Memoir of the Fukui Prefecture Dinosaur Museum*, **1**, 19-28.

Lü Junchang. 2003. A new pterosaur: Beipiaopterus chenianus, gen. et sp. nov. (Reptilia: Pterosauria) from western Liaoning Province of China. Memoir of the Fukui Prefectural Dinosaur Museum, **2**, 153-160.

Lü Junchang and Ji Qiang. 2005. A new Ornithocheirid from the Early Cretaceous of Liaoning Province, China. *Acta Geologica Sinica*, **79**, 157-163.

Lydekker, R. 1901. Pterodactyles. *Nature*, **64**, 645–646. [Review of Seeley's *Dragons of the Air*.]

MacCready, P. 1985. The great pterodactyl project. *Engineering and Science*, **49** (2), 18–24.

Mader, B. and Kellner, A.W.A. 1999. A new anhanguerid pterosaur from the Cretaceous of Morocco. *Boletim do Museu Nacional, Geologia, Nova Série, Rio de Janeiro*, **45**, 1-11.

Maier, G. 2004. *African Dinosaurs Unearthed: The Tendaguru Expeditions* (Life of the Past). Indiana University Press.

Maisey, J.G. 1991. *Santana Fossils: An Illustrated Atlas*. T.F.H. Publications, inc., Neptune City, New Jersey. The Santana Formation pterosaurs, 351-370.

Manabe, M., Barrett, P. and Isaji, S. 2000. A refugium for relics. *Nature*, **404**, 953.

Marsh, O.C. 1876. Notice of new sub-order of Pterosauria. *American Journal of Science*, **1** (3), 507-509.

Marsh, O.C. 1882. The wings of pterodactyles. *American Journal of Science*, **23** (3), 251–256 (also appeared in *Nature*, **25**, 531–533).

Martill, D.M. 1989. The Medusa effect: instantaneous fossilization. *Geology Today*, **5** (6), Nov.-Dec., 201-205.

Martill, D.M. 1993. Fossils of the Santana and Crato Formations, Brazil. *The Palaeontological Association Field Guides to Fossils*, **5**.

Martill, D.M. 1997. From hypothesis to fact in a flight of fantasy: the responsibility of the popular scientific media. *Geology Today*, March/April, 71-73.

Martill, D.M. and Frey, E. 1998. A new pterosaur lagerstätte in N.E. Brazil (Crato Formation; Aptian, Lower Cretaceous): preliminary observations. *Oryctos*, **1**, 79-85.

Martill, D.M. and Unwin, D.M. 1989. Exceptionally well-preserved pterosaur wing membrane from the Cretaceous of Brazil. *Nature*, **340**, 138–140.

Mayr, E. 1942. *Systematics and the Origin of Species*. Columbia University Press, New York.

Mazin, J.M., Hantzpergue, P., La Faurie, G. and Vignaud, P. 1995. Des pistes de ptérosaures dans le Tithonien de Crayssac (Quercy, France). *C. r. Acad. Sci. Paris II Sci. Terre*, **321**, 417–424.

Mazin, J-M., Billon-Bruyat, J-P., Hantzpergue, P. and Lafaurie, G. 2003. Ichnological evidence for quadrupedal locomotion in pterodactyloid pterosaurs: trackways from the Late Jurassic of Crayssac (southwestern France) In: Buffetaut, E. and Mazin, J.-M. (eds), *Evolution and Palaeobiology of Pterosaurs*. Geological Society of London, Special Publication, **217**, 283-296.

McGowan, C. 1999. *A Practical Guide to Vertebrate Mechanics*. Cambridge University Press, Cambridge.

Müller, J. 2003. Early loss and multiple return of the lower temporal arcade in diapsid reptiles. *Naturwissenschaften*, **90**, 473-476.

Murry, P.S., Winkler, D.A. and Jacbs, L.L. 1991. An azhdarchid pterosaur humerus form the Lower Cretaceous Glen Rose Formation of Texas. *Journal of Paleontology*, **65**, 167-170.

Naish, D. 2001. Crocodilians. *Geology Today*, **17**, 71-77.

Nesov, L.A. 1991. [Giant flying reptiles of the family Azhdarchidae: I. Morphology and systematics.] *Bulletin of Leningrad University, Series 7, Geology and Geography*, **2**(14), 14-23.

Nesov, L.A. 1995. Dinosaurs of Northern Eurasia: New Data About Assemblages, Ecology and Paleobiogeography. University of St Petersburg.

Newman, E. 1843. Note on the pterodactyl tribe considered as marsupial bats. *Zoologist*, **1**, 129–131.

Nopsca, F. von. 1916. Zur Körpertemperatur der Pterosaurier. *Zbl. Miner. Geol.*, 418–419.

Nopsca, F. von. 1924. Bemerkungen und Ergänzungen zu G. v. Arthabers Arbeit über Entwicklung und Absterben der Pterosaurier. *Paläont. Z.*, **6**, 80–91.

Norberg, U. 2002. Structure, form, and function of flight in engineering and the living world. *Journal of Morphology*, **252**, 52-81.

Norrel, M.A. 2005. *Unearthing the Dragon: The Great Feathered Dinosaur Discovery*. A Peter N. Névraumont Book, Pi Press, New York.

O'Connor, P.M. 2004. Pulmonary Pneumaticity in the Postcranial Skeleton of Extant Birds: A Case Study Examining Anseriforms. *Journal of Morphology*, **261**,141-161.

Olsen, P.E., Kent, D.V., Sues, H.-D., Koeberl, C., Huber, H., Montanari, A., Rainforth, E. C., Fowell, S.J., Szajna, M.J., and Hartline, B.W. 2002, Ascent of dinosaurs linked to an iridium anomaly at the Triassic-Jurassic boundary. *Science*, **296**, 1305-1307.

Ostrom, J.H. 1978. The osteology of *Compsognathus longipes* Wagner. *Zitteliana*, **4**, 73-118.

Owen, R. 1846. On the supposed fossil bones of birds from the Wealden. *Quaterly Journal of the Geological Society of London London*, **2**, 96–102.

Owen, R. 1874. Monograph on the Fossil Reptilia of the Mesozoic Formations. *Monographs of the Palaeontographical Society*, 1-14.

Padian, K. 1983. A functional analysis of flying and walking in pterosaurs. *Paleobiology*, **9**, 218–239.

Padian, K. 1983. Osteology and functional morphology of *Dimorphodon macronyx* (Buckland) (Pterosauria: Rhamphorhynchoidea) based on new material in the Yale Peabody Museum. *Postilla*, **189**, 1-44.

Padian, K. 1984. The origin of pterosaurs. In: Reif, W.-E. and Westphal, F. (eds.), *Proceedings of the 3rd Symposium on Mesozoic Terrestrial Ecosystems*. Attempto, Tübingen, 163-168.

Padian, K. 1985. The origins and aerodynamics of flight in extinct vertebrates. *Palaeontology*, **28**, 413–433.

Padian, K. 1987. The case of the bat-winged pterosaur. Typological taxonomy and the influence of pictorial representation on scientific presentation. In: Czerkas, S.J. and Olson, E.C. (eds.). *Dinosaurs Past and Present*. Natural History Museum of Los Angeles County and University of Washington Press, **II**, 65–81.

Padian, K. 1997. Pterosauromorpha. In: Currie, P.J. and Padian, K. (eds.), *Encyclopedia of Dinosaurs*. Academic Press, 617-618.

Padian, K., Horner, J.R., and Ricqlès, A. de. 2004. Growth in small dinosaurs and pterosaurs: the evolution of archosaurian growth strategies. *Journal of Vertebrate Paleontology*, **24** (4), 555-571.

Padian, K. and Olsen, P. E. 1984. The fossil trackway *Pteraichnus*: not pterosaurian, but crocodilian. *Journal of Paleontology*, **58**, 178–184.

Padian, K. and Rayner, J.M.V. 1993. The wings of pterosaurs. *Am. J. Sci.*, **293-A**, 91–166.

Pennycuick, C.J. 1986. Mechanical constraints on the evolution of flight. In: Padian, K. (ed.), *The origin of birds and the evolution of flight. Memoirs of the California Academy of Sciences*, **8**, 83–98.

Pennycuick, C.J. 1988. On the reconstruction of pterosaurs and their manner of flight, with notes on vortex wakes. *Biol. Rev.* **63**, 209–231.

Pereda-Suberbiola, X., Bardet, N., Jouve, S., Larochene, M., Bouya, B., and Amaghzaz, M. 2003. A new azhdarchid pterosaur from the Late Cretaceous phosphates of Marocco, In: Buffetaut, E. and Mazin, J.-M. (eds.), *Evolution and Palaeobiology of Pterosaurs*. Geological Society of London, Special Publication, **217**, 79-90.

Plieninger, F. 1895. *Campylognathus Zitteli*, ein neuer Flugsaurier aus dem obersten Lias Schwabens. *Paläontographica*, **41**, 193-222.

Plieninger, F. 1901. Beiträge zur Kenntnis der Flugsaurier. *Paläontographica*, **48**, 65-90.

Plieninger, F. 1907. Die Pterosaurier der Juraformation Schwabens. *Paläontographica*, **53**, 209-313.

Ponomarenko, A.G. 1976. A new insect from the Cretaceous of Transbaikalia, a possible parasite of pterosaurs. Paleont. *Journal of Ac. of Scie.*, Moscow, **3**, 339-343.

Ponomarenko, A.G. 1986. [Insects in the early Cretaceous ecosystems of West Mongolia.] *Transactions of the Joint Soviet-Mongolian Palaeontological Expedition*, **28**, 183-201 [in Russian].

Pough, F.H., Andrews, R.M., Cadle, J.E., Savitzky, A.H., and Wells, K.D. 1998. *Herpetology*. Prentice Hall, New Jersey.

Price, L.I. 1953. A presenca de Pterosauria no Cretaceo Superior do Estada da Paraiba. *Divisao de Geologia e Mineralogia Notas Preliminaries e Estudos*, **71**, 1-10.

Prum, O.R. and Brush, A.H. 2003. Which came first, the feather or the bird. *Scientific American*, **288**(3), 84-93.

Quammen, S. 1997. *The Song of the Dodo: Island Biogeography in an Age of Extinction*. Scribner, New York.

Rasnitsyn, A.P., Quicke, D.L.J. (eds.) 2002. *History of Insects*. Kluwer Academic Publishers, Dordrecht, Netherlands.

Rauhut, O.W.M., Martin, T., Ortiz Jaureguizar, T., and Puerta, P. 2002. A Jurassic mammal from South America. *Nature*, **416**, 165-168.

Ricqlès, A. de, Padian, K., Horner, J. R. and Francillon-Vieillot, H. 2000. Palaeohistology of the bones of pterosaurs (Reptilia: Archosauria): antomy, ontogeny, and biomechanical implications. *Zoological Journal of the Linnean Society*, **129**, 349-385.

Ride, W.D.L., Cogger, H.G., Dupuis, C., Kraus, O., Minelli, A., Thompson, F.C., and Tubbs, P.K. 1999. *International Code of Zoological Nomenclature*, Fourth Edition. International Trust for Zoological Nomenclature, c/o The Natural Hisatroy Museum, London.

Ryabinin, A.N. 1948. [Remarks on a flying reptile from the Jurassic of the Karatau.] *Transactions of the Palaeontological Institute*, **15**, 86-93 [in Russian].

Salée, A. 1928. L'exemplaire de Louvain de *Dorygnathus banthensis* Theodori sp. *Mémoires de L'Institut de Géologique de L'Université de Louvain*, **4**, 289-344.

Sanchez, T.M. 1973. Redescription del craneo y mandibulas de *Pterodaustro guinazui* Bonaparte (Pterodactyloidea, Pterodaustriidae). *Ameghiniana*, **10**, 313-325.

Sayão, J.M. and Kellner, A.W.A. 2001. New data on the pterosaur fauna from Tendaguru (Tanzania), Upper Jurassic, Africa. *Journal of Vertebrate Paleontology*, **21** (suppl. to nb. 3), 97A.

Schmidt-Nielsen, K. 1984. *Scaling: Why Is Animal Size So Important?* Cambridge University Press, Cambridge.

Seeley, H.G. 1901. *Dragons of the Air: An Account of ExtinctFlying Reptiles*. Methuen, London.

Senter, P. 2003. New information on cranial and dental features of the Triassic archosauriform reptile *Euparkeria capensis*. *Palaeontology*, **46**, 613-621.

Senter, P. 2004. Phylogeny of Drepanosauridae (Reptilia: Diapsida). *Journal of Systematic Palaeontology*, **2**, 257-268.

Sharov, A.G. 1971. [New flying reptiles from the Mesozoic of Kazakhstan and Kirghizia.] *Transactions of the Palaeontological Institute*, **130**, 104-113.

Short, G.H. 1914. Wing adjustments of pterodactyls. *Aeronautical Journal* **18**, 336–343.

Silvio, R. 2004. New data on the anatomy and biogeography of the Drepanosauridae. In: Renesto, S. and Pagganoni, A. (eds.), *A symposium honouring the 30th anniversary of the discovery of* Eudimorphodon. Bergamo 2003. *Riv. Mus. civ. Sc. Mat. "E. Caffi" Bergamo*, **22**, 65-66.

Smith, A.B. 1994. *Systematics and the Fossil Record: Documenting Evolutionary Patterns*. Blackwell Scientific, Oxford.

Soemmerring, S.T. von. 1812. Über einen *Ornithocephalus* oder über das unbekannten Thier der Vorwelt, dessen Fossiles Gerippe Collini im 5. Bande der Actorum Academiae Theodoro-Palatinae nebst einer Abbildung in natürlicher Grösse im Jahre 1784 beschrieb, und welches Gerippe sich gegenwärtig in der Naturalien-Sammlung der königlichen Akademie der Wissenschaften zu München befindet. *Denkschr. k. bayer. Akad. Wiss. math.-phys. Kl.*, **3**, 89–158.

Steele, L. 2005. *Studies on pterosaur bone histology*. Ph.D. Thesis, University of Portsmouth.

Stein, R.S. 1975. Dynamic analysis of *Pteranodon ingens*: a reptilian adaptation for flight. *Journal of Paleontology* **49**, 534-548.

Stein, R.S. 1976. Aerodynamics of the long pterosaur wing. *Science*, **191**, 898–899.

Stieler, C. 1922. Neuer Rekonstruktionsversuch eines liassischen Flugsauriers. *Naturw. Wschr. NF*, **21**, 273–280.

Stokes, W.L. 1954. Pterodactyl tracks from the Morrison Formation near Carrizo Mountains, Arizona. *Bull. Geol. Soc. Am.*, **65**, 1309.

Stokes, W.L. 1957. Pterodactyl tracks from the Morrison Formation. *Journal of Paleontology*, **31**, 952–954.

Sues, H.-D. and Averianov, A. 2004. Dinosaurs from the Upper Cretaceous (Turonian) of Dzharakuduk, Kyzylkum Desert, Uzbekistan. *Journal of Vertebrate Paleontology*, **24** (suppl. to **3**), 119A-120A.

Theodori, K. von. 1830. Knochen von *Pterodactylus* aus der Liasformation von Banz. *Not. Geb. Natur Heilk.*, **29**, 101–103.

Thompson, M.B. and Speake, B.K. 2004. Egg morphology and composition. In: Deeming, D.C. (ed.), *Reptilian Incubation: Environment, Evolution and Behaviour*. Nottingham University Press, Nottingham, 45-74.

Tischlinger, H. 2002. Der Eichstätter *Archaeopteryx* im langwelligen UV-Licht. *Archaeopteryx*, **20**, 21-38.

Tischlinger, H. 2003. Professor Goldfuß und sein 'Dickschnabel' aus dem Altmühljura. *Globulus - Beitr. Natur- u. kulturwiss. Ges.*, **10**, 95-104.

Tischlinger, H. and Frey, E. 2001. Mit Haut und Haar! Flugsaurier-Neufunde mit spektakulärer Weichteilerhaltung. *Fossilien*, **3** (1), 151-158.

Tischlinger, H. and Frey, E. 2002. Ein *Rhamphorhynchus* (Pterosauria, Reptilia) mit ungewöhnlicher Flughauterhaltung aus dem Solnhofener Plattenkalk. *Archaeopteryx*, **20**, 1-20.

Unwin, D.M. 1987. Pterosaur Locomotion – joggers or waddlers? *Nature*, **327**, 13-14.

Unwin, D.M. 1988. A new pterosaur from the Kimmeridge Clay of Kimmeridge, Dorset. *Proc. Dorset Nat. Hist. Archaeol. Soc.*, **109**, 150-153.

Unwin, D.M. 1995. Preliminary results of a phylogenetic analysis of the Pterosauria (Diapsida: Archosauria). In: Sun, A. and Wang, Y. (eds.), *Sixth Symposium on Mesozoic Terrestrial Ecosystems and Biota*. Beijing, China Ocean Press, 69–72.

Unwin, D.M. 1996. The fossil record of Middle Jurassic pterosaurs. In: Morales, M. (ed.), *The Continental Jurassic*. Museum of Northern Arizona Bulletin, **60**, 291-304.

Unwin, D.M. 1997. Pterosaur tracks and the terrestrial ability of pterosaurs. *Lethaia*, **29**, 373-386.

Unwin, D.M. 2001. An overview of the pterosaur assemblage from the Cambridge Greensand (Cretaceous) of Eastern England. *Mitteilungen Museum für Naturkunde Berlin, Geowissenschaftlichen Reihe*, **4**, 189-222.

Unwin, D.M. 2002. On the systematic relationships of *Cearadactylus atrox*, an enigmatic Early Cretaceous pterosaur from the Santana Formation of Brazil. *Mitteilungen Museum für Naturkunde Berlin, Geowissenschaftlichen Reihe*, **4**, 237-261.

Unwin, D.M. 2003. On the phylogeny and evolutionary history of pterosaurs: In: Buffetaut, E. and Mazin, J.-M. (eds.), *Evolution and Palaeobiology of Pterosaurs*. Geological Society Special Publication, **217**, 139-190.

Unwin, D.M. 2003. Smart-winged pterosaurs. *Nature*, **425**, 910-911.

Unwin, D.M. 2004. *Eudimorphodon* and the early history of pterosaurs. In: Renesto, S. and Pagganoni, A. (eds.), A symposium honouring the 30th anniversary of the discovery of Eudimorphodon, Bergamo 2003. *Riv. Mus. civ. Sc. Mat. "E. Caffi" Bergamo*, **22**, 41-48.

Unwin, D.M., Alifanov, V.R., and Benton, M.J. 2000. Enigmatic small reptiles from the Middle-Late Triassic of Kirgizstan. In: Benton, M.J., Shishkin, M.A., Unwin, D.M., and Kurochkin, E.N. (eds.), *The Age of Dinosaurs in Russia and Mongolia*. Cambridge University Press, Cambridge.

Unwin, D.M. and Bakhurina, N.N. 1994. *Sordes pilosus* and the nature of the pterosaur flight apparatus. *Nature*, **371**, 62-64.

Unwin, D.M. and Heinrich, W.-D. 1999. On a pterosaur jaw remain from the Late Jurassic of Tendaguru, East Africa. *Mitteilungen Museum für Naturkunde Berlin, Geowissenschaftlichen Reihe*, **2**, 121-134.

Unwin, D.M. and Lü Junchang. 1997. On *Zhejiangopterus* and the relationship of pterodactyloid pterosaurs. *Historical Biology*, **12**, 199-910.

Unwin, D.M., Lü, Junchang and Bakhurina, N.N. 2000. On the systematic and stratigraphic significance of pterosaurs from the Lower Cretaceous Yixian Formation (Jehol Group) of Liaoning, China. *Mitteilungen Museum für Naturkunde Berlin, Geowissenschaftlichen Reihe*, **3**, 181–206.

Wagler, J.G. 1830. *Natürliches System der Amphibien*. München, Stuttgart, Tübingen.

Wanderer, K. 1908. *Rhamphorhynchus Gemmingi* H. v. Meyer. Ein Exemplar mit teilweise erhaltener Flughaut an dem Kgl. Mineralog.-Geol. Museum zu Dresden. *Paläontographica*, **55**, 195–216.

Wang Xiao-Lin and Lü Junchang. 2001. Discovery of a pterodactyloid pterosaur fron the Yixian Formation of western Liaoning, China. Chinese Science Bulletin, **45**, 447-454.

Wang Xiao-Lin and Zhou Zhonghe. 2003. A new pterosaur (Pterodactyloidea, Tapejaridae) from the Early Cretaceous Jiufotang Formation of western Liaoning, China and its implications for biostratigraphy, Chinese Science Bulletin, **48**(1), 16-23.

Wang Xiao-Lin and Zhou Zhonghe. 2003. Two new pterodactyloid pterosaurs from the Early Cretaceous Jiufotang Formation of western Liaoning, China. *Vertebrata PalAsiatica*, **41**, 34-41.

Wang Xiao-Lin and Zhou Zhonghe. 2004. Pterosaur embryo from the Early Cretaceous. *Nature*, **429**, 621.

Wang Xiao-Lin, Zhou Zhonghe, Zhang Fucheng, and Xu Xing. 2002. A nearly complete articulated rhamphorhynchoid pterosaur with exceptionally well-preserved wing membranes and "hairs" from Inner Mongolia, Northeast China. *Chinese Science Bulletin*, **47**(3), 226-232.

Weimerskirsch, H., Chastel, O. Barbraud, C., and Tostain, O. 2003. Frigate birds ride high on thermals. *Nature*, **421**, 333-334.

Weishampel, D.B., Dodson, P., and Osmólska, H. 2004. *The Dinosauria*, Second Edition. University of California Press, Berkeley.

Wellnhofer, P. 1970. Die Pterodactyloidea (Pterosauria) der Oberjura Plattenkalke Süddeutschlands. *Bayerische Akademie der Wissenschaften, mathematisch-naturwissenschaftliche Klasse, Abhandlungen*, **141**, 1-133.

Wellnhofer, P. 1974. *Campylognathoides liasicus* (Quenstedt), an Upper Liassic pterosaur from Holzmaden. The Pittsburgh specimen. *Annals of the Carnegie Museum*, **45**, 5-34.

Wellnhofer, P. 1975. Die Rhamphorhynchoidea (Pterosauria) der Oberjura-Plattenkalke Süddeutschlands. *Palaeontographica* A, **148**, 1-33, 132-186, **149**, 1-30.

Wellnhofer, P. 1978. Pterosauria. In: Wellnhofer, P. (ed.), *Handbuch der Paläoherpetologie*. Gustav Fischer Verlag, Stuttgart, **19**, 1-82.

Wellnhofer, P. 1985. Neue Pterosaurier aus der Santana Formation (Apt) der Chapada do Araripe, Brasilien. *Palaeontographica* A, **187**, 105-182.

Wellnhofer, P. 1987. Die Flughaut von *Pterodactylus* (Reptilia, Pterosauria) am Beispiel des Wiener Examplares von *Pterodactylus kochi* (Wagner). *Annln Nathist. Mus. Wien*, **88A**, 149–162.

Wellnhofer, P. 1987. New crested pterosaurs from the Lower Cretaceous of Brazil. *Mitteilungen der Bayerischen Staatssammlung für Paläontologie und historische Geologie*, **27**, 175-186.

Wellnhofer, P. 1988. Terrestrial locomotion in pterosaurs. *Historical Biology*, **1**, 3–16.

Wellnhofer, P. 1991. *The Illustrated Encyclopedia of Pterosaurs*. Salamander Books, London.

Wellnhofer, P. and Buffetaut, E. 1999. Pterosaur remains from the Cretaceous of Morocco. *Palaontologische Zeitschrift*, **73**, 133-142.

Wellnhofer, P. and Kellner, A.W.A. 1991. The skull of *Tapejara wellnhoferi* Kellner (Reptilia: Pterosauria) from the Lower Cretaceous Santana Formation of the Araripe Basin, Northeastern Brazil. *Mitteilungen der Bayerischen Staatssammlung für Paläontologie und historische Geologie*, **31**, 89-106.

Welman, J. 1995. *Euparkeria* and the origin of birds. *South African Journal of Science*, **91**, 533-537.

Wiffen, J. and Molnar, R.E. 1988. First pterosaur from New Zealand. *Alcheringa*, **12**, 53–59.

Wild, R. 1973. Die Triasfauna der Tessiner Kalkalpen. XXIII. *Tanystropheus longobardicus* (Bassani) (Neue Ergebnisse). *Abh. Schweiz. Paläont. Ges.*, **95**, 1–162.

Wild, R. 1978. Die Flugsaurier (Reptilia, Pterosauria) aus der Oberen Trias von Cene bei Bergamo. *Bollettino della Società Paleontologica Italiana*, **17**, 176-256.

Wild, R. 1984. A new pterosaur (Reptilia: Pterosauria) from the Upper Triassic (Norian) of Friuli, Italy. *Gortania - Atti del Museo Friulano di Storia Naturale*, **5**, 45-62.

Wild, R. 1990. Ein Flugsaurierrest (Reptilia, Pterosauria) aus der Unterkreide (Hauterive) von Hannover (Niedersachsen). *Neues Jahrbuch für Geologie und Paläontologie Abhandlung*, **181**, 241–254.

Wild, R. 1993. Holzmaden. In: Briggs, D.E.G. and Crowther, P.R. (eds.), *Palaeobiology: A Synthesis*. Blackwell Scientific Publications, London, 282-285.

Wild, R. 1993. A juvenile specimen of *Eudimorphodon ranzii* Zambelli (Reptilia, Pterosauria) from the Upper Triassic (Norian) of Bergamo. *Rivista del Museo civico di Scienze Naturali "E. Caffi" Bergamo*, **16**, 95-120.

Wilkinson, M. 2003. *Flight of the ornithocheirid pterosaurs*. Ph.D. Thesis, University of Cambridge.

Winston, J.E. 2000. *Describing Species: Practical Taxonomic Procedure for Biologists*. Columbia University Press, New York.

Witmer, L.M. 1997. The evolution of the antorbital cavity of archosaurs: a study in soft-tissue reconstruction in the fossil record with an analysis of the function of pneumaticity. *Memoirs of the Society of Vertebrate Paleontology, Journal of Vertebrate Paleontology*, **17** (Suppl. to **1**),1–73.

Witmer, L.M. 2001. Nostril position in dinosaurs and other vertebrates and its significance for nasal function. *Science*, **293**, 850–853.

Witmer, L.M., Chatterjee, S., Franzosa, J., and Rowe, T. 2003. Neuroanatomy of flying reptiles and implications for flight, posture and behavior. *Nature*, **425**, 950-953.

Wright, J.L. and Lockley, M.G. 1999. Tracking Late Jurassic pterosaurs on land and sea. *Journal of Vertebrate Paleontology*, **19**, 85, 86A.

Wright, J.L., Unwin, D.M., Lockley, M.G. and Rainforth, E.C. 1997. Pterosaur tracks from the Purbeck Limestone Formation of Dorset, England. *Proceedings of the Geologists' Association*, **108**, 39-48.

Wu Qicheng (ed.). 2002. *Fossil Treasures from Liaonin*. Geological Publishing House, China.

Yalden, D.W. 1985. Forelimb function in *Archaeopteryx*. In: Hecht, M.K., Ostrom, J.H., Viohl, G., and Wellnhofer, P. (eds.), *The Beginning of Birds*. Eichstät, 91-97.

Yaoming Hu, Jin Meng, Yuanqing Wang, and Chuankui Li. 2005. Large Mesozoic mammals fed on young dinosaurs. *Nature* **433**, 149-152.

Young Chung-Chien (Yang Zhong-jian). 1964. On a new pterosaurian from Sinkiang, China. *Vertebrata Palasiatica*, **8**, 221-255.

Young Chung-Chien (Yang Zhong-jian). 1973. [Wuerho pterosaurs]. Special Publication of the Institute of Vertebrate Palaeontology and Palaeoanthropology, Academia Sinica, **11**, 18-34.

Zambelli, R. 1973. *Eudimorphodon ranzii* gen. nov., sp. nov., uno pterosauro triassico (nota preliminare). *Rendiconti Scienca Instituto Lombardo* B, **107**, 27-32.

Zhou Zhonge and Zhang Fucheng. 2004. A precocial avian embryo from the Lower Cretaceous of China. *Science*, **306**, 653.

Zittel, K.A. 1882. Über Flugsaurier aus dem lithographischen Schiefer Bayerns. *Paläontographica*, **29**, 47-80.

ACKNOWLEDGMENTS

Many of my friends and colleagues have spent much of their valuable time discussing pterosaurs with me, in conversation, as part of formal discussions at meetings, via email and the World Wide Web and through books and papers. I thank the following individuals for broadening and enriching my understanding of these extraordinary animals: Sasha Averianov, Natasha Bakhurina, Natalie Bardet, Chris Bennett, Michael Benton, Jean-Paul Billon-Bruyat, Nils Bonde, Eric Buffetaut, Cai Zhenquan, Ken Carpenter, Sankar Chatterjee, Anusuya Chinsamy, Luis Chiappe, Per Chistiansen, Leon Claessens, Jane Clarke, Laura Codorniú, Julio Company, Phil Currie, Fabio Dalla Vecchia, Charles Deeming, Dong Zhiming, Danny Elvidge, Greg Erickson, Steve Etches, Michael Fastnacht, Dino Frey, David Gillette, Bev Halstead, Grant Hazlehurst, Wolf-Dieter Heinrich, Donald Henderson, David Hone, Jim Hopson, Stafford Howse, Karl Hirsch, Koo-Geun Hwang, Farish Jenkins, Ji Chiang, Ji Shu-an, Alex Kellner, Andrew Kitchener, Wann Langston, Yuong-Nam Lee, Guiseppe Leonardi, Martin Lockley, Lü Junchang, Richard Kemp, Makoto Manabe, Todd Marshall, David Martill, Jean-Michel Mazin, Andrew Milner, Ralph Molnar, Darren Naish, Lev Nesov, Kevin Padian, Pat O'Connor, John Ostrom, Colin Pennycuick, Xavier Pereda-Suberbiola, Perle Altangerel, David Peters, Emma Rainforth, Ollie Rauhut, Jeremy Rayner, Jurgen Riess, J. I. Ruiz-Omeñaca (Ome), Lorna Steele, Tony Thulborn, Helmut Tischlinger, Wang Xiao-lin, Peter Wellnhofer, Rupert Wild, Matt Wilkinson, Larry Witmer, Jo Wright, Derek Yalden, Rocco Zambelli, and Zhou Zhonghe.

I am extremely grateful to all those who provided the various materials that have gone into the illustrations, thereby considerably enhancing the visual appearance of this work. In particular, I would like to give special thanks to Helmut Tischlinger, for generously giving me unrestricted access to his superb photographs, and to Peter Wellnhofer, Dino Frey and Wang Xiao-Lin for sending numerous pictures often at very short notice.

My thanks also to the following, all of whom generously provided illustrations: Chris Bennett, Eric Buffetaut, Cai Zhenquan, Ken Carpenter, Sankar Chatterjee, Luis Chiappe, Per Chistiansen, Jim Clark, Fabio Dalla Vecchia, Michael Fastnacht, Dino Frey, Donald Henderson, Jose Carlos Garcia-Ramos, Koo-Geun Hwang, Alex Kellner, Martin Lockley, Lü Junchang, Makoto Manabe, David Martill, Jean-Michel Mazin, Lorna Steele, Larry Witmer and Jo Wright.

Rodrigo Solar Gijon and Adriana Lopez-Arabrello kindly provided translations of Spanish texts, Natasha Bakhurina held my hand as I struggled to read Russian texts, Olga Bakhurina Unwin assisted with translation of French texts and Lü Junchang very kindly made translations of several Chinese texts.

I am very grateful indeed to Jason Dunlop, David Martill and David Lazarus for their heroic efforts in reading the entire text, grappling with my arcane phraseology and tactfully removing my worst excesses. Natasha Bakhurina, Charlie Deeming, Jesus Marugan, Matt Wilkinson, Donald Henderson, Thorsten Bergmann, and the absolutely irrepressible Nizar Ibrahim all read and commented on various chapters.

The diagrams and drawings were prepared by as fine a set of illustrators as any one could wish for: Natasha Bakhurina, Jan Müller-Edzards, Jorg-Peter Mendau and Elke Siebert. Todd ("Dude") Marshall produced an excellent series of color paintings and was a joy to work with.

Special thanks go to three people. Dave "woof-woof" Martill lent me money, his house and huge amounts of his time. On the other hand, his cat fleas got quite a lot of my blood, so I think we are about even, apart from the jar of mustard. Bev Halstead, my Ph.D. supervisor, gave me the chance to work on pterosaurs and, perhaps even more importantly, taught me how to think. Sadly, he was killed in an accident shortly after seeing my doctoral work through to its completion—I miss him still. Natasha Bakhurina helped in just about every way possible and is practically the only person in the world who actually went out to look for pterosaurs—and found them. What more could a husband want?

Peter N. Névraumont made this book happen. Without his pleas, demands, threats and incredible patience, it would still be lurking at the back of the hard drive on my computer. If any credit is due to anyone on this project it is Peter. My heartfelt thanks to you, Peter, for dragging this book out of me inch by inch—it was the only way. My thanks also to Stephen Morrow at Pi Press for his forbearance and Jean Christensen for taking my tangled syntax and turning it into readable text.

INDEX

Page numbers followed by *n* indicate endnotes.